Studies in Systems, Decision and Control

Volume 348

Series Editor

Janusz Kacprzyk, Systems Research Institute, Polish Academy of Sciences, Warsaw, Poland

The series "Studies in Systems, Decision and Control" (SSDC) covers both new developments and advances, as well as the state of the art, in the various areas of broadly perceived systems, decision making and control–quickly, up to date and with a high quality. The intent is to cover the theory, applications, and perspectives on the state of the art and future developments relevant to systems, decision making, control, complex processes and related areas, as embedded in the fields of engineering, computer science, physics, economics, social and life sciences, as well as the paradigms and methodologies behind them. The series contains monographs, textbooks, lecture notes and edited volumes in systems, decision making and control spanning the areas of Cyber-Physical Systems, Autonomous Systems, Sensor Networks, Control Systems, Energy Systems, Automotive Systems, Biological Systems, Vehicular Networking and Connected Vehicles, Aerospace Systems, Automation, Manufacturing, Smart Grids, Nonlinear Systems, Power Systems, Robotics, Social Systems, Economic Systems and other. Of particular value to both the contributors and the readership are the short publication timeframe and the world-wide distribution and exposure which enable both a wide and rapid dissemination of research output.

Indexed by SCOPUS, DBLP, WTI Frankfurt eG, zbMATH, SCImago.

All books published in the series are submitted for consideration in Web of Science.

More information about this series at http://www.springer.com/series/13304

Ibrahim Arpaci · Mostafa Al-Emran ·
Mohammed A. Al-Sharafi · Gonçalo Marques
Editors

Emerging Technologies During the Era of COVID-19 Pandemic

 Springer

Editors
Ibrahim Arpaci
Tokat Gaziosmanpasa University
Tokat, Turkey

Mostafa Al-Emran ⓘ
The British University in Dubai
Dubai, United Arab Emirates

Mohammed A. Al-Sharafi
Universiti Malaysia Kelantan
Kelantan, Malaysia

Gonçalo Marques ⓘ
Polytechnic of Coimbra
Rua General Santos Costa, Portugal

ISSN 2198-4182   ISSN 2198-4190   (electronic)
Studies in Systems, Decision and Control
ISBN 978-3-030-67715-2   ISBN 978-3-030-67716-9   (eBook)
https://doi.org/10.1007/978-3-030-67716-9

This Springer imprint is published by the registered company Springer Nature Switzerland AG
The registered company address is: Gewerbestrasse 11, 6330 Cham, Switzerland

Preface

The pandemic resulted from the novel coronavirus (COVID-19) has dramatically affected the society and individuals in many ways around the world. While emerging technologies have been widely used in the past, the appearance of the COVID-19 pandemic requires the use of these technologies across many sectors. In health care, these technologies could be used in detecting, identifying, and predicting the COVID-19 cases, controlling and preventing its spread, developing vaccines, and discovering and testing new drugs, among many others. Concerning the educational sector, several technologies were evolved with the purpose of delivering distance learning to students at "anytime anywhere" settings. In addition, surveillance and control technologies were used to monitor and track individuals to detect those with COVID-19 symptoms. It is essential to understand how these technologies were used in managing the pandemic in order to determine future needs and research directions. These issues are the core of this edited book. Further, this edited book attempts to cover empirical and review studies that mainly focus on the use of emerging technologies and their impact on health care, education, and society.

This book has several implications. For research, it assists scholars in determining the current challenges and constraints in using technologies during the era of COVID-19 pandemic and uncover new opportunities for future research in the domain. Further, it helps postgraduate students to gain an insight into the recent techniques and applications of the emerging technologies. For health care, the results of the chapters are expected to assist in identifying, detecting, and predicting the COVID-19 cases. It is also expected that the derived results will assist in using emerging technologies for drug discovery and vaccine developments. For education, the results are believed to help the policymakers in formulating and developing their policies and procedures for using the emerging technologies in delivering distance learning during the COVID-19 pandemic and other similar crises. For society, the results are expected to assist organizations and individuals in understanding the issues associated with the use of emerging technologies and the factors affecting the adoption of such technologies during the era of COVID-19.

This book is intended to present the state-of-the-art research studies concerning the use of recent applications and techniques during the era of COVID-19 pandemic. It was able to attract 99 submissions from different countries across the world. Out of the 99 submissions, we accepted 23 submissions, representing an acceptance rate of 23.2%. Each submission was reviewed by at least two reviewers, who have expertise in the related submitted paper. The evaluation criteria include several issues, such as correctness, originality, technical strength, significance, quality of presentation, interest, and relevance to the book scope. The chapters of this book provide a collection of high-quality research works that address broad challenges, constraints, and opportunities in both theoretical and application aspects of several technologies. The chapters of this book are published in *Studies in Systems, Decision and Control Series* by Springer, which has a high SJR impact.

We acknowledge all those who contributed to the production of this edited book. We would also like to express our gratitude to the referees for their valuable feedback and suggestions. Without them, it would not be possible for us to maintain the high-quality and the success of the "Emerging Technologies During the Era of COVID-19 Pandemic" edited book. Therefore, on the next page, we list the reviewers along with their affiliations as a recognition of their time and efforts.

Tokat, Turkey   Ibrahim Arpaci
Dubai, United Arab Emirates   Mostafa Al-Emran
Kelantan, Malaysia   Mohammed A. Al-Sharafi
Coimbra, Portugal   Gonçalo Marques
March 2021

List of Reviewers

Ajay Gadicha, Sant Gadge Baba Amravati University, India
Ali Q. Al-Shetwi, Universiti Tenaga Nasional, Malaysia
Fadi Herzallah, Palestine Technical University, Palestine
Ibrahim Mahariq, American University of the Middle East, Kuwait
Kasım Karataş, Karamanoğlu Mehmetbey University, Turkey
Khaled Shaalan, The British University in Dubai, United Arab Emirates
Nidal Drissi, UAE University, United Arab Emirates
Noor Al-Qaysi, Universiti Pendidikan Sultan Idris, Malaysia
Rajmohan Seetharaman, Lokmanya Tilak Municipal General Hospital and Medical College, India
Ritambhara Parashar, Jaipur Engineering College and Research Centre, India
Salome Oniani, Ilia State University, Georgia
Shadi Alshehabi, University of Turkish Aeronautical Association, Turkey
Yojna Arora, Amity University, India

Contents

A Survey of Using Machine Learning Algorithms During the COVID-19 Pandemic

Mostafa Al-Emran⬡, Mohammed N. Al-Kabi,
and Gonçalo Marques⬡

Abstract The emergence of novel coronavirus (COVID-19) is considered a worldwide pandemic. In response to this pandemic and following the recent developments in artificial intelligence (AI) techniques, the literature witnessed an abundant amount of machine learning applications on COVID-19. To understand these applications, this study aims to provide an early review of the articles published on the employment of machine learning algorithms in predicting the COVID-19 infections, survival rates of patients, vaccine development, and drug discovery. While machine learning has had a more significant impact on healthcare, the analysis of the current review suggests that the use of machine learning is still in its early stages in fighting the COVID-19. Its practical application is hindered by the unavailability of large amounts of data. Other challenges, constraints, and future directions are also discussed.

Keywords Coronavirus · COVID-19 · Artificial intelligence · Machine learning · Infections · Vaccine development · Drug discovery

M. Al-Emran (✉)
Faculty of Engineering & IT, The British University in Dubai, Dubai, UAE
e-mail: mustafa.n.alemran@gmail.com

M. N. Al-Kabi
Department of Information Technology, Al Buraimi University College, Al Buraimi, Oman
e-mail: mohammed@buc.edu.om

G. Marques
Polytechnic of Coimbra, ESTGOH, Rua General Santos Costa,
3400-124 Oliveira do Hospital, Portugal
e-mail: goncalosantosmarques@gmail.com

I. Arpaci et al. (eds.), *Emerging Technologies During the Era of COVID-19 Pandemic*, Studies in Systems, Decision and Control 348,
https://doi.org/10.1007/978-3-030-67716-9_1

1

1   Introduction

A novel coronavirus (COVID-19) has appeared in Wuhan, China, since December 2019 and has spread to many countries in a short period [1]. In March 2020, the World Health Organization (WHO) has declared the outbreak of COVID-19 as a worldwide pandemic. Therefore, a global response is required to provide healthcare systems to work against this crisis [2]. The delivery of such systems requires the availability of new technologies, such as artificial intelligence (AI), machine learning, and the Internet of Things (IoT), to combat the new epidemic [3]. AI techniques have gradually made a paradigm shift in the healthcare sector [4]. With the recent developments of AI applications, these applications have been employed in fields that are formerly believed to be the fields of only human experts [5]. The emergence of these innovative techniques will be the key in enhancing the identification, prevention, and prediction of COVID-19 cases [6]. In this regard, various expectations have been raised by the scientific community concerning the role of AI techniques in improving the treatment and diagnosis of COVID-19 infections [7].

Healthcare organizations are in an urgent need for decision support systems that could help them in handling the pandemic and suggesting proper solutions to avoid the outbreak [3]. Machine learning, as one of the well-known applications of AI, has been extensively applied on several COVID-19 datasets [8]. Since its emergence in December 2019, the literature witnessed an abundant amount of machine learning applications on COVID-19. This has been perceived through the analysis and prediction of the current and potential future patients [3]. Tracking the confirmed, recovered, and death cases is another application of machine learning. These applications can also be extended to the development of vaccines [3] and drug discovery [9]. To understand these applications, the main contribution of this study is to provide an early review of the articles published on the employment of machine learning algorithms in predicting the COVID-19 infections, survival rates of patients, vaccine development, and drug discovery. The current study also attempts to provide the main limitations and challenges of the existing techniques.

2   COVID-19 Infection Prediction

A vast amount of research studies were conducted for applying the machine learning algorithms to predict the infections of COVID-19. Through the use of a mobile-based web survey, Rao and Vazquez [10] proposed a machine learning algorithm to accelerate the prediction and identification of COVID-19 infected cases. Besides, Maghdid et al. [11] have applied machine learning algorithms for identifying and predicting the preliminary stage of some COVID-19 symptoms based on smartphone sensors (microphone, camera, temperature, and inertial). Further, Metsky et al. [12] developed machine learning algorithms to design nucleic acid detection assays. This approach has detected 67 viral species and subspecies through which the COVID-19 virus is one of them. In addition, Qi et al. [13] designed a machine learning-based CT radiomics model for predicting the period of residence in the hospital for those who are suffering from pneumonia associated with COVID-19 based on a dataset of 52 patients with laboratory-confirmed cases. The designed model is based on two algorithms, including Random Forest and Linear Regression. Moreover, Yu et al. [14] developed a supervised decision-tree classifier based on several features, including CT images, Anal-swab or Urine specimens, measurements of Throat-swab, and chest radiography. The study aimed to predict the COVID-19 pediatric cases using a dataset of 105 infected children.

Chest computed tomography (CT) scans are considered one of the essential techniques to evaluate the severity of COVID-19 [15]. Therefore, several machine learning-based studies were conducted by applying deep learning techniques on CT scan images to predict the infection of COVID-19. For example, Tang et al. [15] applied the Random Forest model on chest CT images of 176 patients by classifying them into severe/non-severe COVID-19 cases. Besides, Zheng et al. [16] applied a supervised deep learning model on a dataset of 540 patients by analyzing their CT scans to predict the COVID-19 infections. Further, Gozes et al. [17] applied a deep learning approach on a dataset of 157 patients by adopting their CT scans and classifying them into COVID-19 infected cases and non-infected cases. Additionally, Li et al. [18] applied a deep learning approach on the CT scans of 3322 patients for identifying the infected COVID-19 cases. Moreover, Xu et al. [19] employed two CNN three-dimensional classification models using a deep learning approach for screening 618 CT images to measure the probability of COVID-19 infections. In addition, Narin et al. [20] applied three convolutional neural network models on 100 Chest X-ray images to detect the infected patients of COVID-19 pneumonia.

For predicting the development of COVID-19 infections to make better decision-making, several studies were also carried out. In that, Fong et al. [21] used GROOMS methodology in conjunction with the Composite Monte-Carlo simulator (CMC) to improve the deep learning network and fuzzy rule induction to predict the development of COVID-19 infections. In the same vein, Jia et al. [22] applied three mathematical models (i.e., Logistic, Bertalanffy, and Gompertz) to predict the evolution of COVID-19 infected cases. Besides, Qiang et al. [23] employed three encoding algorithms for screening the spike protein features to predict the infection risk and monitor the evolution of COVID-19. In addition, Poole [24] suggested the employment of machine learning for building models that use big data to predict and monitor the spread of COVID-19 and seasonal flu. It has been suggested that there is a correlation between climatological temperatures, latitude, and the spread of COVID-19. Further, Bai et al. [25] used two methods, namely deep learning and multivariate logistic regression to compare between the patients' data at the admission stage and hospital residence and predict the progression of the COVID-19 disease.

3   Survival Prediction of COVID-19 Patients

In addition to employing the machine learning algorithms in identifying and pre-dicting the infection of COVID-19 cases, these algorithms have also been used in predicting the survival rates of COVID-19 patients. For instance, Yan et al. [26] proposed a prognostic prediction model based on the XGBoost machine learning algorithm to determine the crucial predictive biomarkers of disease severity that could be used to predict the survival of COVID-19 infected patients using a dataset of 2799 patients. In the same vein, Yan et al. [27] adopted the XGBoost supervised classifier to predict the survival of severe COVID-19 patients using a larger dataset of 3000 patients. Besides, Yan et al. [28] employed the XGBoost machine learning-based prognostic model to predict the survival rates of COVID-19 cases using a dataset of blood samples of 404 infected patients.

4   Vaccine Development

Several data centers and research labs reported that they are employing AI tech-niques to look for a vaccine against COVID-19 [29]. In that, some studies reported the use of machine learning in vaccine development. For instance, Ong et al. [30] have reviewed the current status of coronavirus vaccine development and applied the "Vaxign RV" and "Vaxign-ML" techniques to predict the COVID-19 protein candidates for developing the vaccine. In addition, Prachar et al. [31] tested 19 epitope-HLA-binding prediction techniques and used them for the purpose of vaccine development for COVID-19. Despite the efforts made in employing machine learning algorithms for vaccine development, it is suggested that it is not very likely that the vaccine would be available soon [29].

5   Drug Discovery

Several studies were conducted with the potential of using AI techniques in screening the existing drugs and expediting the process of antiviral development to assist in treating the COVID-19. Magar et al. [32] developed a machine learning model to discover the antibodies that potentially inhibit the COVID-19. Besides, Patankar [33] trained the Long Short-Term Memory (LSTM) model to screen 310,000 drug-like compounds from the ZINC database to inhibit the RNA Dependent RNA Polymerase for COVID-19. Further, Tang et al. [34] developed an advanced deep Q-learning network with the fragment-based drug design (ADQN-FBDD) to produce effective drugs against COVID-19.

6 Critical Reflections

It is apparent that while several machine learning studies were carried out to combat the COVID-19, the use of machine learning is still limited in providing more efficient results. This is a surprising outcome as machine learning has had a more significant impact on Medicine, which has led to a paradigm shift in the healthcare sector [35]. Machine learning algorithms could be used to provide attribute prediction (e.g., infection prediction, survival prediction), which although possible, has not been perceived in the reviewed literature. Unlike the other domains, the emergence of COVID-19 and its impact on humanity requires that the patients' data be available to the public. This would allow researchers to analyze these data and generate efficient results that serve the healthcare sector. To this end, a large number of COVID-19 patients' attributes need to be available in order to determine the interrelationships among these features and how this affects the infection or survival rate of patients.

It is argued that AI still did not prove its efficiency concerning COVID-19 [29]. This argument is also supported by the current investigation with regard to machine learning. This stems from several reasons. First, machine learning requires a massive amount of data to train and test the prediction models. Unlike other diseases, there is still inadequate data that can be used to predict and track its outbreak. Second, it has been observed from the existing literature that most of the machine learning studies tend to determine and predict COVID-19 infections with small datasets. The use of small datasets might lead, in some scenarios, into possible biased or unreliable results. Therefore, generalizing the results should be treated with caution. Third, most of the published studies on using machine learning in the COVID-19 pandemic have not been peer-reviewed and tend to use the Chinese datasets [29]. This, in turn, raises some concerns, such as accuracy and reliability.

AI techniques were praised for their potential in contributing to the discovery of new drugs [29]. However, there was a minimal number of studies that contributed to the discovery of drugs and the development of vaccines concerning the use of machine learning in COVID-19. Even the conducted studies did not evaluate the efficiency of the drugs or vaccines on the patients' clinical features. It was hoped that machine learning would provide more interesting patterns. In that, machine learning could be used to determine the relationship between the developed drugs and their impact on patients' treatment using their clinical features. Machine learning could also be used to test the efficiency of the proposed vaccines and their effect on future infections. These issues open up new opportunities for future studies and encourage scholars to uncover new areas of research.

7 Conclusion and Future Insights

The widespread use of machine learning in relevance to COVID-19 is evident through the reviewed studies. However, the effectiveness of machine learning was not perceived concerning the prediction of COVID-19 infections and patients' survival rates. Most of the conducted studies tended to use simple classification algorithms with minimal attempts to take advantage of more recent algorithms. These outcomes open the door for future research and encourage scholars to work on more advanced techniques, such as developing new algorithms or improving the existing ones. In addition, the limited access to patients' data hinders the use of machine learning in an appropriate manner. Hence, the availability of patients' clinical features is another challenge to the application of machine learning algorithms. In order to fight the pandemic, it is essential that authorities take special care of handling the COVID-19 data and communicating them to the public [29].

In summary, AI techniques in general and machine learning, in particular, have the potential to combat the COVID-19 and relevant crisis. However, it can be seen from the rapid review of the existing studies that the use of machine learning in fighting the pandemic is still in its early stages. This outcome is also supported by the conclusions drawn in previous studies [29, 36], which reported that AI techniques are still at their preliminary stages in working against the COVID-19. While the use of machine learning is still limited, the current pandemic may expedite its application for more advanced issues related to the prediction of COVID-19 infections, survival rates of patients, and vaccine and drug developments.

References

1. Zhu, N., et al.: A novel coronavirus from patients with pneumonia in China, 2019. N. Engl. J. Med. (2020). https://doi.org/10.1056/NEJMoa2001017
2. Belfiore, M.P., et al.: Artificial intelligence to codify lung CT in Covid-19 patients. Radiol. Medica (2020). https://doi.org/10.1007/s11547-020-01195-x
3. Vaishya, R., Javaid, M., Khan, I.H., Haleem, A.: Artificial intelligence (AI) applications for COVID-19 pandemic. Diabetes Metab. Syndr. Clin. Res. Rev. (2020). https://doi.org/10.1016/j.dsx.2020.04.012
4. Albahri, A.S., et al.: Role of biological data mining and machine learning techniques in detecting and diagnosing the novel coronavirus (COVID-19): a systematic review. J. Med. Syst. 44(7). (2020). https://doi.org/10.1007/s10916-020-01582-x
5. Yu, K.H., Beam, A.L., Kohane, I.S.: Artificial intelligence in healthcare. Nat. Biomed. Eng. 2(10), 719–731 (2018). https://doi.org/10.1038/s41551-018-0305-z
6. Long, J.B., Ehrenfeld, J.M.: The role of augmented intelligence (AI) in detecting and preventing the spread of novel coronavirus. J. Med. Syst. (2020). https://doi.org/10.1007/s10916-020-1536-6
7. Neri, E., Miele, V., Coppola, F., Grassi, R.: Use of CT and artificial intelligence in suspected or COVID-19 positive patients: statement of the Italian Society of Medical and Interventional Radiology. Radiol. Medica (2020). https://doi.org/10.1007/s11547-020-01197-9

8. Arpaci, I., et al.: Analysis of twitter data using evolutionary clustering during the COVID-19 pandemic. Comput. Mater. Contin. **65**(1), 193–203 (2020). https://doi.org/10.32604/cmc.2020.011489
9. Ahuja, A.S., Reddy, V.P., Marques, O.: Artificial intelligence and COVID-19: a multidisciplinary approach. Integr. Med. Res. (2020)
10. Rao, A.S.R.S., Vazquez, J.A.: Identification of COVID-19 can be quicker through artificial intelligence framework using a mobile phone-based survey in the populations when cities/towns are under quarantine. Infect. Control Hosp. Epidemiol. (2020). https://doi.org/10.1017/ice.2020.61
11. Maghdid, H.S., Ghafoor, K.Z., Sadiq, A.S., Curran, K., Rabie, K.: A novel AI-enabled framework to diagnose coronavirus COVID 19 using smartphone embedded sensors: design study (2020)
12. Metsky, H.C., Freije, C.A., Kosoko-Thoroddsen, T.-S.F., Sabeti, P.C., Myhrvold, C.: CRISPR-based COVID-19 surveillance using a genomically-comprehensive machine learning approach. bioRxiv (2020). https://doi.org/10.1101/2020.02.26.967026
13. Qi, X., et al.: Machine learning-based CT radiomics model for predicting hospital stay in patients with pneumonia associated with SARS-CoV-2 infection: a multicenter study. medRxiv (2020)
14. Yu, H., et al.: Data-driven discovery of clinical routes for severity detection of COVID-19 pediatric cases. medRxiv (2020)
15. Tang, Z., et al.: Severity assessment of coronavirus disease 2019 (COVID-19) using quantitative features from chest CT images (2020)
16. Zheng, C., et al.: Deep learning-based detection for COVID-19 from chest CT using weak label. medRxiv (2020)
17. Gozes, O., et al.: Rapid AI development cycle for the coronavirus (COVID-19) pandemic: initial results for automated detection & patient monitoring using deep learning CT image analysis (2020)
18. Li, L., et al.: Artificial intelligence distinguishes COVID-19 from community acquired pneumonia on chest CT. Radiology, 200905 (2020). https://doi.org/10.1148/radiol.2020200905
19. Xu, X., et al.: Deep learning system to screen coronavirus disease 2019 pneumonia (2020)
20. Narin, A., Kaya, C., Pamuk, Z.: Automatic detection of coronavirus disease (COVID-19) using X-ray images and deep convolutional neural networks (2020)
21. Fong, S.J., Li, G., Dey, N., Crespo, R.G., Herrera-Viedma, E.: Composite monte carlo decision making under high uncertainty of novel coronavirus epidemic using hybridized deep learning and fuzzy rule induction. Appl. Soft Comput. **93**, 1–14 (2020)
22. Jia, L., Li, K., Jiang, Y., Guo, X., Zhao, T.: Prediction and analysis of coronavirus disease 2019 (2020)
23. Qiang, X.-L., Xu, P., Fang, G., Liu, W.-B., Kou, Z.: Using the spike protein feature to predict infection risk and monitor the evolutionary dynamic of coronavirus. Infect. Dis. Poverty **9**(1), 1–8 (2020). https://doi.org/10.1186/s40249-020-00649-8
24. Poole, L.: Seasonal influences on the spread of SARS-CoV-2 (COVID19), causality, and forecastabililty (3-15-2020). SSRN Electron. J. (2020). https://doi.org/10.2139/ssrn.3554746
25. Bai, X., et al.: Predicting COVID-19 malignant progression with AI techniques. medRxiv (2020)
26. Yan, L., et al.: Prediction of criticality in patients with severe Covid-19 infection using three clinical features: a machine learning-based prognostic model with clinical data in Wuhan. medrxiv.org (2020)
27. Yan, L., et al.: Prediction of survival for severe Covid-19 patients with three clinical features: development of a machine learning-based prognostic model with clinical data in Wuhan. medRxiv (2020). https://doi.org/10.1101/2020.02.27.20028027
28. Yan, L., et al.: A machine learning-based model for survival prediction in patients with severe COVID-19 infection (2020). https://doi.org/10.1101/2020.02.27.20028027

29. Naudé, W.: Artificial intelligence versus COVID-19: limitations, constraints and pitfalls. Ai Soc. 1–5 (2020)
30. Ong, E., Wong, M.U., Huffman, A., He, Y.: COVID-19 coronavirus vaccine design using reverse vaccinology and machine learning. bioRxiv (2020)
31. Prachar, M., et al.: COVID-19 vaccine candidates: prediction and validation of 174 SARS-CoV-2 epitopes. bioRxiv (2020). https://doi.org/10.1101/2020.03.20.000794
32. Magar, R., Yadav, P., Farimani, A.B.: Potential neutralizing antibodies discovered for novel corona virus using machine learning (2020)
33. Patankar, S.: Deep learning-based computational drug discovery to inhibit the RNA dependent RNA polymerase: application to SARS-CoV and COVID-19 (2020). doi: https://doi.org/10.31219/osf.io/6kpbg
34. Tang, B., He, F., Liu, D., Fang, M., Wu, Z., Xu, D.: AI-aided design of novel targeted covalent inhibitors against SARS-CoV-2. bioRxiv (2020)
35. Rajkomar, A., Dean, J., Kohane, I.: Machine learning in medicine. N. Engl. J. Med. (2019). https://doi.org/10.1056/NEJMra1814259
36. Bullock, J., Pham, K.H., Lam, C.S.N., Luengo-Oroz, M.: Mapping the landscape of artificial intelligence applications against COVID-19 (2020). arXiv preprint: arXiv:2003.11336

Deep Learning Techniques and COVID-19 Drug Discovery: Fundamentals, State-of-the-Art and Future Directions

Mohammad Behdad Jamshidi, Ali Lalbakhsh, Jakub Talla, Zdeněk Peroutka, Sobhan Roshani, Vaclav Matousek, Saeed Roshani, Mirhamed Mirmozafari, Zahra Malek, Luigi La Spada, Asal Sabet, Mojgan Dehghani, Morteza Jamshidi, Mohammad Mahdi Honari, Farimah Hadjilooei, Alireza Jamshidi, Pedram Lalbakhsh, Hamed Hashemi-Dezaki, Sahar Ahmadi, and Saeedeh Lotfi

Abstract The world is in a frustrating situation, which is exacerbating due to the time-consuming process of the COVID-19 vaccine design and production. This chapter provides a comprehensive investigation of fundamentals, state-of-the-art and some perspectives to speed up the process of the design, optimization and

M. B. Jamshidi (✉) · J. Talla · Z. Peroutka · H. Hashemi-Dezaki
Research and Innovation Centre for Electrical Engineering (RICE), University of West Bohemia, 301 00 Pilsen, Czech Republic
e-mail: Jamshidi@fel.zcu.cz

M. B. Jamshidi · J. Talla · Z. Peroutka
Department of Power Electronics And Machines (KEV), University of West Bohemia, 301 00 Pilsen, Czech Republic

A. Lalbakhsh
School of Engineering, Macquarie University, Sydney, NSW 2109, Australia

S. Roshani · S. Roshani
Department of Electrical Engineering, Islamic Azad University, Kermanshah Branch, 1477893855 Kermanshah, Iran

V. Matousek
Department of Computer Science and Engineering (KIV), University of West Bohemia, 301 00 Pilsen, Czech Republic

M. Mirmozafari
Department of Electrical and Computer Engineering, University of Wisconsin–Madison, Madison, WI 53706, USA

Z. Malek
Medical Sciences Research Center, Faculty of Medicine, Tehran Medical Sciences Branch, IAU, 1477893855 Tehran, Iran

L. La Spada
School of Engineering and the Built Environment, Edinburgh Napier University, Edinburgh EH11 4DY, UK

© The Author(s), under exclusive license to Springer Nature Switzerland AG 2021
I. Arpaci et al. (eds.), *Emerging Technologies During the Era of COVID-19 Pandemic*, Studies in Systems, Decision and Control 348,
https://doi.org/10.1007/978-3-030-67716-9_2

production of the medicine for COVID-19 based on Deep Learning (DL) methods. The proposed platforms are able to be used as predictors to forecast antigens during the infection disregarding their abundance and immunogenicity with no requirement of growing the pathogen in vitro. First, we briefly survey the latest achievements and fundamentals of some DL methodologies, including Deep Boltzmann Machines (DBM), Restricted Boltzmann Machine (RBM), Deep Belief Network (DBN), Hopfield network and Long Short-Term Memory(LSTM). These techniques help us to reach an integrated approach for drug development by non-conventional antigens. We then propose several DL-based platforms to utilize for future applications regarding the latest publications and medical reports. Considering the evolving date on COVID-19 and its ever-changing nature, we believe this survey can give readers some useful ideas and directions to understand the application of Artificial Intelligence (AI) to accelerate the vaccine design not only for COVID-19 but also for many different diseases or viruses.

Keywords Artificial intelligence · Artificial neural network · Bioinformatics · COVID-19 · Deep learning · Drug discovery

A. Sabet
Department of Pharmacy, Irma Lerma Rangel College of Pharmacy, Texas A&M University, Kingsville, TX 78363, USA

M. Dehghani
Department of Physics and Astronomy, Louisiana State University, Baton Rouge, LA 70803, USA

M. Jamshidi
Young Researchers and Elite Club, Islamic Azad University, Kermanshah Branch, Kermanshah 1477893855, Iran

M. M. Honari
Electrical and Mechanical Engineering Department, University of Alberta, Edmonton, AB T6G1H9, Canada

F. Hadjilooei
Department of Radiation Oncology, Cancer Institute, Tehran University of Medical Sciences, Tehran 1416753955, Iran

A. Jamshidi
Dentistry School, Babol University of Medical Sciences, Babol 4717647745, Iran

P. Lalbakhsh
Department of English Language and Literature, Razi University, Kermanshah 6714414971, Iran

S. Ahmadi
Department of Technologies and Measurement (KET), University of West Bohemia, 301 00 Pilsen, Czech Republic

S. Lotfi
Department of Theoretical Electrical Engineering (KTE), University of West Bohemia, 301 00 Pilsen, Czech Republic

1 Introduction

Although the second decade of the third millennium was expected to start with exciting signs of progress in medicine development and some digital technologies to resolve issues associated with major clinical problems and diseases [1]; COVID-19outbreak and its consequent pandemic and concerns regarding its future bounce back highlighted the need for speed in developing new treatments and designing intervention protocols [2].However, history has proved that the gravest health challenges can lead to the greatest opportunities for the development of novel treatments, medicines and vaccines. In this particular case where accurate but cheap diagnostic tests are extremely necessary, digital technology, Artificial Intelligence (AI) and Deep Learning (DL) can serve with their potentials in detecting and diagnosing COVID-19 [2, 3]. A novel feature learning method consists of various hidden layers of representation [4]. Recently, a number of researches have proved that Machine Learning (ML) can be considered an appropriate technique to solve complex problems or analyze the big data [5, 6]. Hence, we aim to review some potential methods to direct future research in the case of technology and drug discovery.

In this chapter, we try to introduce reliable techniques regarding the most recent achievements in the area of AI, DL with a concentration on COVID-19 and its different problems. Although we presented some DL methods for diagnosis and treatment to combat COVID-19 in our previous research [3], in this research, we try to develop those platforms that are useful to accelerate the process of drug discovery. Generally, the ideas mentioned in the last research has been developed and extended to drug discovery goals. In Sect. 2, some fundamentals about bioinformatics and drug discovery have been demonstrated to review the ability and potential of these modern approaches to developing drugs and vaccines. In Sect. 3, we explain DL-based drug discovery and state-of-the-art within it. In Sect. 4, the proposed drug discovery strategy is proposed. The proposed algorithm in this section is useful for finding and developing drugs for COVID-19-related diseases. In Sect. 5, some platforms based on the most recent publications for COVID-19 are considered to demonstrate the effectiveness of the algorisms, for instance, prioritizing the vaccine design approaches, prediction of antibody in infants through their mother and prediction of change in patients receiving the drug. These structures are important to give many ideas to apply these methods to combat COVID-19. Section 6 and Sect. 7 present discussion and conclusion, respectively.

2 Bioinformatics and Drug Discovery

On the one hand, bioinformatics can be considered a powerful tool to accelerate drug discovery. Considering the current situation, using bioinformatics method is essential to conquering the problem. On the other hand, tools such as binding simulations, modeling of proteins and computational chemistry, are usually used within the related fields, they should be used cautiously because the conformational space is complex and entropic contributions from the surrounding solvent are strong [7, 8]. Besides, emphasizing the conventional methods to find the drugs are not effective. Because traditional approaches lead to squandering the time and money for Pharmaceutical companies, based on an investigation [9], conventional drug discovery is an expensive process, and it can take approximately 12–16 years. Figure 1 illustrates the long process of drug discovery in a conventional form, including six stages taking several years to reach the approved drugs [9]. Needless to say, to combat COVID-19 or similar viruses which may spread in the future, we need to use a more effective and faster process to decrease the problems associated with such viruses. Fortunately, from the 1960s onward medicinal chemistry has successfully applied different forms of AI to the design compounds and wherever training models have been done through the use of labelled training supervised learning has been applied extensively [10].In contrast, the present models such as the ones in the industry of drug discovery software are capable of forecasting simple physicochemical features and can use simple mechanisms to predict pharmacokinetic properties of new compounds with relatively high degrees of precision [11]. They are not optimal when it comes to complex biological properties such as drug efficacy and the related side effects [12]. The gravest current challenge of biomedicine is developing new drugs that could be used in fighting against diseases. To overcome this challenge bioinformatics and cheminformatics have started to utilize computational methods in the last three decades to come to a deeper understanding of the molecular mechanisms and find a vantage point to maximize available options for disease treatment [13]. Recognition and characterization of applicable T and B-cell epitopes, which generates the epitopic vaccine to combat SARS-COV-2

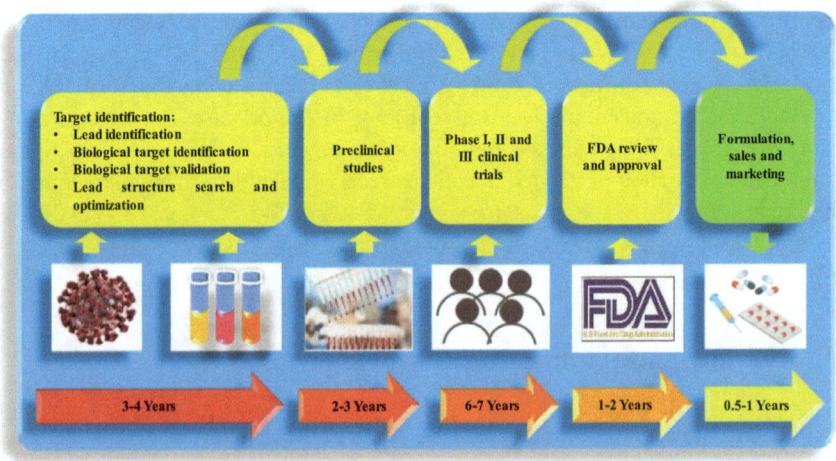

Fig. 1 The process of design and development of drug and vaccine with a conventional approach

could have been possible through the use of immunoinformatics [14]. Advanced computational abilities create the opportunity for conceptual progress of this sort to be accessed while analysis of large data sets of ligands is done at the same time. Therefore, dynamic virtual screening can be utilized as a systematic tool to find new potential drugs [15]. Taking advantage of available large data sets resulting from high throughput experiments with gene expression profiles, AI can play an important role in repurposing drugs [16].

3 DL-Based Drug Discovery: State-of-the-Art

In this section, we take a glance at some most recent accomplishments in Deep Neural Networks (DNN), which can be used in different stages of drug discovery. While conventional approaches in developing vaccines are cumbersome and time-consuming, they are not desirably efficient because there is not only the risk for failure when pathogens are impossible to be cultivated under laboratory conditions, but there are also arrays of antigens with no guarantee to provide immunity against the targeted disease. In contrast, AI-based approaches to vaccine development do not suffer from these weaknesses and provide researchers with the opportunity to identify novel antigen vaccine candidates [17]. Moreover, ML technologies can be introduced to supply chains to establish and consolidate more intelligent supervision [18]. Using various study designs and statistical methods to complete clinical trials in a scientifically effective manner is of great importance when it comes to bringing COVID-19 under control [19]. A useful tool for designing safer medicines with more efficacy seems to be coupling MD simulations with AI approaches as an integral part of a general pharmacological model of drug action [15]. ML methods, including ANN, have a long history of application to predict compound activities and other similar complex variables in both medicine and engineering [20, 21]. There is a collection of layers in ANNs, and they function in a way that one output layer is the next layer input until the list of the outputs is exhausted, and an output layer that could predict a property is reached. A matrix that reflects the weights of connections between the layers represents pairs of connected layers to make ANNs amenable to matrix operations. Recently, ANNs have played a significant role in applications to discover potential drugs, especially the deep versions of these networks [22]. Figure 2 demonstrates a conventional DL method consisting of an input layer, hidden layers, and an output layer. This structure can be used as a simple technique for drug discovery goals as a reliable predictor [23]. DL methods are most importantly employed to address problems in the process of activity prediction [21].

The concepts of D Land ANNs were first introduced in the 1980s [24]. Having a lot of hidden layers, Deep Neural Networks (DNNs), as a major interest of information technology companies, was developed as a solution to overcome challenges such as those in speech recognition [25]. ANN were both designed to mimic neuron excitation in the human brain through analogizing the activation of a binary logic

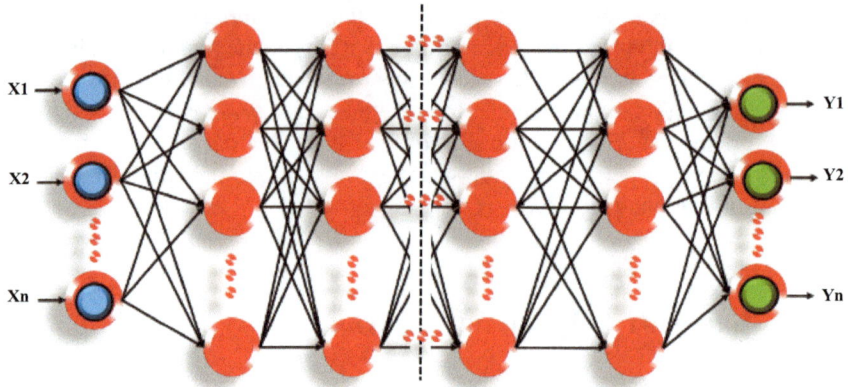

Input Layer   Hidden Layer 1 Hidden Layer 2 Hidden Layer N-1 Hidden Layer N Output Layer

Fig. 2 Deep conventional ANN used as a predictor for drug discovery, including several hidden layers that are helpful to analyze the complex data

gate in the networks [26]. An instance of this case that is used extensively for the prediction of properties is the quantitative structure-activity relationship (QSAR) methodology. In this scenario, for example, log P, solubility and bioactivity for given chemical structures are predicted [27]. If the objective, in QSAR ranking applications, is the identification of compounds with higher activities, the desired ranking goal is to have compounds with greater activities in a higher rank than compounds which have lower activities [27]. Entering an era in which clinical trial failures are minimized to the lowest rate and the process of drug development could be completed in a faster, cheaper and more effective way can be realized if proper methods of AI are employed [28]. As such, extracting knowledge from six categories of data known as proteomics, microarrays, genomics, biological systems, data mining, and text mining is possible through the application of ML techniques [29]. Additionally, advanced feature learning enables DL to achieve high accuracy in identification when the training set contains a huge bulk of data [26].Due to fast happening and increasing time complexity which is a result of the network architecture complication, more advanced programming skills and hardware technology are needed if DL methods are to be feasible and effective [26]. The output of such AI-based methods comprises designing drug in de novo and selecting the best structure based on experimental tests, both of which are achievable through modelling and quantum chemistry [30]. The idea of employing AI in the process of drug development is now a part of reality and not a dream anymore [31]. AI-related computational algorithms have achieved such levels of advancement that computer-based inference engines are now capable of reaching unprecedently deep conclusions [32]. In the process of fighting COVID-19 as a devastating and life-threatening disease, it is important to understand how COVID-19 recognizes the host cell. Moreover, it is impossible to effectively monitor and predict the manner in which the infection spreads during an epidemic if the required epidemiological data are not available [33]. However, there are other data sources that focus mainly on aggregated case counts in every geographic location [34]. A Deep Boltzmann Machines (DBM) is a structured model that works based on or is adapted to the theory of probability and consists of several layers of variables that are random and mostly latent [35, 36]. DBM shave been utilized as effective methods for drug discovery. Figure 3 shows a DBM which has been used in several pieces of research for drug discovery [37]. DBMs are a type of generative models which are able to be used for feature learning techniques. These classifiers are good options to be used as classifiers for COVID-19 vaccine methods. This model is also a good tool for extracting a unified representation that fuses modalities together. It is also very useful in cases such as hyperspectral imagery where human expertise is limited. DBM is also capable of approximating the features of prior knowledge bases samples without having any information available from the labels [38], which makes it a preferable candidate over others in assisting scientists in finding the best way to develop vaccines [39]. Figure 4 demonstrates a Deep Belief Network (DBN), which are used in many complex applications. DBM is an extremely efficient hierarchical generative model for extracting features that are capable of describing highly variant functions and discovering the manifold of the features [40]. The DBN, which was proposed in 2006 was trained to maximize the likelihood of its training data [41]. The DBN is a powerful multilayer generative model in which layers encode statistical dependencies among the units in the layer below them. The advantage of DBN over training methods of traditional deep models such as multilayer perceptron is that DBN can rely on a special unsupervised pre-training procedure to prevent over-fitting to the training set via [4, 40].

Furthermore, Fig. 5 depicts a Hopfield neural network (Hopfield net) structure which is used for predictions problems and drug discovery. Hopfield net is a type of Recurrent ANNs, and it is an appropriate technique for the content type in the

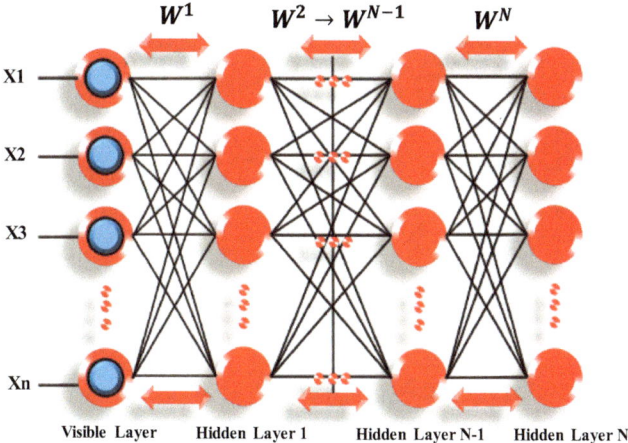

Fig. 3 The structure of Deep Boltzmann Machine

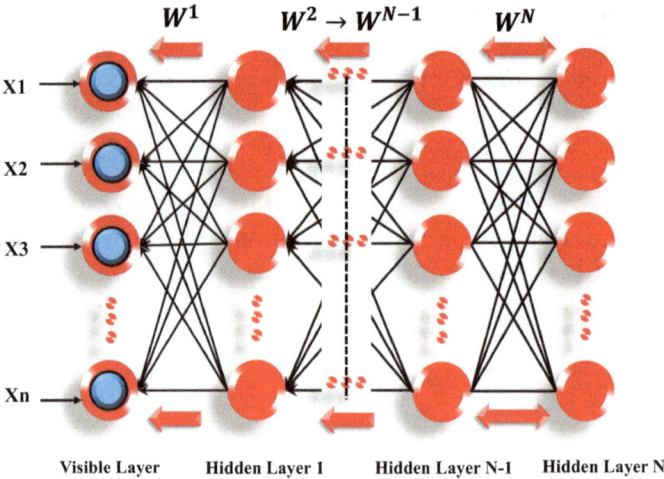

Fig. 4 The structure of Deep Belief Network

address memories and also it an effective method to solve optimization problems [42]. Although it is an old technique, it has been used in many different types of bioinformatics and biological problems [43, 44]. A type of generative stochastic ANNs called Restricted Boltzmann Machine (RBM) is able to be trained through a probability distribution by inputs. That is why it is an appropriate option to be applied in drug discovery and bioinformatics problems [13, 45]. The structure of an RBM is illustrated in Fig. 6. Sangari and Sethares [46] was the first to develop RBM, a special traditional DBM, that can be formed by hierarchically connecting the neurons of DBM. RBMs have been effectively utilized to model distributions over binary-valued data without any connections between hidden layers [36].

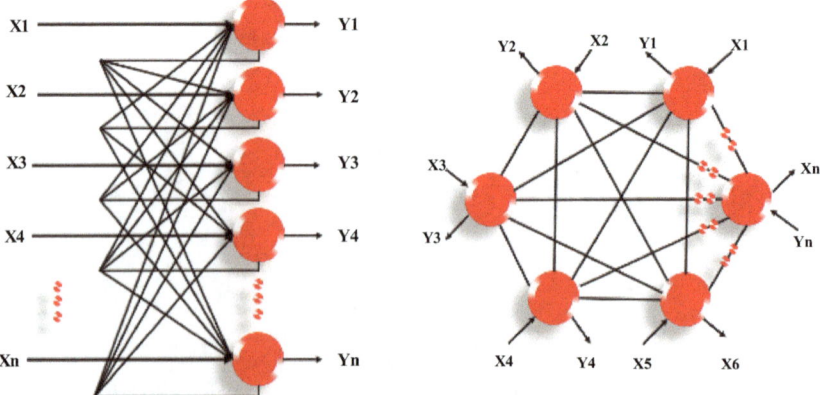

Fig. 5 Hopfield neural network structure

Fig. 6 Restricted Boltzmann
Machine (RBM)

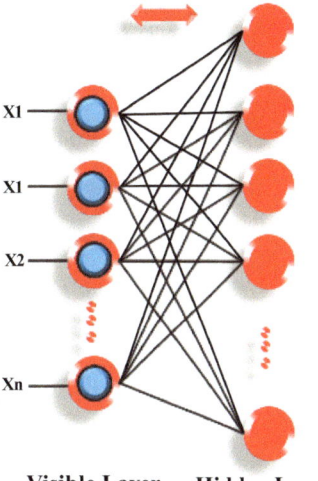

Visible Layer   **Hidden Layer**

The model serves well when a unified representation that fuses modalities together is to be extracted. The key point is learning a joint density model over the space of multimodal inputs. Table 1 depicts useful information about some powerful techniques based on AI, which can be considered to develop the AI-based platforms for drug discovery. In addition, some most recent publications about COVID-19 and technology have been presented in this table.

Furthermore, Table 2 gives information about the strong publications about AI and drug discovery. Although numerous methods have recently been presented to speed up the drug discovery process, we here try to review those approaches that play a crucial role in drug discovery powered by AI, particularly in the terms applicability, reliability and generalizability. Because not only do such methodologies have to be practical enough to save time, but they also must be flexible to merge with other conventional methods.

Training each RBM for modeling samples from the previous RBM's posterior distribution increases a variational lower bound on the likelihood of the DBM, which can be used as an acceptable approach for initializing the joint model [48]. In 1986 Paul Smolensky invented RBMS prototype, which was called Harmonium [59]. However, its prominence was not achieved until Geoffrey Hinton, and his collaborators equipped them with fast learning algorithms they invented in the mid-2000. Long Short-Term Memory (LSTM) is a type of RNN utilized in the area of DL [60]. This network has feedback connections between its layers, unlike another type of standard feedforward neural networks. LSTM is an efficient ANN to solve many different kinds of problems in drug discovery [52, 53, 61, 62].The structure of an LSTM is shown in Fig. 7.

Table 1 The most important AI-based techniques that are infrastructure for drug discovery and some research about COVID-19 and technology

Ref., Author(s) and Year	Publication	Methodology	The purpose of the study	Outcomes
[1] Ting et al. (2020)	Nature Medicine	AI-based methods	Survey applications for COVID-19	Describing challenges of AI methods to combat COVID-19
[47] Arpaci et al. (2020)	CMC-Computers Materials & Continua	Evolutionary clustering	Analyzing big data from Twitter	Some useful recommendation to combat COVID-19
[36] Salakhutdinov and Hinton (2009)	Artificial Intelligence and Statistics	DBM	Introducing DBM	Clarification the effectiveness of DBM
[37] Taherkhani et al. (2018)	Neurocomputing	DBM	Forecasting	Description of DBM for prediction
[39] Hess et al. (2017)	Bioinformatics	Partitioned learning in DBM	Application for SNP data	Description of DBM for SNP data
[40] Zhang and Wu (2012)	IEEE Transactions on Audio, Speech, and Language Processing	DBN	Comprehensive research to use DBNs	Showing abilities of DBN to solve complex problems
[41] Hinton et al. (2006)	Neural Computation	DBN	Speeding up calculation with DBN	Presenting a fast learning algorithm
[42] Maetschke and Ragan (2014)	Bioinformatics	Hopfield networks	Using Hopfield as attractors	Classification of cancer subtypes
[43] Conforte et al. (2020)	Frontiers in Genetics	Hopfield networks	Application of Hopfield for medicine	Modelling basins of attraction in breast cancer
[44] Al-Maitah (2020)	Neural Computing Applications	Hopfield networks	Using Hopfield for genetic diseases	Some image processing techniques powered by Hopfield
[3] Jamshidi et al. (2020)	IEEE Access	DL-based techniques	Diagnosis of COVID-19 by DL	Demonstrating several DL platforms
[48] Kim et al. (2020)	Journal of Neuroscience Methods	RBM	Using RBM techniques in medicine.	Analyzing MRI by RBMs
[49] Vogelstein et al. (2018)	Nature Methods	Neuro-based computing	Review on big data and calculation	Presenting an open-source platform

(continued)

Table 1 (continued)

Ref., Author(s) and Year	Publication	Methodology	The purpose of the study	Outcomes
[50] Beam and Kohane (2018)	Jama	Big data and ML	Review on applications of ML methods	Presenting some fruitful recommendations
[51] Arpaci et al. (2020)	Personality and Individual Differences	confirmatory factor analyses	Analysis of phobia associated with COVID-19	A practical instrument namely, corona phobia scale

Table 2 The approaches powered by AI, ML and DL to accelerate drug discovery; such methods have appropriate potential to find drugs for COVID-19 related disease

Ref., Author (s) and Year	Publication	Methodology	The purpose of the study	Outcomes
[2] Ton et al. (2020)	Molecular Informatics	Deep Docking (DD)	Finding potential ligands	Forecasting a large number of purchasable molecules quickly
[8] Li and Robson (2000)	Drugs and the Pharmaceutical Sciences	Bioinformatics	Survey of computational methods in molecular design	Description of several methods based on bioinformatics for drug discovery
[11] Zhu (2020)	Annual Review of Pharmacology and Toxicology	Big data and AI	Review on novel data mining for drug discovery	Demonstrating novel drug development and optimization
[13] Rifaioglu et al. (2019)	Briefings in Bioinformatics	DL in silico drug discovery	Review on new machine intelligence methods	illustrating the original DL-based method for drug discovery
[15] Díaz et al. (2019)	Rise of Machines in Medicine	Molecular Dynamics Simulations and ML	Accelerating drug discovery	Classification of ligands and identified functional receptor motifs successfully
[16] Zhavoronkov (2018)	Mol. Pharmaceutics	A novel DL method	Improving the productivity in drug discovery	
[17] Chen et al. (2019)	Nature Biotechnology	An integrated DL method	Prediction of antigen	Prediction of HLA class II antigen presentation
[21] Chen et al. (2018)	Drug Discovery Today	DL methodologies	Survey on the effectiveness of the methods	Future of DL in drug discovery
[22] Ghasemi et al. (2018)	Drug Discovery Today	DL in QSAR	A review of DL algorithms	Explaining some drawbacks of the DL methods

(continued)

Table 2 (continued)

Ref., Author (s) and Year	Publication	Methodology	The purpose of the study	Outcomes
[24] Gawehn et al. (2016)	Molecular Informatics	RBM and CNN in drug discovery	An overview of DL methods	Analyzing the understudied methods for drug discovery goals
[28] Fleming (2018)	Nature	AI and drug discovery	An analysis of the effectiveness	Reflecting the challenges associated with drug discovery
[29] Larranaga et al. (2006)	Briefings in Bioinformatics	ML in bioinformatics	Evaluation of The ML methods in biological knowledge	Introducing some practical methods
[31] Smalley (2017)	Nature Biotechnology	AI techniques in drug discovery	Overview of different aspects of AI in drug discovery	Some useful analysis
[45] Stephenson et al. (2019)	Current Drug Metabolism	ML techniques	Survey on drug discovery	Analysis of ML methods for drug discovery goals
[52] Cai et al. (2020)	Journal of Medicinal Chemistry	LSTM	Evaluation of ML in drug discovery	A comprehensive analysis of transfer Learning for Drug Discovery
[53] Baskin (2020)	The Expert Opinion on Drug Discovery	LSTM	Review on DL methods in drug discovery	Presenting some useful ideas about DL and drug discovery
[54] Yildirim et al. (2016(	Frontiers in Pharmacology	Novel approaches in drug discovery	A review of challenges associated with big data	Analyzing the potential methods to improve drug discovery
[55] Walls et al. (2020)	Cell	Infrastructures for the COVID-19 drug discovery	Survey of fundamentals for COVID-19 drug discovery	Fundamental methods for COVID-19 drug discovery
[56] Shoichet (2004)	Nature	Technology-based drug discover	An overview of challenges using technology for drug discovery	Presenting some recommendations
[57] Zhang et al. (2017)	Drug Discovery Today	DL and ML approaches	A review on progress from ML to DL	Analyzing the pros and cons of both approaches
[58] Wang et al. (2018)	Scientific Reports	DL techniques	Possibility of DL-based method for the design of protein	Showing acceptable results

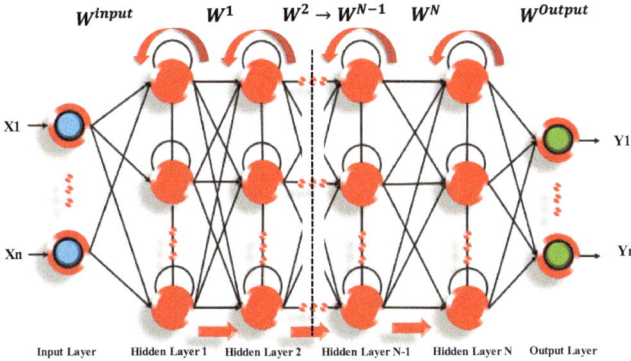

Fig. 7 Long Short-Term Memory (LSTM) structure

4   COVID-19 Drug Discovery Strategy: A Viewpoint

Discovering a natural or artificial molecule that could battle a protein target implicated in disease is hard and makes Drug discovery a difficult process to complete. Even if millions of molecules are screened the hit rates from screening may be very low. Despite a large number of drugs such as anti-influenza drugs which are present in clinical trials, none could be presented as a practical drug to cure COVID-19 complications. By combining the structural bioinformatics and molecular modeling docking in [63], researchers have predicted the COVID-19 spike binding site to the cell-surface receptor (Glucose Regulated Protein 78 (GRP78)) as the main driving force for host cell recognition. The SARS spike has been used to model spike protein of the COVID-19. The structural alignments and sequence demonstrated that moreover to its cyclic nature, there are four regions, in which similarities to sequence and physicochemical similarities to the cyclic Pep42 do exist. To test the four spike regions that fit tightly in the GRP78 Substrate Binding Domain β (SBDβ), Protein-protein docking was performed [63].

We employed DL approaches to present a comprehensive strategy for drug discovery of COVID-19. The proposed DL-based method for drug discovery of COVID-19 complications is demonstrated in Fig. 8. As observed in Fig. 8, a drug could be prepared through the introduced DL-based algorithm in 10 layers, commencing from the input data layer to the approved drug layer. The proposed strategy in this study seems to be prominently more efficient than previous ones at least in terms of drug designing for COVID-19; however, it is imperative that the presented methods are investigated along with thorough and close consultant with FDA guidelines. The first layer that contains primary data and pharmaceutical knowledge is the layer of input data, and the data collected from different sources as a Big Data Repository (BDR) is used for building this layer.

BDR's function is to support the publication and discovery of biomedical data [49]. BDR, as a set of data services, can be used to store and publish biomedical data

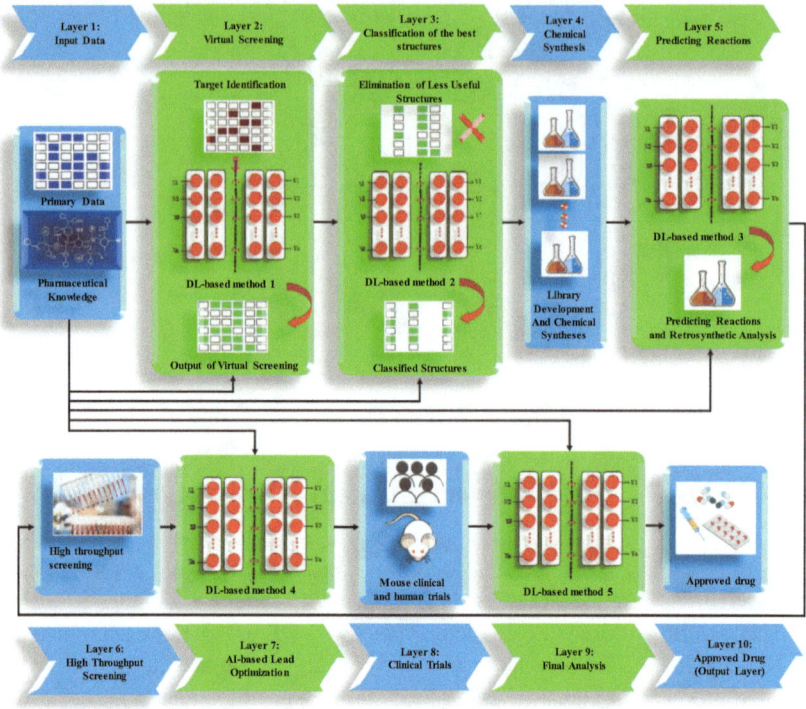

Fig. 8 Schematic diagram the proposed DL-powered strategy of COVID-19 drug discovery

from a host of domains such as neuroimaging, proteomics, and genomics. Using big data facilitates experimentation as well as new knowledge creation and transparency [54]. In health care contexts the sources of data are mainly hospitals and clinics, health insurance companies, pharmaceutical and medical device R&D and physicians with various specialties; however, patient's behavior and sentiments, and population and public health data could contribute as well. Even data coming from genomic sources and large scale phenotyping efforts may be available to utilize [64]. Virtual screening that includes a supervised DL method responsible for data analysis and identifying the best matching between target identification and the inputs according to virtual screening takes place in the second layer. Virtual screening is employed to search libraries of small molecules during the process of drug discovery to identify structures with a greater possibility of binding to a drug target such as a protein receptor or enzyme [56]. The third layer, classification of the best structure, is a layer in which a Hopfield network classifies the best outputs of the second layer. Recognition and classification of given grain samples are done by a Discrete Hopfield Network, which is a type of Auto-associative NN and is capable of learning and storing the data in the form of weights. Chemical syntheses and prediction of the reactions happen in the third and fourth layer. Being an expensive and time-consuming drug synthesis as to deal with omnipresent ethical disputes and limited outcomes [65]. In addition to that, a reaction predictor is a DL-based method to reaction prediction operating at the level of elementary reactions. Such design choice is desirable because they reflect the process and the ways that human experts contemplate and understand chemical reactions. Layer sixth and layer seventh are responsible for high throughput screening and lead optimization, respectively. While clinical trials and final analysis are done in order by layer eighth and layer ninth. Finally, the drug will be approved through the layer tenth.

5 Future Directions

The first example to discuss here is related to prioritizing vaccine strategies. Reference [55] presents some significant approaches and candidate antigens to provide effective vaccines against SARS-CoV-2. Since it follows previous studies that focus on prevention and control of seasonal influenza vaccines, the entire virus particle-based preparation of vaccines that consists of attenuated and inactivated virus vaccines is advisable [66]. As [55] demonstrated, we can determine the extent of the effectiveness of the vaccine against SARS-CoV-2 with high precision [67–72]. The DBM illustrated in Fig. 9 can be used to classify the best possible strategy with all of the limitations taken into account. In the process of classification or regression, the main job is predicting the outcome associated with a particular individual while a feature vector that describes the individual is provided. Individuals are grouped together in clustering according to the properties they share. In feature selection, however, the task is the selection of features that play important roles in the prediction of outcome for an individual [40].

Estimating antibody in infants by their mother is another example that we discuss here. Reference [63] illustrates antibodies in infants born to mothers with COVID-19. In the serum of all 6 infants, the antibodies were identified, two of whom had IgM and IgG concentrations more than the normal level (<10 AU/mL).

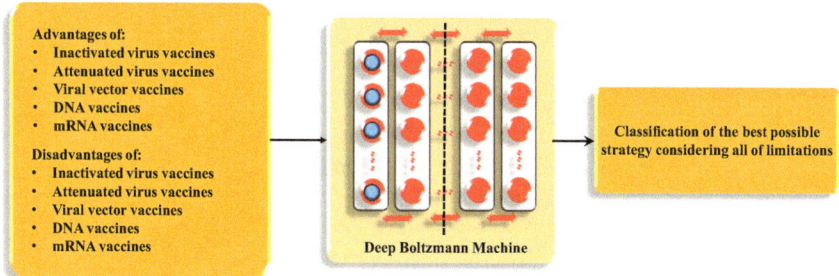

Fig. 9 The proposed developed strategy for classification of the best strategies to create COVID-19 powered by DBM network

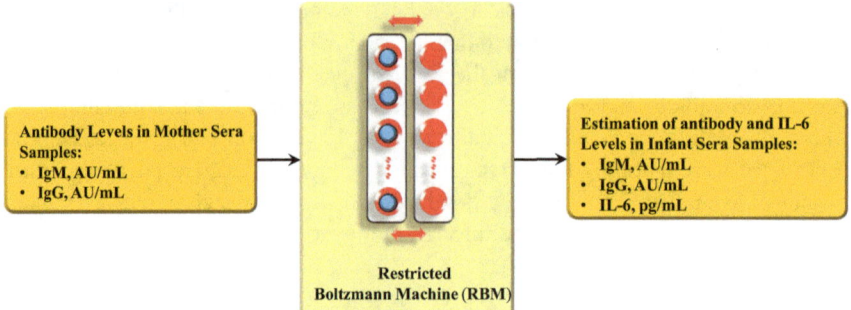

Fig. 10 An AI-equipped method to estimate antibodies and IL-6 levels in infant sera samples

Also, their neonatal blood sera samples demonstrated virus-specific antibodies. Five of the infants had elevated IgG concentrations [63]. To estimate antibody and IL-6 levels in infant sera samples based on [63], a Restricted Boltzmann Machine (RBM) is proposed. To learn the proposed network, as illustrated in Fig. 10, it is possible to use values of IgG and IgM antibodies in the mothers' sera samples.

RBM is a generative stochastic ANN that can learn a probability distribution over the network inputs. In [63], the selected samples, six mothers infected by COVID-19, are limited values for learning the introduced method; therefore, it is advisable that more samples are selected to implement this method. A developed strategy is presented here so that conventional networks could use the DL method. Predictions change after patients receive the drug. Improvement in the patients' clinical status in [73] followed an administration of convalescent plasma that contains neutralizing antibody. Since the limited sample size and study design prevent the expression of a definitive judgment on the extent of this treatment's effectiveness, evaluation in clinical trials are required. Hence for [73], an LSTM approach is recommended to predict changes in those patients who receive Convalescent Plasma Transfusion. LSTM is an Artificial Recurrent Neural Network (RNN) architecture that is adopted in the field of DL. Contrary to standard feed-forward neural networks, LSTM has feedback connections and can process single data points as well as the entire sequences of data. The proposed LSTM-powered technique is illustrated in Fig. 11. As is shown in this figure, the clinical characteristics comprise IL-6, pg/mL (normal range, 0–7), length of hospital stay, cycle threshold, PAO2/FIO2 ratio, SOFA score. Moreover, the inputs to achieve such results can be characteristics and antibody titer of Convalescent a plasma donor like donated plasma volume, the interval between symptom onset and discharge, interval between discharge and neutralizing antibody titer RBD-specific IgG ELISA titer, plasma donation and RBD-specific IgM ELISA titer.

A. An integrated Drug and Vaccine Strategy for Future Disease

Figure 12 illustrates the concept of an intelligent approach which can be used as a DL-based method to find the drugs and vaccine. This algorithm can be generalized for finding drugs or vaccines for any disease and virus in the future. In this way,

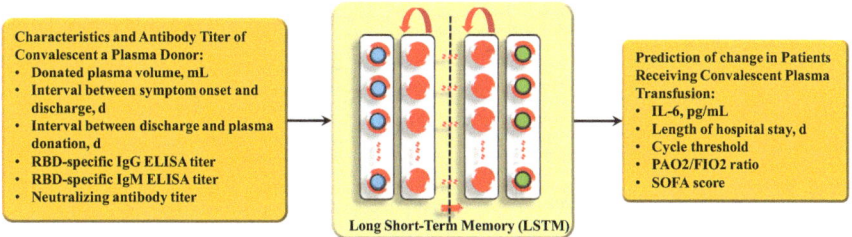

Fig. 11 The proposed LSTM-powered technique for the prediction of change in patients receiving the drug

when a disease is targeted, new compounds with desired and functional activities could be identified by active learning algorithms. Targeting COVID-19 and using novel structural features and network architecture alongside conventional approaches, we applied DL in computational drug and vaccine design in the present study. However, instead of limiting ourselves to a specific drug or method, we extended our study to develop analytical methods and demonstrate several DL networks for the discovery of drugs. Therefore, we aimed at providing an overview of DL approaches to discover drugs. The following section that illustrates DL drug development is followed by three sections present an explanation for the proposed vaccine development strategy, suggest a drug discovery strategy, and present the conclusion in order. DL has been demonstrably efficient and accurate in predicting drug properties and proposing drug candidates, which has been reliable in outlining possible toxicity risks [74]. The spike glycoprotein of SARS-COV-2 was a specific target because of its role in shaping the virus's characteristic protruding crown [75]. The vaccine component was eventually modelled in the SPARKS-X server [76]. A catalogue of all protein antigens, which can be expressed by pathogen at any time is immediately provided by the genome sequence. In this approach initiated by the genomic sequence, antigens that have the potentials to be the best vaccine candidates are predicted based on computer analysis. However, this approach is naive in the sense that it cannot determine if any of the potential antigen candidates ends in the provision of protective immunity without information regarding the abundance of the antigen or its immunogenicity in the period of infection or when expressed in vitro [77]. DL approaches are apparently more preferable to be used when raw high-dimensional data exist because when compared to conventional ANN, they describe the vanishing effects of gradients [41, 78]. Machine intelligence is now more efficient in drug discovery because of the result of data-driven and power-driven computational studies [57]. As recent research show, DL algorithm is now able to use retinal photographs to detect diabetic retinopathy, and they do this demonstrating equal or even greater sensitivity than ophthalmologists. The diagnosis procedure in this model which is void of human intervention, is learnt from raw pixels of the images and in the absence of ophthalmologists who could annotate pictures with their correct diagnosis [50]. DNNs' ability to learn features from simple inputs data, such as atom types and coordinates makes them more desirable

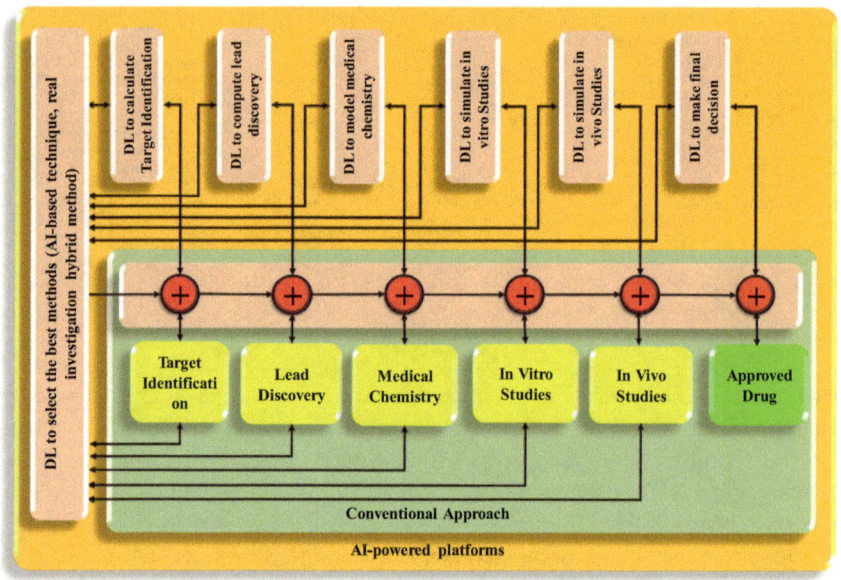

Fig. 12 Future drug discovery algorithm empowered with DL-based techniques

for application. Although technical details vary and network architecture and data representations change depending on the type of application, there is just one significant requirement of using deep ANN, and that is the existence and availability of large sets of data [58]. Applying a DNN on the Merck Kaggle challenge dataset through the use of a large number of 2D topological descriptors, Dahl et al.; witnessed that in comparison to the standard Random Forest (RF) method DNN is of slightly better performance in 13 of the total 15 targets [79]. A comprehensive multi-task DNN model that there is Tox21 challenge on a dataset comprising 12,000 compounds for 12 high-throughput toxicity assays was reported in [80]. An interesting attempt that originated a novel theory and attracted a lot of researches and pharmaceutical companies was Hinton et al. [41] research in 2006 in which they introduced the deep belief networks that facilitated the construction of nets that consist many hidden layers.

6 Discussion

Despite the full-paced attempts of companies whose aim is to develop a vaccine against coronavirus, the most optimistic news for the public considers a period of 1–1.5 years until such an aim is achieved. It is hard to come into terms with such a long interval when a look is taken at the numbers of ever-increasing patients and death tolls caused by COVID-19; however, new technologies developed in recent years, as well as the converging international efforts to find a way out of the pandemic, provide us with a firm ground of optimism regarding an ultimate solution to get away from COVID-19. Exacerbating the frustrating situation put the whole world in danger leading to recession and the risk of putting lives at risk. While attaining ideal solutions might not be a possibility in the short-run, we cannot afford losing sight of the present possibilities available to us in battling against COVID-19 [81]. All attempts to develop a vaccine for the disease are currently passing through primary clinical trials to ensure that they do not cause further risks for the receivers, while, in the meantime, more and more patients have to face the risk of hospitalization and even death. Therefore, the use of DL-based methods to find effective drugs and vaccines for COVID-19 could have a great impact on the fight against the disease spread. The present study aligns with this aim and suggests such models enhance approaches that could realize the objectives of the study. The comprehensive approach illustrated in Fig. 9 provides us with a powerful tool to improve conventional and recent intelligent techniques. The suggested algorithm in this study (Fig. 12) adds more value to such straightforward techniques in the sense that it decreases the amount of time needed in conventional and AI-based models to ensure safety and precision. One important feature of these techniques is that predicting vaccines or drugs' caused reaction in the body is quite possible prior to its application to animals and humans. Another significant feature is to analyze the results of trails by DBN. Various descriptors successfully used in this network to ensure the cost-effectiveness of ligand-based virtual screening in chemical data bases demonstrated that similarity searching was performed as expected. Vaccines cannot be made available to the public until they have obtained official approval for the competent authorities. Besides, there is always the probability of diseases disappearing by themselves or fresh medicine or drugs being released to the market for disease treatment. In either case, manufacturers could be at risk of losing their resources which in such case they have to suffer the loss of precious resources that could be employed and used elsewhere and for another disease. The suggested models in this study make it possible to review a large number of drug discovery methods and merge them with the effective DL-based techniques that demonstrate a well-matched AI approach to not only save resources of the pharmaceutical industry but also create a synergistic approach that could contribute to better quality and fine tune the distribution system performance.

7   Conclusion

Amidst the outbreak of COVID-19 and in the absence of a definite medical treatment and working vaccine to fight and possibly stop the fast-moving pandemic, we need to rely on the massive and rapid shift in behavior to bring the disease under control. Utilizing novel intelligent approaches based on Deep Learning (DL) and Artificial Intelligence (AI) can help us through speeding the process of design and optimization any potential drugs and vaccines. This chapter proposed a survey of fundamentals and state-of-the-art to use DL-based methods for drug discovery goals. The proposed methods included Deep Boltzmann Machines (DBM), Hopfield network, Restricted Boltzmann Machine (RBM), Long Short-Term Memory (LSTM), and Deep Belief Network (DBN). Particular emphasis was placed on proposing some DL-based platforms for applications of DL in different aspects of drug discovery and describing a comprehensive approach consisting of several AI-based layers to find the best results in the shortest possible time.

Acknowledgements This research has been supported by the Ministry of Education, Youth and Sports of the Czech Republic under the project OP VVV Electrical Engineering Technologies with High-Level of Embedded Intelligence CZ.02.1.01/0.0/0.0/18_069/0009855.

References

1. Ting, D.S.W., Carin, L., Dzau, V., Wong, T.Y.: Digital technology and COVID-19. Nat. Med. **26**(4), 459–461 (2020)
2. Ton, AT, Gentile, F., Hsing, M., Ban, F., Cherkasov, A.: Rapid identification of potential inhibitors of SARS-CoV-2 main protease by deep docking of 1.3 billion compounds. Mol. Inform. (2020)
3. Jamshidi, M.B., et al.: Artificial Intelligence and COVID-19: deep learning approaches for diagnosis and treatment. IEEE Acc. **8**, 109581–109595 (2020)
4. Hinton, G.E., Salakhutdinov, R.R.: Reducing the dimensionality of data with neural networks. Science **313**(5786), 504–507 (2006)
5. Arpaci, I., Al-Emran, M., Al-Sharafi, M.A., Shaalan, K.: A novel approach for predicting the adoption of smartwatches using machine learning algorithms. In: Recent Advances in Intelligent Systems and Smart Applications, pp. 185–195. Springer (2020)
6. Arpaci, I.: A hybrid modeling approach for predicting the educational use of mobile cloud computing services in higher education. Comput. Hum. Behav. **90**, 181–187 (2019)
7. Robson, B., McBurney, R.: The role of information, bioinformatics and genomics. In: Drug Discovery and Development, pp. 77–94. Elsevier (2013)
8. Li, J., Robson, B.J.D.: Bioinformatics and computational chemistry in molecular design: recent advances and their applications, vol. 101, pp. 285–307 (2000)
9. Arodola, O.A., Soliman, M.E.: Quantum mechanics implementation in drug-design workflows: does it really help? Drug Des. Dev. Ther. **11**, 2551 (2017)
10. Pohl, R., Gilman, R., Miller, G.A., Pachucki, K.: Muonic hydrogen and the proton radius puzzle. Annu. Rev. Nucl. Part. Sci. **63**, 175–204 (2013)
11. Zhu, H.: Big data and artificial intelligence modeling for drug discovery. Annu. Rev. Pharmacol. Toxicol. **60**, 573–589 (2020)
12. Zhu, H., et al.: t4 report: supporting read-across using biological data. Altex **33**(2), 167 (2016)
13. Rifaioglu, A.S., Atas, H., Martin, M.J., Cetin-Atalay, R., Atalay, V., Doğan, T.: Recent applications of deep learning and machine intelligence on in silico drug discovery: methods, tools and databases. Brief. Bioinform. **20**(5), 1878–1912 (2019)
14. Yin, D., et al.: A novel multi-epitope recombined protein for diagnosis of human brucellosis. BMC Infect. Dis. **16**(1), 219 (2016)
15. Díaz, Ó., Dalton, J.A., Giraldo, J.: Artificial intelligence: a novel approach for drug discovery. Trends Pharmacol. Sci. **40**(8), 550–551 (2019)
16. Zhavoronkov, A.: Artificial intelligence for drug discovery, biomarker development, and generation of novel chemistry. ACS Publications (2018)

17. Chen, B., et al.: Predicting HLA class II antigen presentation through integrated deep learning. Nat. Biotechnol. **37**(11), 1332–1343 (2019)
18. Kim, Y., et al.: A computational framework for deep learning-based epitope prediction by using structure and sequence information. In: 2019 IEEE International Conference on Bioinformatics and Biomedicine (BIBM), pp. 1208–1210. IEEE (2019)
19. Zhu, R.-F., Gao, R.-L., Robert, S.-H., Gao, J.-P., Yang, S.-G., Zhu, C.: Systematic review of the registered clinical trials of coronavirus diseases 2019 (COVID-19). medRxiv (2020)
20. Jamshidi, M.B., Jamshidi, M., Rostami, S.: An intelligent approach for nonlinear system identification of a li-ion battery. In: 2017 IEEE 2nd International Conference on Automatic Control and Intelligent Systems (I2CACIS), pp. 98–103. IEEE (2017)
21. Chen, H., Engkvist, O., Wang, Y., Olivecrona, M., Blaschke, T.: The rise of deep learning in drug discovery. Drug Discov. Today **23**(6), 1241–1250 (2018)
22. Ghasemi, F., Mehridehnavi, A., Perez-Garrido, A., Perez-Sanchez, H.: Neural network and deep-learning algorithms used in QSAR studies: merits and drawbacks. Drug Discov. Today **23**(10), 1784–1790 (2018)
23. Sunseri, J., King, J.E., Francoeur, P.G., Koes, D.R.: Convolutional neural network scoring and minimization in the D3R 2017 community challenge. J. Comput.-Aided Mol. Des. **33**(1), 19–34 (2019)
24. Gawehn, E., Hiss, J.A., Schneider, G.: Deep learning in drug discovery. Mol. Inform. **35**(1), 3–14 (2016)
25. Hinton, G., et al.: Deep neural networks for acoustic modeling in speech recognition: the shared views of four research groups. IEEE Sig. Process. Mag. **29**(6), 82–97 (2012)
26. Jing, Y., Bian, Y., Hu, Z., Wang, L., Xie, X.-Q.S.: Deep learning for drug design: an artificial intelligence paradigm for drug discovery in the big data era. AAPS J **20**(3), 58 (2018)
27. Sellwood, M.A., Ahmed, M., Segler, M.H., Brown, N.: Artificial intelligence in drug discovery. Future Science (2018)
28. Fleming, N.: How artificial intelligence is changing drug discovery. Nature **557**(7706), S55–S55 (2018)
29. Larranaga, P., et al.: Machine learning in bioinformatics. Brief. Bioinform. **7**(1), 86–112 (2006)
30. Cherkasov, A., et al.: QSAR modeling: where have you been? where are you going to? J. Med. Chem. **57**(12), 4977–5010 (2014)
31. Smalley, E.: AI-powered drug discovery captures pharma interest. Nat. Biotechnol. **35**, 604–605 (2017)
32. Ye, D.: Artificial intelligence and deep learning application in evaluating the descendants of Tubo Mgar Stong Btsan and social development. In: Data Processing Techniques and Applications for Cyber-Physical Systems (DPTA 2019), pp. 1869–1876. Springer (2020)
33. Xu, B., et al.: Epidemiological data from the COVID-19 outbreak, real-time case information. Sci. Data **7**(1), 1–6 (2020)
34. Dong, E., Du, H., Gardner, L.: An interactive web-based dashboard to track COVID-19 in real time. Lancet. Infect. Dis. **20**(5), 533–534 (2020)
35. Wang, S., He, M., Gao, Z., He, S., Ji, Q.: Emotion recognition from thermal infrared images using deep Boltzmann machine. Front. Comput. Sci. **8**(4), 609–618 (2014)
36. Salakhutdinov, R., Hinton, G.: Deep Boltzmann machines. In: Artificial Intelligence and Statistics, pp. 448–455 (2009)
37. Taherkhani, A., Cosma, G., McGinnity, T.M.: Deep-FS: a feature selection algorithm for deep Boltzmann machines. Neurocomputing **322**, 22–37 (2018)
38. Alberici, D., Barra, A., Contucci, P., Mingione, E.: Annealing and replica-symmetry in deep Boltzmann machines. J. Stat. Phys., 1–13 (2020)
39. Hess, M., Lenz, S., Blätte, T.J., Bullinger, L., Binder, H.: Partitioned learning of deep Boltzmann machines for SNP data. Bioinformatics **33**(20), 3173–3180 (2017)
40. Zhang, X.-L., Wu, J.: Deep belief networks based voice activity detection. IEEE Trans. Audio Speech Lang. Process. **21**(4), 697–710 (2012)

41. Hinton, G.E., Osindero, S., Teh, Y.-W.: A fast learning algorithm for deep belief nets. Neural Comput. **18**(7), 1527–1554 (2006)
42. Maetschke, S.R., Ragan, M.A.J.B.: Characterizing cancer subtypes as attractors of Hopfield networks **30**(9), 1273–1279 (2014)
43. Conforte, A.J., Alves, L., Coelho, F.C., Carels, N., Silva, F.A.B.D.: Modeling basins of attraction for breast cancer using Hopfield networks. Front. Genet. **11**, 314 (2020)
44. Al-Maitah, M.: Analyzing genetic diseases using multimedia processing techniques associative decision tree-based learning and Hopfield dynamic neural networks from medical images. Neural Comput. Appl. **32**(3), 791–803 (2020)
45. Stephenson, N., et al.: Survey of machine learning techniques in drug discovery. Curr. Drug Metab. **20**(3), 185–193 (2019)
46. Sangari, A., Sethares, W.: Paper texture classification via multi-scale restricted Boltzman machines. In: 2014 48th Asilomar Conference on Signals, Systems and Computers, pp. 482–486. IEEE (2014)
47. Arpaci, I., et al.: Analysis of twitter data using evolutionary clustering during the COVID-19 Pandemic. CMC—Comput. Mater. Continua. **65**(1), 193–203 (2020)
48. Kim, H.-C., Jang, H., Lee, J.-H.: Test–retest reliability of spatial patterns from resting-state functional MRI using the restricted Boltzmann machine and hierarchically organized spatial patterns from the deep belief network. J. Neurosci. Meth. **330**, 108451 (2020)
49. Vogelstein, J.T., et al.: A community-developed open-source computational ecosystem for big neuro data. Nat. Meth. **15**(11), 846–847 (2018)
50. Beam, A.L., Kohane, I.S.: Big data and machine learning in health care. JAMA **319**(13), 1317–1318 (2018)
51. Arpaci, I., Karataş, K., Baloğlu, M.J.P., Differences, I.: The development and initial tests for the psychometric properties of the COVID-19 phobia scale (C19P-S), p. 110108 (2020)
52. Cai, C., et al.: Transfer learning for drug discovery. J. Med. Chem. **63**(16), 8683–8694 (2020)
53. Baskin, I.I.: The power of deep learning to ligand-based novel drug discovery. Expert Opin. Drug Discov.,1–10 (2020)
54. Yildirim, O., Gottwald, M., Schüler, P., Michel, M.C.: Opportunities and challenges for drug development: public–private partnerships, adaptive designs and big data. Front. Pharmacol. **7**, 461 (2016)
55. Walls, A.C., Park, Y.-J., Tortorici, M.A., Wall, A., McGuire, A.T., Veesler, D.: Structure, function, and antigenicity of the SARS-CoV-2 spike glycoprotein. Cell (2020)
56. Shoichet, B.K.: Virtual screening of chemical libraries. Nature **432**(7019), 862–865 (2004)
57. Zhang, L., Tan, J., Han, D., Zhu, H.: From machine learning to deep learning: progress in machine intelligence for rational drug discovery. Drug Discov. Today **22**(11), 1680–1685 (2017)
58. Wang, J., Cao, H., Zhang, J.Z., Qi, Y.: Computational protein design with deep learning neural networks. Sci. Rep. **8**(1), 1–9 (2018)
59. Smolensky, P.: Information processing in dynamical systems: foundations of harmony theory. Department of Computer Science, Colorado University at Boulder (1986)
60. Kim, H.-C., Jang, H., Lee, J.-H.: Test–retest reliability of spatial patterns from resting-state functional MRI using the restricted Boltzmann machine and hierarchically organized spatial patterns from the deep belief network. J. Neurosci. Meth. **330**, 108451 (2020)
61. Karim, M.R., Cochez, M., Jares, J.B., Uddin, M., Beyan, O., Decker, S.: Drug-drug interaction prediction based on knowledge graph embeddings and convolutional-LSTM network. In: Proceedings of the 10th ACM International Conference on Bioinformatics, Computational Biology and Health Informatics, pp. 113–123 (2019)
62. Conover, M., Staples, M., Si, D., Sun, M., Cao, R.: AngularQA: protein model quality assessment with LSTM networks. Comput. Math. Biophys. **7**(1), 1–9 (2019)
63. Ibrahim, I.M., Abdelmalek, D.H., Elshahat, M.E., Elfiky, A.A.: COVID-19 spike-host cell receptor GRP78 binding site prediction. J. Infect. (2020)

64. Kumarakulasinghe, N.B., et al.: EGFR kinase inhibitors and gastric acid suppressants in EGFR-mutant NSCLC: a retrospective database analysis of potential drug interaction. Oncotarget **7**(51), 85542 (2016)
65. Carreira-Perpignan, M.: HGE on contrastive divergence learning. In: Proceedings of the International Conference on Artificial Intelligence and Statistics, vol. 10 (2005)
66. Grohskopf, L.A., Sokolow, L.Z., Broder, K.R., Walter, E.B., Fry, A.M., Jernigan, D.B.: Prevention and control of seasonal influenza with vaccines: recommendations of the Advisory Committee on Immunization Practices—United States, 2018–19 influenza season. MMWR Recommendations and Rep. **67**(3), 1 (2018)
67. Graham, R.L., Donaldson, E.F., Baric, R.S.: A decade after SARS: strategies for controlling emerging coronaviruses. Nat. Rev. Microbiol. **11**(12), 836–848 (2013)
68. Amanat, F., Krammer, F.: SARS-CoV-2 vaccines: status report. Immunity (2020)
69. Ahmed, S.F., Quadeer, A.A., McKay, M.R.: Preliminary identification of potential vaccine targets for the COVID-19 coronavirus (SARS-CoV-2) based on SARS-CoV immunological studies. Viruses **12**(3), 254 (2020)
70. Hu, F., Jiang, J., Yin, P.: Prediction of potential commercially inhibitors against SARS-CoV-2 by multi-task deep model (2020). arXiv preprint: arXiv:2003.00728
71. Du, L., He, Y., Zhou, Y., Liu, S., Zheng, B.-J., Jiang, S.: The spike protein of SARS-CoV—a target for vaccine and therapeutic development. Nat. Rev. Microbiol. **7**(3), 226–236 (2009)
72. Roper, R.L., Rehm, K.E.: SARS vaccines: where are we? Expert Rev. Vaccines **8**(7), 887–898 (2009)
73. Shen, C., et al.: Treatment of 5 critically ill patients with COVID-19 with convalescent plasma. JAMA **323**(16), 1582–1589 (2020)
74. Hughes, J.P., Rees, S., Kalindjian, S.B., Philpott, K.L.: Principles of early drug discovery. Br. J. Pharmacol. **162**(6), 1239–1249 (2011)
75. Chung, M., et al.: CT imaging features of 2019 novel coronavirus (2019-nCoV). Radiology **295**(1), 202–207 (2020)
76. Yang, Y., Faraggi, E., Zhao, H., Zhou, Y.: Improving protein fold recognition and template-based modeling by employing probabilistic-based matching between predicted one-dimensional structural properties of query and corresponding native properties of templates. Bioinformatics **27**(15), 2076–2082 (2011)
77. Rappuoli, R.: Reverse vaccinology. Curr. Opin. Microbiol. **3**(5), 445–450 (2000)
78. Hassan, A., Mahmood, A.J.I.A.: Convolutional recurrent deep learning model for sentence classification. IEEE Acc. **6**, 13949–13957 (2018)
79. Ma, J., Sheridan, R.P., Liaw, A., Dahl, G.E., Svetnik, V.: Deep neural nets as a method for quantitative structure–activity relationships. J. Chem. Inf. Model. **55**(2), 263–274 (2015)
80. Unterthiner, T., Mayr, A., Klambauer, G., Hochreiter, S.: Toxicity prediction using deep learning (2015). arXiv preprint: arXiv:1503.01445
81. Weinstein, M.C., Freedberg, K.A., Hyle, E.P., Paltiel, A.D.: Waiting for certainty on Covid-19 antibody tests—at what cost?. N. Engl. J. Med. (2020)

Covid-19 Detection Using Advanced CNN and X-rays

Basudeba Behera, Nitish Kumar, Mukesh Ranjan Mahato, and Ajay Kumar

Abstract As the Covid-19 pandemic surfaced worldwide, various newer technologies came ahead to help humanity to survive and live a better life. While medical science worked on vaccine development, Artificial intelligence and advanced computing helped in segregating millions of drugs to counter this novel coronavirus disease. Deep learning also came up with an alternative detection model. In this work, we present the detection method with computer vision using advanced CNN (Convolutional Neural Network) and architecture used is VGGNet. Since the virus is a new variant of the coronavirus also named as SARS-2 (Severe Acute Respiratory Syndrome Coronavirus 2) by WHO (World Health Organization), requires a different technique for testing and detection. The detection procedure used to take too much time to confirm the infection. The techniques utilized to this point of time is limited and which creates an impossible type of situation to test a huge crowd of the population like India. So, it's high time to come up with an alternative system for the Covid-19 Tests. Hence this paper came up with detection and partial confirmation of Covid-19 using X-ray images by applying a deep learning Algorithm. By enhancing the images and classifying the features we can easily differentiate between Covid-19 affected people and Normal people X-ray images of the chest. This pandemic highly affects our lungs, X-ray images may be one of the viable options.

Keywords Convolutional neural network · Covid-19 · Deep neural network · Pandemic · SARs-2 · Vaccine development

B. Behera (✉) · N. Kumar · M. R. Mahato · A. Kumar
Department of Electronics and Communication Engineering, National Institute of
Technology Jamshedpur, Jamshedpur, Jharkhand 831014, India
e-mail: basudeb.ece@nitjsr.ac.in

© The Author(s), under exclusive license to Springer Nature Switzerland AG 2021
I. Arpaci et al. (eds.), *Emerging Technologies During the Era of COVID-19
Pandemic*, Studies in Systems, Decision and Control 348,
https://doi.org/10.1007/978-3-030-67716-9_3

1 Introduction

The consequences of the spread of Covid-19 in the world are death tolls, lockdown, loss of economy, the uncertainty of future but certain preventions such as social distancing, face covering, and diagnosis of the cases results in slowing down the spread [1]. Currently, the Techniques available for detecting Covid-19 is very costly, and time taking besides its production is limited to certain countries. Since the development of new testing kits and using new tech may take a lot of time and thus it will be fatal for all of us as the virus is contagious [2]. So, there has to be a technique that is easily available and it can tell about the initial development of symptoms and not fully but partial detection of this pandemic virus. As every feature of an image can be extracted using a Deep Neural network and convolutional neural net (also called ConvNet) [3] which can enhance the features beyond the human eyes and can easily classify the images of different categories after having trained over large images of ImageNet [4]. Due to these developments, now-a-days various emotions can be identified in Google photos. Today Google has already worked on the system which identifies you out of millions of images. Facebook's face recognition system has become so advanced that Duty guards are being replaced with automatic surveillance systems.

So, after several months of the pandemic, there is enough data available on this pandemic to train and analyze for the detection of the infection using image processing techniques. As the X-ray images of people who are affected and normal people are available to overcloud. If it can be trained to a machine to learn how a normal X-ray image of the chest looks and how affected X-rays? Li et. al. [5] presented on the introduction of KELM (Kernel Extreme Learning Machines), a better classifier than softmax so the performance of CNN could be promoted. Thus, if a machine once trained to identify it could easily tell that certain X-ray images are affected and further testing can be done to verify. This method has the possibility of detecting disease through the proposed method.

2 Previous Works

Not many works about Covid-19 have been proposed yet, but many works around image classification can be found out. Emotion recognition software is elsewhere. And these work on the training of images in large amounts. Jmour et al. [6] presented their paper and they talked about convolutional neural networks and classification of images which talked about transfer learning technique also called fine-tuning, introducing reusing of pre-trained models On ImageNet for great accuracy while building the new model. Supe et al. [7] worked on image processing for medical Images where how clearly, they explained about the processing of images in detection of tumors, detection of internal organ failures, if they could clearly understand the internal damages why can't we use these X-rays techniques

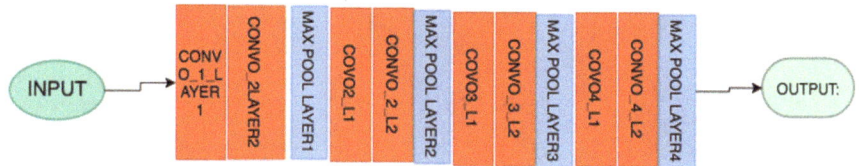

Fig. 1 Architecture of VGGNet

to detect abnormality in the lungs? There is a need to just implement the details. Besides that, we know that Covid-19 affects mostly our lungs.

Sultana et al. [8] worked on the article of advancement in convolutional neural networks where they wrote how ConvNet has advanced from LeNet-5 to the latest SENet model. Today the work on these technologies is such that no details of the image are left behind. They drove the conclusion that Today's era of googleNet, CPU based implementation trained using DistBelief [9] comes from VGGNet as shown in Fig. 1 developed by Krizhevsky [10] which is a deeper configuration of AlexNet, this is a concept we have used in this work.

Once we train the images our system can be fruitful with the main objective of the proposed research work is to use the Neural Networks and convolutional networks to achieve the following goals:

- To find an alternative Testing method for Covid-19 +ve people which is feasible and easily available.
- To replace the old time-consuming method with an automatic system that could be automated and utilizing the image classification methods.

3 Theory System Architecture

To understand how our system works it needs to know the basics behind the Image classification and ConvNet. Here it starts with how computers read an image as shown in Fig. 2. Overview of the computer vision Model will clear out all the doubts and it will clearly understand, how this can be possible.

This work has come with computer vision techniques to classify normal and COVID affected X-ray images. To understand the working of the proposed method, it needs to dive deeper into how Convolutional Neural Networks help in classifying the images by feature extraction. Simple raw images can be classified by a simple Neural network and fully connected layer but when it comes to RGB images simple Neural Net fails and it needs convolutions to build on these images as shown in Fig. 3.

CNN the neuron in a layer will be only connected with the only few of the neurons in the next layer unlike all the neurons connected in the fully connected layer so CNN is preferred.

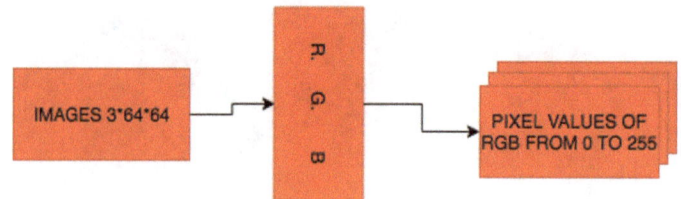

Fig. 2 Image reading pattern of computer

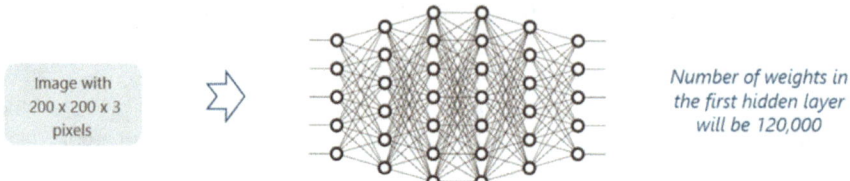

Fig. 3 Very high number of weights in the first hidden layer

CNN works by putting the convolution filters to extract features. Each filter is like a feature and it runs over the image to extract each feature out of the image and then we apply the pooling layer to compress the image besides keeping the important features in the image. It compares the image piece by piece which is very obvious from the Fig. 4 below. It gets better at seeing similar things by roughly matching the pieces at different locations in the image.

After having the normalized image with the input layer, we apply convolution, relu, and pooling layer in multiple amounts and stack it to have the image with less size but enhanced features. This stack of layers is applied to fully connected layers which can be seen in Fig. 5 below.

Feature extraction is an important part as classification can be done with any kind of image be it raw images or self-centered images. This is called feature extraction. Only a neural network once applied the image will make up the model

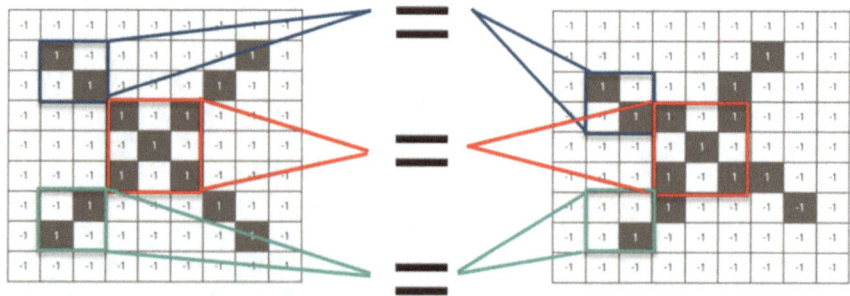

Fig. 4 Working on the image piece by piece

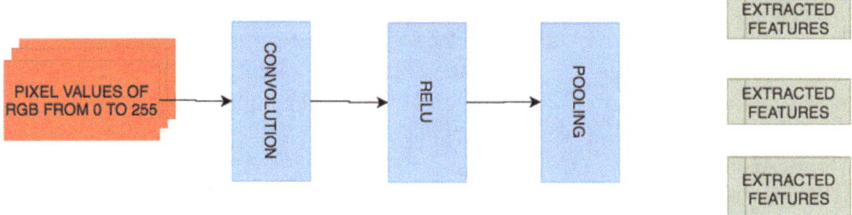

Fig. 5 Feature extraction

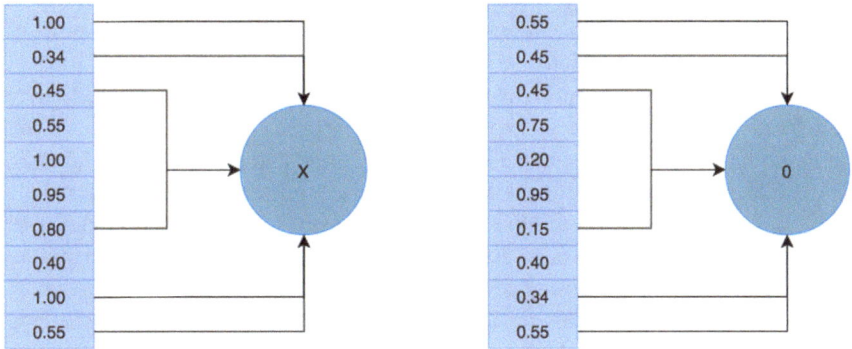

Fig. 6 Fully connected layer applied after feature extraction

but will not be accurate in the case of complex images. ConvNet can classify simple to complex image sets. Thus, this becomes a very useful part of our program.

Once the matrix of the given input layer image is supplied to the fully connected layer. This layer with SoftMax function classifies the image based on the probability reference as shown in Fig. 6.

4 Proposed Model Building

The proposed model building consists of the following steps.

- Load/Process the dataset.
- Normalize the dataset/encode it.
- Derive the ImageNet inbuilt model.
- Make the deep fully connected layer.
- Train the CNN-DNN model.
- Test the CNN-DNN.
- Here evaluates the model to proceed to
- Observation section.

In the proposed model we load the dataset from a GitHub repository [11] of Dr. Cohen where data is constantly updated with the help of various scientists and doctors who are constantly working ahead to find a solution to this pandemic. In this dataset, we take away 25 X-ray images of affected people and 25 normal people X-ray images. This link also contains a dataset for SARS, MERs outbreak in the past is as available [12]. In the given code below, we specify the path to the dataset and output path after the model building. It can well observe the affected dataset in Fig. 7.

```
# construct the argument parser and parse the arguments
ap = argparse.ArgumentParser()
ap.add_argument("-c", "--covid", required=True,
    help="path to base directory for COVID-19 dataset")
ap.add_argument("-o", "--output", required=True,
    help="path to directory where 'normal' images will be stored")
args = vars(ap.parse_args())
```

Now we are ready to preprocess the dataset. Initially, we will capture each image path from the dataset directory and then we will extract the class level from each path as either COVID or normal. Then load the image and convert it into RGB and 224 × 224 pixels to be fit for the convolutional neural network input layer. Once this is done, we will go on to Normalize our intensity to [0, 1] besides converting our label and data into NumPy arrays. Here you can view the above-mentioned process performed below in the code.

Continuing with our preprocessing process once we bring the label to [0, 1] we work with a hot encoder which will encode our label to 1 for COVID+ affected people X-ray images and 0 for normal X-ray people. Besides we will split our dataset into 80% of the dataset for testing purposes while 20% for testing or validating the model which is as follows in the code.

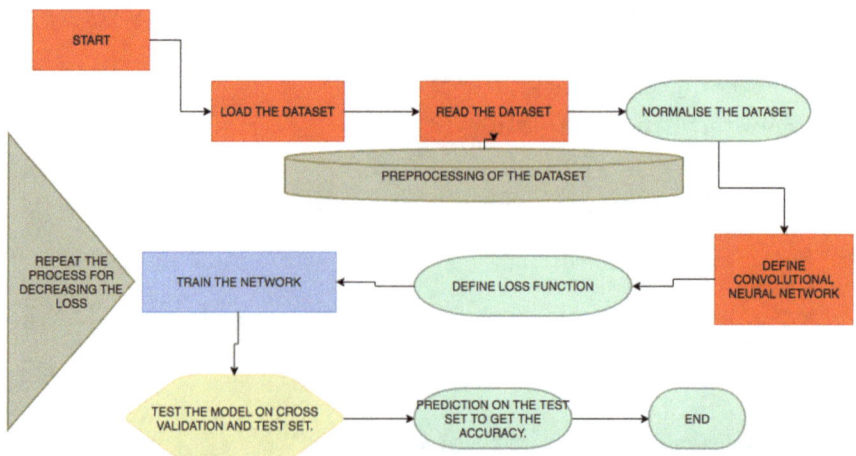

Fig. 7 Model building

```
# grab the list of images in our dataset directory, then initialize
# the list of data (i.e., images) and class images
print("[INFO] loading images...")
imagePaths = list(paths.list_images(args["dataset"]))
data = []
labels = []
# loop over the image paths
for imagePath in imagePaths:
    # extract the class label from the filename
    label = imagePath.split(os.path.sep)[-2]
    # load the image, swap color channels, and resize it to be a fixed
    # 224x224 pixels while ignoring aspect ratio
    image = cv2.imread(imagePath)
    image = cv2.cvtColor(image, cv2.COLOR_BGR2RGB)
    image = cv2.resize(image, (224, 224))
    # update the data and labels lists, respectively
```

```
# perform one-hot encoding on the labels
lb = LabelBinarizer()
labels = lb.fit_transform(labels)
labels = to_categorical(labels)
# partition the data into training and testing splits using 80% of
# the data for training and the remaining 20% for testing
(trainX, testX, trainY, testY) = train_test_split(data, labels,
                                  test_size=0.20, stratify=labels, random_state=42)
# initialize the training data augmentation object
trainAug = ImageDataGenerator(
                    rotation_range=15,
                    fill_mode="nearest")
```

After preprocessing we move ahead with training our dataset. As we have very few datasets available so the model may not be accurate. That's why we will take the help of already heavy models from ImageNet (which is a collection of huge image datasets having millions of images). The VGG16 (Fig. 1) model is trained from there we derive the ImageNet model without the fully connected layer as in the code then we go ahead with building the fully connected layer as POOL $\Rightarrow$ FC = SOFTMAX. Then the final model has a base model and on top of that, we have a fully connected layer besides we will freeze the weights of the convolutional layer so that only the fully connected layer gets trained here in our training.

```
# load the VGG16 network, ensuring the head FC layer sets are left
# off
baseModel = VGG16(weights="imagenet", include_top=False,
                  input_tensor=Input(shape=(224, 224, 3)))
# construct the head of the model that will be placed on top of the
# the base model
headModel = baseModel.output
headModel = AveragePooling2D(pool_size=(4, 4))(headModel)
headModel = Flatten(name="flatten")(headModel)
headModel = Dense(64, activation="relu")(headModel)
headModel = Dropout(0.5)(headModel)
headModel = Dense(2, activation="softmax")(headModel)
# place the head FC model on top of the base model (this will become
# the actual model we will train)
model = Model(inputs=baseModel.input, outputs=headModel)
# loop over all layers in the base model and freeze them so they will
# #not# be updated during the first training process
for layer in baseModel.layers:
    layer.trainable = False
```

Note: VGG16 model which is trained from ImageNet architecture works on the convolutional layer which is well capable of extracting the features of the images, which is a very important part of the classifying the X-ray images correctly.

After the model building is done, it needs to run the training process without any delay besides it will compile the model using binary cross-entropy key ingredients for the classification process. To simplify it, the process converts our dataset into a list of numerical values but the main work based on the probability of classification is done by the optimizer.

```
[INFO] compiling model...
[INFO] training head...
Epoch 1/25
5/5 [==============================] - 20s 4s/step - loss: 0.7169 - accuracy: 0.6000 - val_loss: 0.6590 - val_accuracy: 0.5000
Epoch 2/25
5/5 [==============================] - 0s 86ms/step - loss: 0.8088 - accuracy: 0.4250 - val_loss: 0.6112 - val_accuracy: 0.9000
Epoch 3/25
5/5 [==============================] - 0s 99ms/step - loss: 0.6809 - accuracy: 0.5500 - val_loss: 0.6054 - val_accuracy: 0.5000
Epoch 4/25
5/5 [==============================] - 1s 100ms/step - loss: 0.6723 - accuracy: 0.6000 - val_loss: 0.5771 - val_accuracy: 0.6000
...
Epoch 22/25
5/5 [==============================] - 0s 99ms/step - loss: 0.3271 - accuracy: 0.9250 - val_loss: 0.2902 - val_accuracy: 0.9000
Epoch 23/25
5/5 [==============================] - 0s 99ms/step - loss: 0.3634 - accuracy: 0.9250 - val_loss: 0.2690 - val_accuracy: 0.9000
Epoch 24/25
5/5 [==============================] - 27s 5s/step - loss: 0.3175 - accuracy: 0.9250 - val_loss: 0.2395 - val_accuracy: 0.9000
Epoch 25/25
5/5 [==============================] - 1s 101ms/step - loss: 0.3655 - accuracy: 0.8250 - val_loss: 0.2522 - val_accuracy: 0.9000
[INFO] evaluating network...
```

Fig. 8 Training in progress

```
# compile our model
print("[INFO] compiling model...")
opt = Adam(lr=INIT_LR, decay=INIT_LR / EPOCHS)
model.compile(loss="binary_crossentropy", optimizer=opt,
                metrics=["accuracy"])
# train the head of the network
print("[INFO] training head...")
H = model.fit_generator(
                trainAug.flow(trainX, trainY, batch_size=BS),
                steps_per_epoch=len(trainX) // BS,
                validation_data=(testX, testY),
                validation_steps=len(testX) // BS,
                epochs=EPOCHS)
```

Figure 8 shows a screenshot of the training process. It is observed that the accuracy of around 90% is achieved. Let's get that validation done in the result and observation section Fig. 9.

	precision	recall	f1-score	support
covid	0.83	1.00	0.91	5
normal	1.00	0.80	0.89	5
accuracy			0.90	10
macro avg	0.92	0.90	0.90	10
weighted avg	0.92	0.90	0.90	10

```
[[5 0]
 [1 4]]
acc: 0.9000
sensitivity: 1.0000
specificity: 0.8000
[INFO] saving COVID-19 detector model...
```

Fig. 9 Classification table for evaluation metrics

5   Results and Observation

After training the model, it needs to be validated, besides it will use confusion matrix as a tool to predict the accuracy, sensitivity, and specificity as shown in the code below and results follows:

```
# compute the confusion matrix and and use it to derive the raw
# accuracy, sensitivity, and specificity
cm = confusion_matrix(testY.argmax(axis=1), predIdxs)
total = sum(sum(cm))
acc = (cm[0, 0] + cm[1, 1]) / total
sensitivity = cm[0, 0] / (cm[0, 0] + cm[0, 1])
specificity = cm[1, 1] / (cm[1, 0] + cm[1, 1])
# show the confusion matrix, accuracy, sensitivity, and specificity
print(cm)
print("acc: {:.4f}".format(acc))
print("sensitivity: {:.4f}".format(sensitivity))
print("specificity: {:.4f}".format(specificity))
```

Here in the result, it can well observe that with the help of the confusion matrix, the accuracy of around 90% with a sensitivity of 1.000 and specificity around 0.800 is achieved. Let's go on to plotting the results now to visualize the outcome in a much better way below is the included classification table for the references.

With the help of pyplot in python, we can easily see the plot of Covid-19 detection using an X-ray. The plot is as per the training which took place. Let's describe the plot in detail below.

```
# plot the training loss and accuracy
N = EPOCHS
plt.style.use("ggplot")
plt.figure()
plt.plot(np.arange(0, N), H.history["loss"], label="train_loss")
plt.plot(np.arange(0, N), H.history["val_loss"], label="val_loss")
plt.plot(np.arange(0, N), H.history["accuracy"], label="train_acc")
plt.plot(np.arange(0, N), H.history["val_accuracy"], label="val_acc")
plt.title("Training Loss and Accuracy on COVID-19 Dataset")
plt.xlabel("Epoch #")
plt.ylabel("Loss/Accuracy")
plt.legend(loc="lower left")
plt.savefig(args["plot"])
```

From the plot, we can observe that training loss and validation loss both are decreasing with the number of epochs, moreover both training and testing sets are synchronous.

Training accuracy, in the beginning, was random for a few epochs but later training accuracy increased heavily and validation accuracy constantly increased with several iterations. The best part is that in the end having 25 epochs testing and training is no different. Figure 10 can show that red line training loss is decreasing. While working on the model if it needs to go with further training or not that is all determined by these kinds of plot. It helped us in generalizing our model. The model works well with training and testing this can be well seen from the plot in Fig. 10. The accuracy of the model can be verified from the evaluation metric table Fig. 9 and the ROC curve in Fig. 11.

Fig. 10 Results of training
loss versus epochs

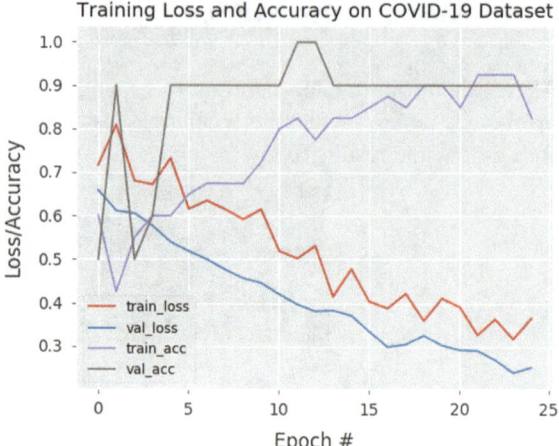

Fig. 11 Auc-ROC curve
FPR versus TPR

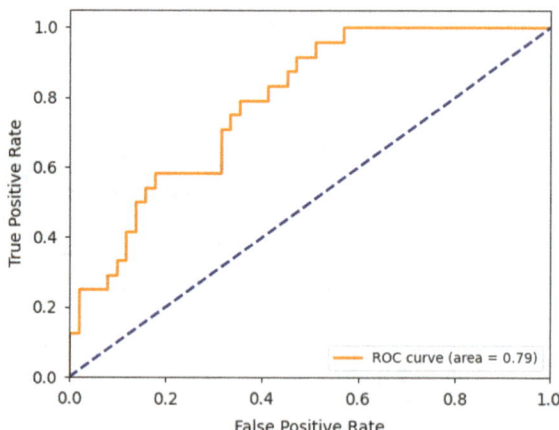

6   Conclusion

The research work focuses on convolutional neural networks to classify the images
after extracting the deep features of the X-rays. Through this work, it is demonstrated to find out an effective method to detect the virus quickly as compared to the
testing kit. As Covid-19 affects our epithelial cells, lungs, and effective regions of
the chest, so, X-ray can be the most viable and easily available method as of now to
detect the coronavirus. Although a very well qualified medical professional can
easily tell the difference by manually looking at the X-ray images. But again, that is
what we lack because doctors are at risk so there is a dearth need for automation,
and here is this we propose. We have achieved an accuracy of 90% but as we are
not medical practitioners nor professionals so we don't guarantee that it.

References

1. Arpaci, I., Karataş, K., Baloğlu, M.: The development and initial tests for the psychometric properties of the COVID-19 Phobia Scale (C19P-S). Pers. Individ. Dif. **164**(April), 110108 (2020)
2. Arpaci, I., et al.: Analysis of twitter data using evolutionary clustering during the COVID-19 pandemic. Comput. Mater. Contin. **65**(1), 193–204 (2020)
3. Semwal, V.B., Singha, J., Sharma, P.K., Chauhan, A., Behera, B.: An optimized feature selection technique based on incremental feature analysis for bio-metric gait data classification. Multimed. Tools Appl., Dec., p. 18 (2016)
4. Tarasenko, A.O., Yakimov, Y.V., Soloviev, V.N.: Convolutional neural networks for image classification. CEUR Workshop Proc. **2546**, 101–114 (2019)
5. Li, Z., Zhu, X., Wang, L., Guo, P.: Image classification using convolutional neural networks and kernel extreme learning machines. Proceedings of International Conference on Image Processing ICIP, pp. 3009–3013 (2018)
6. Jmour, N., Zayen, S., Abdelkrim, A.: Convolutional neural networks for image classification. In: 2018 International Conference on Advanced Systems and Electric Technologies IC_ASET 2018, pp. 397–402 (2018)
7. Supe, P.V., Bhagat, K.S., Chaudhari, J.P.: Image processing for medical image analysis: a review. Int. J. Futur. Revolut. Comput. Sci. Commun. Eng. **4**(12), 105–108 (2018)
8. Sultana, F., Sufian, A., Dutta, P.: Advancements in image classification using convolutional neural network. In: Proceedings 2018 4th IEEE International Conference on Research in Computational Intelligence and Communication Networks (ICRCICN 2018), pp. 122–129 (2018)
9. Dean, J., et al.: Large scale distributed deep networks. (1975), 1–9 (1975)
10. Krizhevsky, A., Sutskever, I., Hinton, G.E.: ImageNet classification with deep convolutional neural networks. Commun. ACM **60**(6), 84–90 (2017)
11. GitHub,: Covid-chestxray-dataset, GitHub. https://github.com/ieee8023/covid-chestxray-dataset (2020). Accessed 17 Aug 2020
12. Grover, R.: Colab research. https://colab.research.google.com/drive/1esbpDOorf7DQJV8GXWON24c-EQrSKOit (2003). Accessed 17 Aug 2020

Integration of Deep Learning Machine Models with Conventional Diagnostic Tools in Medical Image Analysis for Detection and Diagnosis of Novel Coronavirus (COVID-19)

Lakshmi Narasimha Gunturu⊙ and **Girirajasekhar Dornadula**⊙

Abstract Novel coronavirus 19 (COVID-19) had made the lives of humans in dilemma since its outburst from Wuhan, China on December 31, 2019, and is declared as a pandemic by the World Health Organization (WHO). Its spread had been very rapid among the people which is a real task for the authorities to identify the persons infected with viruses and isolate them to prevent virus transmission. At present, various diagnostic aids are used in health care centres to screen the public with the symptoms of COVID-19. However, the use of conventional medical aids is of no use in early detection because of the rapid transmission rate. Therefore, researches are exploring Artificial Intelligence (AI) incorporated techniques that will aid physicians in the early detection of COVID-19 and one such method is Deep Learning Machine Models (DLM). This chapter aims to review research studies available in PubMed database with the Keywords COVID-19, Diagnosis, Deep learning models, and Medical Imaging till August 10, 2020 related to COVID-19 and DLM in medical image analysis. By evaluating 13 full text articles we calculated F1 scores for possible studies and concluded that studies with Computed Tomography (CT) had better diagnostic accuracy when compared to chest X-ray (CXR).

Keywords Artificial intelligence · COVID-19 · Deep learning machine models (DLM) · Diagnostic aids · Early screening

L. N. Gunturu (✉) · G. Dornadula
Department of Pharmacy Practice, Annamacharya College of Pharmacy, Rajampet, Andhra Pradesh, India
e-mail: gunturunarasimha007@gmail.com

G. Dornadula
e-mail: giriraj.pharma@gmail.com

© The Author(s), under exclusive license to Springer Nature Switzerland AG 2021
I. Arpaci et al. (eds.), *Emerging Technologies During the Era of COVID-19 Pandemic*, Studies in Systems, Decision and Control 348,
https://doi.org/10.1007/978-3-030-67716-9_4

1   Introduction

In December 2019 Wuhan provenance many people were infected with symptoms of fever, cough, and shortness of breath and are represented as severe unexplained pneumonia. Later this spread to many countries made the situation worse and resulted in deaths. In February 2020 World Health Organization (WHO) declared this condition as a pandemic and named the new virus as Novel Coronavirus 2019 (COVID-19) [1]. As there are no approved drugs in use for this new virus only treatment option is to reduce the disease severity with the existing drugs like Chloroquine, Remdesivir, and Fapilavir, etc. The prevention and control model for this pandemic is mentioned in Fig. 1:

1. Early detection, reporting, and isolation of infected persons.
2. Providing adequate necessary resources to all infected patients.
3. Hospitalize all the suspected and virus-positive patients.

As per Fig. 1, the biggest worry is the early detection, reporting, and isolation of the infected patients. Because the virus is spreading at a rapid pace so that existing medical facilities and medical aids are not enough to perform adequate viral testing. According to Liu et al. the virus, basic reproduction number (R0) is determined to be 2–3 which is high when compared to the Severe Acute Respiratory Syndrome Virus (SARS) [2]. Due to its high transmission rate existing medical tools like Computed Tomography (CT), reverse transcription-polymerase chain reaction (RT-PCR), X-ray, and sputum test, etc. are not enough for early detection of COVID-19 cases also sometimes false negatives are being resulted. So researches

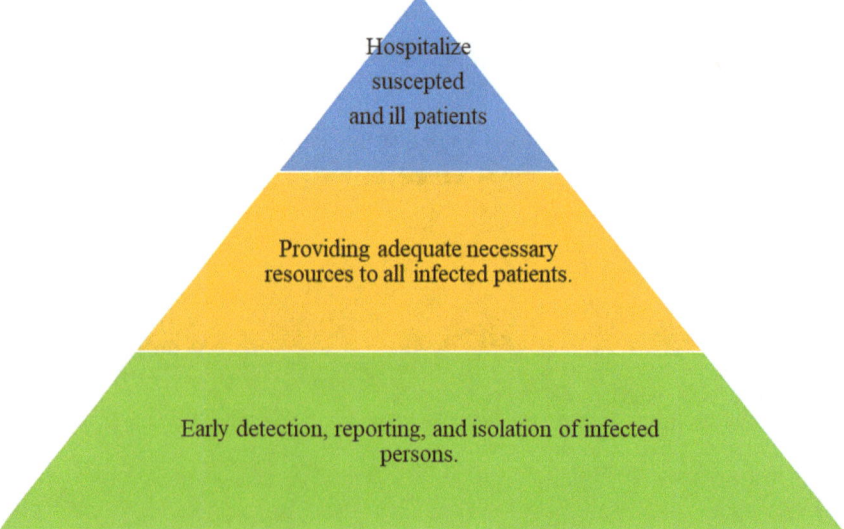

Fig. 1 Illustration of the prevention model for control of corona pandemic

are looking at advanced technologies like Artificial Intelligence to incorporate into diagnostic aids for rapid detection of viral screening. One of such techniques is Deep Learning Machine models (DLM) is a technique of AI that teaches computers to perform what naturally comes to humans. DLM is also used in areas like agriculture, mobile, and wireless networking, Internet of things, bio-informatics, and health management systems. One such application more frequent in use during the COVID-19 pandemic is lung image analysis models of corona infected patients. Because they will automate the characteristics of the symptoms of COVID-19 and also able to distinguish between the non-COVID-19 and COVID-19 persons in a very quick time. DLM performs various tasks such as classification, clustering, regression, image reconstruction, lesion detection, and segmentation. All these functions are performed by the development of paradigms like:

1. Support Vector Machine
2. Random forest
3. Backpropagation network
4. Neuro-fuzzy inference systems
5. Convolutional neural networks [3].

Normally interpretation of X-rays needs technical skill and also time taking. So radiologists use a deep learning model that will automate the analysis of X-rays and reduce the time taking for interpretation and speed up the diagnostic process. Here a group of X-rays is analyzed by the use of a Machine learning model which is equipped with all the necessary sets of instructions to perform its functions. After the analysis of images by the machine learning programs, these are evaluated by the radiologist to provide the final diagnosis [4].

2 Role of Medical Imaging in COVID-19 Detection

Basically, the techniques used in the detection and diagnosis of COVID-19 are Computed Tomography (CT), reverse transcription-polymerase chain reaction (RT-PCR), X-ray, and sputum test. Of these techniques, the use of CT is controversial because it cannot able to diagnose the patients with COVID-19 and other lung-related problems especially pneumonia. Therefore, radiological societies did not recommend CT as a screening tool for the detection of COVID-19. But the study of Simpson et al. predicted the use of CT screening as a tool for the diagnosis of COVID-19 and reported four categories of CT findings related to the COVID-19 [5]. A study of Mahmood et al. that included 12,270 patients reported that CT had gain both sensitivity and specificity and can be used as a screening tool for the detection of COVID-19 patients at the earliest and prevent the spread of infection [6]. In the study of Pereira et al. reported the chest X-ray (CXR) is cheaper, easier, and universal [7]. In yet another study by Dong et al. reported that CT scans can improve the accuracy of diagnosis, detection, and management of COVID-19 condition based on the clinical symptoms of patients [8]. The summary of these studies is represented in Table 1.

3   Materials, Methods and Procedure

3.1   Creating a Systematic Search Strategy

a. Constructing an appropriate question
b. Using the appropriate database(s)
c. Advanced searching in PubMed—MeSH terms and the MeSH database.

PICO (Patients, Intervention, Comparator, Outcomes) table for designing our question as shown in Table 2.

The Table 3 below provides brief descriptions of common databases and sources to search both peer-reviewed and grey literature.

PubMed database was screened with the Keywords COVID-19, Diagnosis, Deep learning models, and Medical Imaging till August 10, 2020. Title and abstract screening were done manually as there are very few articles. Full texts of these screened articles were further screened for possible inclusion and exclusion criteria. Totally 26 studies are obtained of these 13 falls into the category of deep machine

Table 1 Findings of studies related to medical image analysis and COVID-19

Study	Modality	Targeted region	Application	Findings
Simpson et al. [5]	CT	Lung	Diagnosis	CT scan can detect COVID-19 pneumonia and their clinical features
Mahmood et al. [6]	CT	Lung	Diagnosis	CT can be useful as a screening tool without any delay of results when compared to serum assay, CRP, RT-PCR
Pereira et al. [7]	CXR	Lung	Diagnosis	CXR can able to identify the COVID-19 pneumonia cases from other pneumonia-causing pathogens
Dong et al. [8]	Various imaging technologies like CT, CXR, RT-PCR	–	Diagnosis	The Combined use of AI&CT imaging is useful in the rapid diagnosis and prediction of COVID-19

Table 2 PICO for designing a question strategy

Parameters	Included	Excluded
Patients	COVID-19	Non-COVID-19
Intervention	AI and DLM	Other than AI and DLM
Comparator	CXR and CT	Others like serum assays, nucleic acid tests
Outcome	Early Detection	Poor diagnosis results

Table 3 Representing common databases for search strategy

Database and website	Description	What is included	What is excluded
MEDLINE http://www.ncbi.nlm.nih.gov/pubmed/	National Library of Medicine's (NLM) premier bibliographic database PubMed is a free search engine and is maintained by the National Centre for Biotechnology Information (NCBI) at the NLM contains over 19 million citations	Academic journals covering the fields of medicine, the health care system, preclinical sciences Peer-reviewed journals, Deep Learning Machine models, Artificial Intelligence, COVID-19, medical image analysis	Fuzzy systems, Case studies, Pre-prints, Un-authored proofs

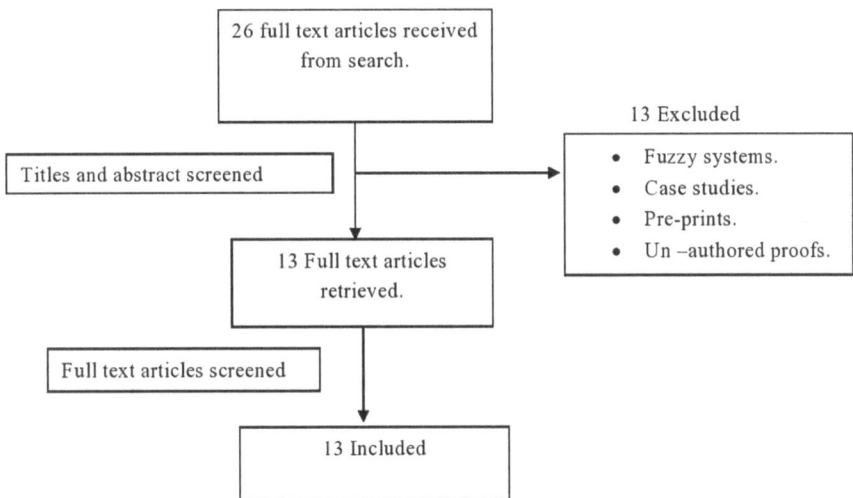

Fig. 2 Flowchart

learning and medical imaging were included. We excluded studies of preprints, Un-authored proofs, case studies, Fuzzy systems, Internet of things, and Robotic learnings. The flowchart of the study is shown in Fig. 2.

4   Research Studies Related to DLM Applications in COVID-19

In the study of Yu et al., they classified the subjects into two groups like 246 severe cases and 483 non-severe cases by using the four pre-trained deep neural networks and chest CT as a tool for the rapid, accurate, automatic tool for screening and follow-up of COVID-19 patients. Their results demonstrated that DenseNet-210 had high accuracy, sensitivity, and specificity for the detection of COVID-19 [9]. Another study of Masot et al. they categorized subjects as 316 training sets and 80 test sets for the identification of pneumonia and COVID-19 using the torso radiographs, VGG 16 based deep learning model. Their results found that VGG 16 based deep learning model had resulted in better sensitivity (0.92%) and specificity (0.85%) [10]. Another study of Mei et al. they categorized the subjects into two groups as positive SARS-Cov-2 and Negative SARS-Cov-2 for the improved detection of patient diagnosis by using three deep learning models. Their findings demonstrated that the Joint model (model 3) had a more specificity of 0.92% when compared to the other two models [11]. In the study of Li et al. they grouped the patients into three categories as COVID-19, Community-acquired Pneumonia, and non-pneumonia to develop an automatic and accurate detection of COVID-19 from other pulmonary diseases by using the COVNet as a deep learning model. Their results demonstrated that it has better sensitivity (90%) and specificity (96%) for COVID-19 [12]. Harmon et al. used a deep learning algorithm for the detection of COVID-19 by using the chest X-rays of patients with both COVID-19 and COVID-19 pneumonia. His study proved that DLM has good accuracy (90.8%), Sensitivity (84%), and Specificity (93%) [13]. Another study of Apostolopoulos and Mpesiana used the Convolutional Neuronal Network models (CNN) to distinguish the X-ray images of common pneumonia, COVID-19, and Non-COVID-19 images. Their study demonstrated the accuracy of 97.82% in the detection of COVID-19 cases with the use of CNN [14]. Another study of Ko et al. demonstrated that the ResNet-50 model has the highest accuracy of 96.97% in the detection of COVID-19 pneumonia [15]. The study of Hurt et al. used the image modality CXR to identify the COVID-19 and pneumonia cases separately using the U-Net model. Their study reported the CXR had better robustness [16]. Study findings of Singh et al. reported the CT scan classifies the COVID-19 positive cases and negative cases accurately using the CNN algorithm [17]. Study findings of Yoo et al. reported that CXR can detect the COVID-19 and other pneumonia diseases with an accuracy of 98% by using the CNN model of deep learning [18]. Another study of Ni et al. used the modality CT of 96 patients integrated with MVP-Net and reported that its results are better when compared to the residential radiologists [19]. Another finding of Rajaraman et al. used the CXR of four different datasets and evaluated the performance of pruned models accuracy as 99% [20]. Findings of Brunese et al. proved that CXR integrated with the VGG-16 algorithm can able to distinguish the health X-rays, Pulmonary disease X-rays, and COVID-19 X-rays [21]. The summary of these studies is represented in Table 4.

Table 4 Studies related to use of DLM models in medical image analysis of COVID-19

Study	Modality	Subjects	Task	Methods	Result
Yu et al. [9]	Chest CT	246 severe cases 483 non-severe cases	Rapid, Accurate automatic tool for severity screening follow up of COVID-19	Deep neural network methods 1. Inception-V3 2. ResNet-50 3. ResNet-101 4. DenseNet-201	DenseNet-210 95.20 Accuracy% 91.87% sensitivity 96.87% specificity
Masot et al. [10]	X-ray	Training set-316 Test set-80	Identification of pneumonia and COVID-19 using torso radiographs	VGG16-Based deep learning model	0.92% sensitivity 0.85% specificity
Mei et al. [11]	CT	419-Positive COVID-19 cases 486-Negative for COVID-19	Improved detection of patients either positive or negative by using AI model	1. Convolutional Neuronal network 2. Machine learning classifiers 3. Joint model	0.86% specificity 0.80 specificity 0.92 specificity
Li et al. [12]	CT	1296-COVID-19 1375-Community-Acquired Pneumonia 1325-non- pneumonia	Detection of COVID-19 from community-acquired pneumonia and other lung diseases	COVNet	90% sensitivity 96% specificity
Harmon et al. [13]	CT	2167 patients both COVID-19 and COVID-19 pneumonia	Evaluate an AI algorithm for detection of COVID-19 on chest CT	Deep learning algorithm	90.8% Accuracy 84% Sensitivity 93% Specificity

(continued)

Table 4 (continued)

Study	Modality	Subjects	Task	Methods	Result
Apostolopoulos and Mpesiana [14]	X-ray	224 images of positive COVID-19 patients. 700 images of common pneumonia. 504 images of Normal conditions.	Evaluation of performance and state of art of CNN model	Transfer learning	97.82% Accurate
Ko et al. [15]	CT	3993 images of chest CT was used	To develop a fast track COVID-19 network to diagnose COVID-19 pneumonia	2D deep learning frame work	ResNet-50 model highest accuracy of 96.97%
Hurt et al. [16]	CXR	5 patients	To identify the COVID-19 and pneumonia cases separately	U-Net	CXR augment the generalizability and robustness of DL
Singh et al. [17]	CT	Datasets were used	Classification of COVID-19 positive cases and COVID-19 negative cases	CNN	CT performed better results in the differentiation of negative and positive cases of COVID-19
Yoo et al. [18]	CXR	Datasets were used	Detect COVID-19, tuberculosis, normal images of CXR	CNN	98% accuracy (Normal vs. Abnormal data) 80% accuracy (TB vs. non-TB data)
Ni et al. [19]	CT	96 patients	Detect COVID-19 pneumonia on CT images	MVP-Net	Algorithm -based findings showed better results compared to residential radiologists
Rajaraman et al. [20]	CXR	Datasets were used	Evaluate the performance of pruned models	Pruned deep learning	99% accuracy AUC 0.9972%
Brunese et al. [21]	CXR	6523 images	Distinguish between healthy, pulmonary disease X rays, COVID-19 X-rays	VGG-16	0.96% accuracy for healthy and pulmonary disease patients 0.98% accuracy for COVID-19 diagnosis

All the studies related to the deep learning models that measure the study weightage are based on these important parameters:

1. AUC (Area Under the receiver operating Characteristic Curve)
2. Accuracy
3. Sensitivity
4. Specificity
5. F1-Score.

5 Discussion

The outbreak of the virus from the Wuhan had made the lives of people pathetic. Existing drugs and diagnostic aids do not cope up with the virus because of rapid spread across the world. Some reports stated the whatever the nucleic acid testing more used in the diagnosis of COVID-19 had an accuracy of about 30–50%. So Governments and Health care organizations advised the CT scans to analyze the pulmonary symptoms of the COVID-19 patients to detect false positives and false negatives accurately.

In this review, we analyzed 13 studies in which they used medical images and datasets of patients to detect the COVID-19 by the integration of deep learning models. F1 score was calculated by using a fx solver a mathematical guide. To conclude the best out of all these studies we calculated F1 scores for 7 studies that included CT or CXR as a diagnostic tool as shown in the Table 5. We could not calculate the F1 scores for the other six studies because the datasets of both the COVID-19 positive and COVID-19 negative patients were represented as a whole. F1 scores are calculated by using the formula

$$F_1 = 2 \cdot p \cdot r / p + r$$

Here

p = precision
r = recall

Table 5 F1 scores of different studies

No.	Study	F1 score
1	Harmon et al.	0.1569
2	Hurt et al.	0.1324
3	Li et al.	0.3243
4	Yu et al.	0.3374
5	Mei et al.	0.4630
6	Masot et al.	0.7980
7	Ni et al.	0.9792

5.1 *Interpretation*

When the **F**1 scores had point 1 then the study had provided better results. If the **F**1 score is 0 score, then the study had achieved the poor results.

Of all the included studies to calculate the **F**1 score, the highest scores were obtained for the study of Ni et al. (0.9792) that was very close to the 1 followed by other studies. Therefore, this review suggested that calculation of **F**1 scores would provide the more suitable metrics for the accuracy of studies that used the DLM in medical image analysis. Our study findings suggest that use of CT serves better in early detection of COVID-19 when compared to CXR as a diagnostic aid. Therefore, the applicability of CT integrated with DLM as a better diagnostic aid to detect and diagnosis the COVID-19 must be addressed with further research studies.

6 Advantages of DLM Applications

1. Rapid screening
2. Segmentation
3. Detection and
4. Classification.

6.1 *Rapid Screening*

The use of the artificial intelligence or its related DLM models is pre-trained and integrates with the mathematical models for the processing and analysis of images automatically without any manual help hence the radiologists finds it easier to interpret the images. This can also minimize the time required for screening and also improve the accuracy of detection when compared to the manual image analysis. Therefore, more number of medical images are screened within less time.

6.2 *Segmentation*

Segmentation is a process of the partitioning of a digital image into multiple segments such as pixels and segments. So that the segmented portions are very easy to analyze. A study of Butt et al. proposed a 3D CNN to segregate many cubes from the CT scan. After the segmentation, it can classify the images into COVID-19 patches, Viral pneumonia images, and other related infections. Their study also achieved a sensitivity of 98.2% and specificity of 92.2% for COVID-19 cases and non-COVID-19 cases [22].

6.3   Detection

Detection is a process associated with a computer vision and can relate to the system that can recognize the presence of the desired object within an image. A study of Hurt et al. proposed a localization system of U-Net trained with 22k medical images that could be interpreted by radiologists for producing the maps of probability. This phenomenon can be applied to the patients of COVID-19 for the detection of images [15].

6.4   Classification

This is a type of supervised learning and can able to identify the specific class to which the data elements belong to. This process was used by Wu et al. and proposed a study about COVID-19 patients. They used the deep learning network models for the fusion of lung images in different views like axial, coronal, and sagittal views and can able to identify the medical images of COVID-19 patients.

7   AI Used Techniques to Prevent the Spread of COVID-19

As per Fig. 3, AI is not only useful in the control of COVID-19 but also it can be useful through indirectly as a Monitoring technology where we can estimate the economic burden facing by the country in these pandemic situations by satellite systems. As COVID-19 is spreading rapidly to people, treating physicians, and health care workers we are now exploring alternate options to look after the needs of infected patients. One such attempt is bringing ROBOTS in hospitals where they can look over the needs of patients. These robots are fixed by AI-based manual programs that an actual human can perform. Another application is the implementation of various AI-based social networking sites like Chatbots. Ex. Canadas COVID-19 Chatbot. These apps can itself diagnose and advise the public depending upon the symptoms and any need to visit the hospital. They can also able to identify the nearby COVID-19 cases in the surroundings within a specific distance. We can also able to calculate the risk of the patient by using the EpiRisk. This requires entering the symptoms of a person and calculating the probability of getting a disease. By the use of technology like social media we can suggest the public regarding the spread of COVID-19 and its precaution measures. This was similar to a study conducted by Arpaci et al. They analyzed the 43 million twitter tweets between March 22 and March 30, 2020 and reported that social media posts may affect human psychology and behavior and can assist the public in decision making and task prioritizing [23]. This can also be used as a platform for the government to communicate easily with the public to prevent the spread of COVID-19.

Fig. 3 Various techniques of
AI to prevent the spread of
COVID-19

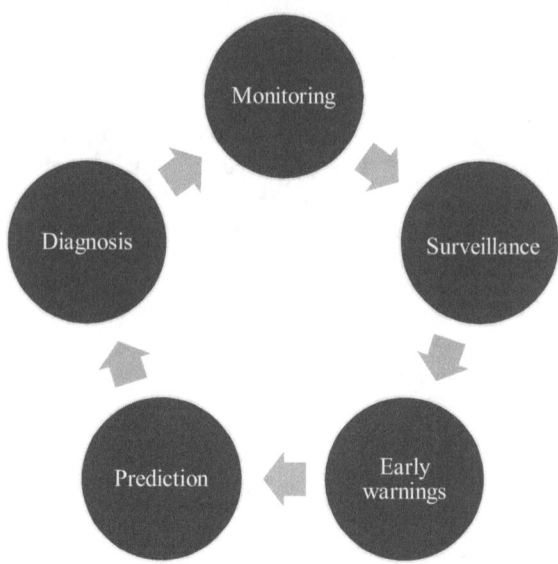

8   Limitations

1. Many of the countries are still at the initial stages of implementing AI-based approaches to combat COVID-19 because of various factors like resources, technology, budget for implementation etc.
2. Another thing that must be mentioned is the size of data sets. Normally if the data or the subjects involved in the study are more we can estimate the results very accurately and precisely but in contrast to this whatever the data sets are available in the healthcare settings are too small to estimate the results accurately.
3. Regulatory approval for the DLM techniques is quite challenging and is poorly addressed currently but it is very important for the healthcare settings to get the regulatory approval to establish the quality assurance and robustness of the study.

9   Conclusion

Existing drugs and treatment options are not in a position to counteract the virus where they can only act as a symptom normalizer. So we need a different approach like AI and advanced technology for early identification of vaccines and to identify the lead molecules. Machine learning is more frequently used in medical settings and recently techniques like DLM are predominant. This review concluded that F1 scores is the accurate measure to estimate the results of study in an efficient manner and CT integrated with DLM models had provided better diagnostic accuracy when compared to X-rays for the detection of COVID-19.

In this chapter, we presented advances in deep learning models in healthcare to combat the COVID-19 are useful for the radiologists as a diagnostic tool to identify the accurate cause with diagnosis to avoid false negatives and false positives and detect the positive cases within less time. Hence we motivate to use the deep learning models instead of the traditional machine learning to address the significant problems efficiently.

Conflicts of Interests None

Funding Nil

Ethical Approval Not applicable

References

1. Sanders, J.M., Monogue, M.L., Jodlowski, T.Z., Cutrell, J.B.: Pharmacologic treatments for coronavirus disease 2019 (COVID-19): a review. JAMA **323**(18), 1824–1836 (2020)
2. Liu, Y., Gayle, A.A., Wilder-Smith, A., Rocklov, J.: The reproductive number of COVID-19 is higher compared to SARS coronavirus. J Travel Med. **27**(2), 1–4 (2020)
3. Narayan Das, N., Kumar, N., Kaur, M., Kumar, V., Singh, D.: Automated deep transfer learning-based approach for detection of COVID-19 infection in chest X-rays. IRBM, 1–6 (2020)
4. Ozturk, T., Talo, M., Yildirim, E.A., Baloglu, U.B., Yildirim, O., Acharya, U.R.: Automated detection of COVID-19 cases using deep neural networks with X-ray images. Comput. Biol. Med. **121**, 1–11 (2020)
5. Simpson, S., Kay, F.U., Abbara, S., Bhalla, S., Chung, J.H., Chung, M.: Radiological society of North America expert consensus statement on reporting chest CT findings related to COVID-19. Endorsed by the society of thoracic radiology, the American College of Radiology, and RSNA. Radio. Cardiothorac Imaging **2**(2), e200152 (2020)
6. Mahmood, A., Gajula, C., Gajula, P.: Covid-19 diagnostic tests: a study of 12,270 patients to determine which test offers the most beneficial results. Surg. Sci. **11**(4), 82–88 (2020)
7. Pereira, R.M., Bertolini, D., Teixeira, L.O., Silla, C.N. Jr., Costa, Y.M.: COVID-19 identification in chest X-ray images on flat and hierarchical classification scenarios. Comput. Meth. Prog. Biomed. **194**, 1–18 (2020)
8. Dong, D., Tang, Z., Wang, S., Hui, H., Gong, L., Lu, Y., et al.: The role of imaging in the detection and management of COVID-19: a review. IEEE Rev. Biomed. Eng. (2020)
9. Yu, Z., Li, X., Sun, H., Wang, J., Zhao, T., Chen, H., et al.: Rapid identification of COVID-19 severity in CT scans through classification of deep features. BioMed. Eng. OnLine **19**(63), 1–13 (2020)
10. Masot, J.C., Perejon, F.L., Morales, M.D., Civit, A.: Deep learning system for COVID-19 diagnosis aid using X-ray pulmonary images. Appl. Sci. **10**(4640), 1–10 (2020)
11. Mei, X., Lee, H., Diao, K., Huang, M., Lin, B., Liu, C., et al.: Artificial intelligence–enabled rapid diagnosis of patients with COVID-19. Nat. Med. **26**, 1224–1228 (2020)

12. Li, L., Qin, L., Xu, Z., Yin, Y., Wang, X., Kong, B., et al.: Artificial intelligence distinguishes COVID-19 from community acquired pneumonia on chest CT. Radiology **296**(2), E65–E72 (2020)

13. Harmon, S.A., Sanford, T.H., Xu, S., Turkbey, E.B., Roth, H., Xu, Z., et al.: Artificial intelligence for the detection of COVID-19 pneumonia on chest CT using multinational datasets. Nat. Commun. **11**(4080), 1–7 (2020)

14. Apostolopoulos, I.D., Mpesiana, T.A.: COVID-19: automatic detection from X-ray images utilizing transfer learning with convolutional neural networks. Phys. Eng. Sci. Med. **43**, 635–640 (2020)

15. Ko, H., Chung, H., Kang, W.S., Kim, K.W., Shin, Y., Kang, S.J., et al.: COVID-19 pneumonia diagnosis using a simple 2D deep learning framework with a single chest CT image: model development and validation. J. Med. Internet Res. **22**(6), e19569 (2020)

16. Hurt, B., Kligerman, S., Hsiao, A.: Deep learning localization of pneumonia: 2019 coronavirus (COVID-19) outbreak. J. Thorac. Imaging **35**(3), W87–W89 (2020)

17. Singh, D., Kumar, V., Vaishali, Kaur, M.: Classification of COVID-19 patients from chest CT images using multi-objective differential evolution–based convolutional neural networks. Eur. J. Clin. Microbiol. Infect. Dis **39**, 1379–1389 (2020)

18. Yoo, S.H., Geng, H., Chiu, T.L., Yu, S.K., Cho, D.C., Heo, J., et al.: Deep learning-based decision-tree classifier for COVID-19 diagnosis from chest X-ray imaging. Front. Med. **7**, 1–8 (2020)

19. Ni, Q., Sun, Z.Y., Qi, L., Chen, W., Yang, Y., Wang, L., et al.: A deep learning approach to characterize 2019 coronavirus disease (COVID-19) pneumonia in chest CT images. Eur. Radiol. 1–11 (2020)

20. Rajaraman, S., Siegelman, J., Alderson, P.O., Folio, L.S., Folio, L.R., Antani, S.K.: Iteratively pruned deep learning ensembles for COVID-19 detection in chest X-rays. IEEE Acc. **8**, 115041–115050 (2020)

21. Brunese, L., Mercaldo, F., Reginelli, A., Santone, A.: Explainable deep learning for pulmonary disease and coronavirus COVID-19 detection from X-rays. Comput. Meth. Prog. Biomed. **196**, 1–11 (2020)

22. Butt, C., Gill, J., Chun, D., Babu, B.A.: Deep learning system to screen coronavirus disease 2019 pneumonia. Appl. Intell. 1–7 (2020)

23. Arpaci, I., Alshehabi, S., Emran, M.A., Khasawneh, M., Mahariq, I., Abdeljawad, T., et al.: Analysis of twitter data using evolutionary clustering during the COVID-19 pandemic. CMC **65**(1), 193–203 (2020)

Intelligent Systems and Novel Coronavirus (COVID-19): A Bibliometric Analysis

Mostafa Al-Emran⑩ and Ibrahim Arpaci

Abstract In late 2019, a novel coronavirus (COVID-19) was determined in Wuhan, China. The newly emerged epidemic has spread rapidly, with an increasing number of confirmed cases worldwide. While intelligent systems have been immensely tested and implemented across a wide range of health problems, the emergence of COVID-19 requires the need to use these systems in detecting, identifying, and preventing its outbreak. By using the bibliometric analysis approach, this research aims to provide a holistic view on the state-of-the-art research concerning intelligent systems and COVID-19 by analyzing the most used keywords, most cited articles and journals, most productive countries and institutions, most cited authors, and the role of intelligent systems during the COVID-19 outbreak. The results indicated that the existing research studies on intelligent systems during the COVID-19 outbreak have mainly concentrated on the use of machine learning algorithms in identifying and diagnosing the potential COVID-19 cases and predicting its extinction time. However, the number of articles published on the role of intelligent systems during COVID-19 pandemic is relatively few, suggesting that research in this field is still in its early stages, and more intensive research is required.

Keywords Intelligent systems · Artificial intelligence · Machine learning · Novel coronavirus · COVID-19

M. Al-Emran (✉)
Faculty of Engineering & IT, The British University in Dubai, Dubai, UAE
e-mail: mustafa.n.alemran@gmail.com

I. Arpaci
Department of Computer Education and Instructional Technology,
Tokat Gaziosmanpasa University, Tokat, Turkey
e-mail: ibrahim.arpaci@gop.edu.tr

1 Introduction

In December 2019, a novel coronavirus (COVID-19) was determined in Wuhan, China [1]. It is imperative to report that the number of confirmed cases of COVID-19 has exceeded those of severe acute respiratory syndrome (SARS) [2]. As of today (June 11, 2020), the number of confirmed cases becomes 7,404,092, and more than 416,598 death cases were recorded worldwide [3]. Both SARS and COVID-19 spread quickly through countries, infect humans and animals, and employ the same mechanism to enter and infect the body cell [2].

While intelligent systems, including artificial intelligence (AI) applications and machine learning algorithms, have been vastly tested and employed across various sectors in general [4–7] and the healthcare sector in specific [8], the newly emerged epidemic requires the need to use these systems in detecting, identifying, and preventing its outbreaks. It is argued that AI techniques would cause a paradigm shift in the healthcare sector, and this might assist the employment of these techniques to the existing COVID-19 outbreaks [2]. While the involvement of public health officials and specialist epidemiologists cannot be substituted, AI techniques can serve to manipulate the rapidly emerging data to support the public health experts in complex decision-making [9]. Besides, AI techniques might help in developing more precise symptom screening to predict the probability of COVID-19 infection [9].

In light of these arguments, the purpose of this research is to provide a holistic view on the state-of-the-art research concerning intelligent systems and novel coronavirus (COVID-19) by analyzing the most used keywords, most cited articles and journals, most productive countries and institutions, most cited authors, and the role of intelligent systems during the COVID-19 outbreaks through the use of a bibliometric analysis approach.

2 Method

This research follows a bibliometric analysis approach for analyzing COVID-19 related literature. Bibliometric analysis is a statistical approach used to determine the intellectual structure and improvement of a scientific domain [10]. The bibliometric analysis is usually used to determine the development of publications over time, the most dominant scholars, the most effective studies, and the subjects related to a particular research domain [11]. To draw a comprehensive picture of the COVID-19 and the intelligent systems/techniques used during its outbreak, the bibliometric analysis has been used in this research through the VOSviewer tool. To analyze the extant literature on COVID-19, the relevant articles were collected from the Web of Science database on 16th March 2020.

3 Results and Discussion

3.1 *Most Used Keywords*

The bibliometric mapping analysis suggested that the most used author keywords in the analyzed studies were "2019-ncov" (f = 9), "coronavirus" (f = 8), and "Wuhan" (f = 3). Figure 1 illustrates the most used author keywords in the analyzed studies. Further, the bibliometric analysis with full counting indicated that "disease" (f = 11), "analysis" (f = 9), "affected bone" (f = 6), "range" (f = 5), "oral cavity" (f = 5), and "ace2" (f = 5) were the most used words in the abstracts of the analyzed studies. Figure 2 illustrates the most used words in the abstracts. These results would assist scholars in finding a sufficient number of COVID-19 articles by referring to the aforementioned keywords rather than simply searching for "COVID-19" or "novel coronavirus".

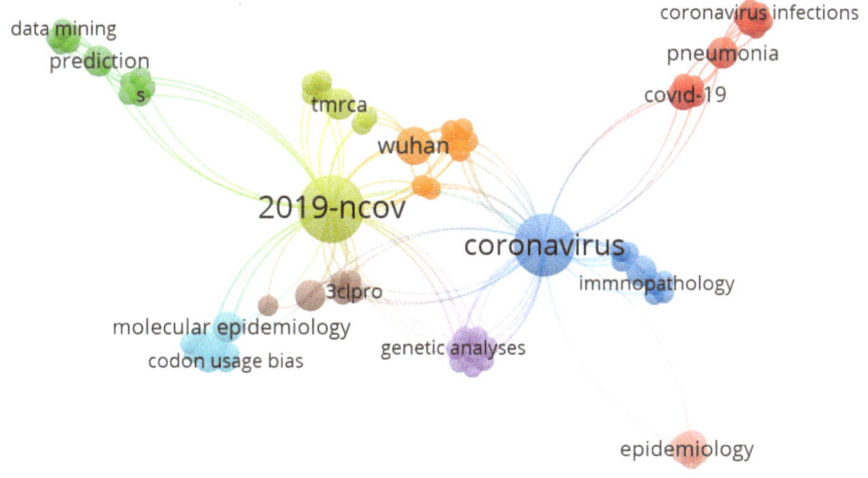

Fig. 1 Most used author keywords

Fig. 2 Most used words in
the abstracts

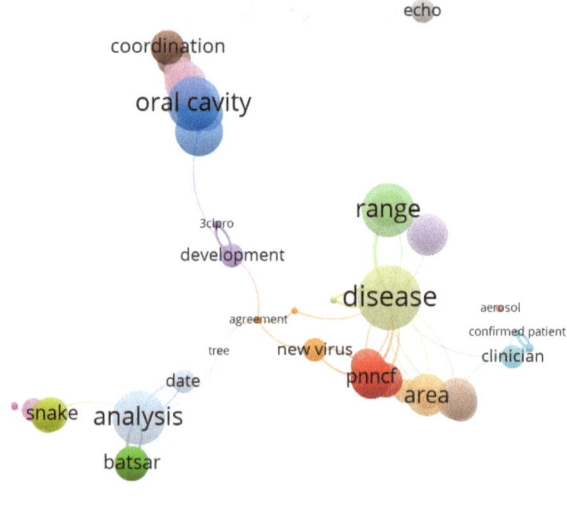

3.2 Most Cited Articles and Journals

As shown in Fig. 3, the bibliometric mapping analysis results showed that studies
conducted by Hui et al. [12] (11 citations), Corman et al. [13] (9 citations), Ji et al.
[14] (9 citations), Wang et al. [15] (6 citations), and The Lancet [16] (4 citations)
were the most cited articles. The increasing number of citations for these articles
stems from the fact in which these articles have provided a comprehensive
understanding of the COVID-19 outbreaks at the time where most researchers are in
need of this information.

Further, the results revealed that "The Lancet" (12 articles, 6 citations),
"Eurosurveillance" (10 articles, 11 citations), and "Journal of Medical Virology" (8
articles, 14 citations) were the most cited journals. The identification of the most
productive journals would assist scholars in finding COVID-19 related articles or
publishing their prospective research studies.

3.3 Most Productive Countries and Institutions

The bibliometric mapping analysis results also indicated that China (25 articles, 42
citations), the USA (14 articles, 15 citations), England (9 articles, 20 citations),
Germany (7 articles, 20 citations), and Italy (7 articles, 11 citations) were the most

Fig. 3 Most cited articles

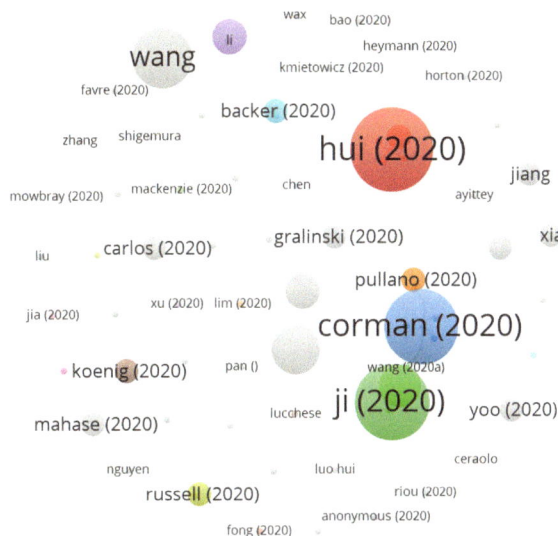

productive countries. These results stem from the fact in which these countries were the most infected environments by COVID-19 [3], and the scholars in these countries are continually working on finding the appropriate drug or vaccine to hinder its outbreaks.

In addition, the Chinese Academy of Sciences (4 articles, 7 citations), Charité University Medicine Berlin (3 articles, 20 citations), Alfaisal University (3 articles, 11 citations), Ningbo University (3 articles, 11 citations), and Wuhan Institute of Bioengineering (3 articles, 11 citations) were found as the most productive institutions in publishing COVID-19 related articles.

3.4 Most Cited Authors

As illustrated in Fig. 4, the bibliometric mapping analysis results suggested that Huang et al. [17] (20 citations), Chan et al. [18] (13 citations), Zhu et al. [1] (10 citations), and Zhou et al. [19] (8 citations) were the most cited co-citation authors. It is imperative to mention that these are the citations received from the Web of Science database. In fact, these citations are much higher in the Scopus database or Google Scholar.

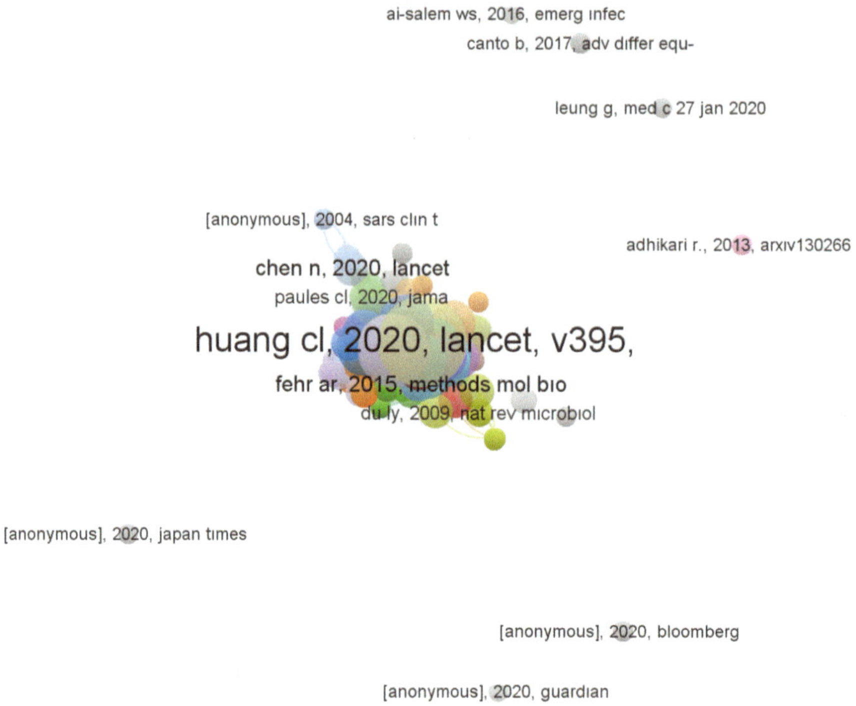

ai-salem ws, 2016, emerg ınfec

canto b, 2017, adv dıffer equ-

leung g, med c 27 jan 2020

[anonymous], 2004, sars clın t

chen n, 2020, lancet

adhikari r., 2013, arxıv130266

paules cl, 2020, jama

huang cl, 2020, lancet, v395,

fehr ar, 2015, methods mol bıo

du ly, 2009, nat rev microbıol

[anonymous], 2020, japan tımes

[anonymous], 2020, bloomberg

[anonymous], 2020, guardıan

Fig. 4 Most cited co-citation authors

3.5 Role of Intelligent Systems During COVID-19 Outbreaks

The applications of intelligent systems, including AI techniques to the current COVID-19, such as predicting the extinction time, potential risk groups, and location of the next outbreak, may cause a paradigm shift in the healthcare sector [2]. There were important studies on the use of intelligent systems in detecting the older versions of coronavirus such as SARS-CoV [20]. For example, Quek et al. [21] used an intelligent medical decision support tool based on fuzzy neural networks to screen the potential SARS-CoV patients. Besides, Yang [22] developed a new approach for building non-orthogonal decision trees for mining protease data in order to examine the specificity of cleavage activity of SARS-CoV and employ the resulted patterns for efficient inhibitor design to fight the virus.

Despite its new outbreaks, there are some completed and ongoing studies on the application of intelligent systems to the current COVID-19 epidemic [9]. In that, two independent research groups reported that they had used AI algorithms to find possible treatments, such as stopping the COVID-19 from replicating in humans' bodies [23]. In another ongoing study, Rao and Vazquez [24] aimed to employ machine learning algorithms to identify potential COVID-19 cases based on travel history data collected via a phone-based survey. The findings might help in early identifying the potential COVID-19 cases and classifying them as high, moderate, minimal, and no risk groups. Hu et al. [25] used AI techniques for predicting the extinction time of COVID-19 across China. In that, clustering algorithms were used to classify nine provinces to build the transmission structure based on the data obtained from WHO. Multiple-step forecasting was conducted to predict the dynamic transmission curves. Their results suggested that the COVID-19 epidemic will be over by the mid of April across China, given that there is no second transmission. In another study, Peng et al. [26] used AI to analyze the COVID-19 diagnosis association index to improve diagnosis accuracy. Four AI techniques (i.e., SRLSR, ARMED, GFS, and RFE) were employed based on 32 diagnosed and 85 undiagnosed cases from Taizhou Hospital, Zhejiang. Their results suggested that the most important attributes in an accurate prediction were Eosinophil count and rate, WBC, Amyloid-A, and COVID-19 RNA.

4 Conclusion

The main purpose of this research is to provide a holistic view on the state-of-the-art research indexed in the Web of Science database with regard to intelligent systems and novel coronavirus (COVID-19) by analyzing the most used keywords, most cited articles and journals, most productive countries and institutions, most cited authors, and the role of intelligent systems during the COVID-19 outbreaks. This has been accomplished through a bibliometric analysis approach.

It can be derived from the results that research on the topic of COVID-19 is still evolving, and tens of publications are getting published every day. The recently published articles on the topic of COVID-19 are significantly contributing to the understanding of the COVID-19 outbreaks at the time where most scholars are intensively working on the problem. However, the number of articles published on the role of intelligent systems during COVID-19 outbreaks is very few, suggesting that research in this field is still in its early stages. Given that the research into the COVID-19 was started in 2020, this also indicates that the role of intelligent systems is still scarce.

The extant research studies on intelligent systems during the COVID-19 outbreaks have mainly concentrated on the use of machine learning algorithms in identifying the potential COVID-19 cases [24], predicting the extinction time of COVID-19 [25], and diagnosing the COVID-19 cases [26]. In spite of all these trials, more intelligent systems through AI techniques are required to provide more insights into the COVID-19 epidemic.

In conclusion, the role of intelligent systems in predicting or identifying the COVID-19 outbreaks is still in its development stage, and more intensive research is required. These results agree with the conclusions drawn in a recent study, which suggested that AI techniques have not yet been impactful against COVID-19 [27]. The need for more research stems from two main reasons. First, there is still much to be examined through machine learning algorithms by predicting the infections using different attributes other than those identified in the previous literature. Second, the growing number of death cases day-by-day increases the possibilities in finding more effective solutions by medical informatics researchers.

It is believed that this study would provide valuable insights into the research published on the topic of COVID-19 and intelligent systems. The conclusions of this bibliometric analysis can assist scholars in understanding the current status of intelligent systems and motivating them to develop more effective systems that could help in detecting, identifying, or preventing the COVID-19 outbreaks.

References

1. Zhu, N., et al.: A novel coronavirus from patients with pneumonia in China, 2019. N. Engl. J. Med. (2020). https://doi.org/10.1056/NEJMoa2001017
2. McCall, B.: COVID-19 and artificial intelligence: protecting health-care workers and curbing the spread. Lancet Digit. Heal. (2020). https://doi.org/10.1016/S2589-7500(20)30054-6
3. Worldometers.: COVID-19 coronavirus outbreak. Worldometers. (2020). https://www.worldometers.info/coronavirus/
4. Arpaci, I., et al.: Analysis of twitter data using evolutionary clustering during the COVID-19 pandemic. Comput. Mater. Contin. **65**(1), 193–203 (2020). https://doi.org/10.32604/cmc.2020.011489
5. Al-Emran, M.: Hierarchical reinforcement learning: a survey. Int. J. Comput. Digit. Syst. **4**(2), 137–143 (2015)
6. Mhamdi, C., Al-Emran, M., Salloum, S.A.: Text mining and analytics: a case study from news channels posts on Facebook. **740** (2018)
7. Zaza, S., Al-Emran, M.: Mining and exploration of credit cards data in UAE. In: Proceedings —2015 5th International Conference on e-Learning, ECONF 2015, pp. 275–279 (2015). https://doi.org/10.1109/econf.2015.57
8. Bian, J., Modave, F.: The rapid growth of intelligent systems in health and health care. Health Inf. J. (2020)
9. Long, J.B., Ehrenfeld, J.M.: The role of augmented intelligence (AI) in detecting and preventing the spread of novel coronavirus. J. Med. Syst. (2020). https://doi.org/10.1007/s10916-020-1536-6
10. Culnan, M.J.: The intellectual development of management information systems, 1972–1982: a co-citation analysis. Manage. Sci. (1986). https://doi.org/10.1287/mnsc.32.2.156
11. Veloutsou, C., Ruiz Mafe, C.: Brands as relationship builders in the virtual world: a bibliometric analysis. Electron. Commer. Res. Appl. (2020). https://doi.org/10.1016/j.elerap.2019.100901
12. Hui, D.S., et al.: The continuing 2019-nCoV epidemic threat of novel coronaviruses to global health—the latest 2019 novel coronavirus outbreak in Wuhan, China. Int. J. Infect. Dis. (2020). https://doi.org/10.1016/j.ijid.2020.01.009
13. Corman, V.M., et al.: Detection of 2019 novel coronavirus (2019-nCoV) by real-time RT-PCR. Euro. Surveill. (2020). https://doi.org/10.2807/1560-7917.ES.2020.25.3.2000045
14. Ji, W., Wang, W., Zhao, X., Zai, J., Li, X.: Cross-species transmission of the newly identified coronavirus 2019-nCoV. J. Med. Virol. (2020). https://doi.org/10.1002/jmv.25682
15. Wang, W., Tang, J., Wei, F.: Updated understanding of the outbreak of 2019 novel coronavirus (2019-nCoV) in Wuhan, China. J. Med. Virol. (2020). https://doi.org/10.1002/jmv.25689

16. The Lancet: Emerging understandings of 2019-nCoV. Lancet (2020). https://doi.org/10.1016/S0140-6736(20)30186-0
17. Huang, C., et al.: Clinical features of patients infected with 2019 novel coronavirus in Wuhan, China. Lancet (2019). https://doi.org/10.1016/S0140-6736(20)30183-5
18. Chan, J.F.W., et al.: A familial cluster of pneumonia associated with the 2019 novel coronavirus indicating person-to-person transmission: a study of a family cluster. Lancet (2020). https://doi.org/10.1016/S0140-6736(20)30154-9
19. Zhou, P., et al.: A pneumonia outbreak associated with a new coronavirus of probable bat origin. Nature (2020). https://doi.org/10.1038/s41586-020-2012-7
20. Eysenbach, G.: SARS and population health technology. J. Med. Internet Res. (2003). https://doi.org/10.2196/jmir.5.2.e14
21. Quek, C., Irawan, W., Ng, E.Y.K.: A novel brain-inspired neural cognitive approach to SARS thermal image analysis. Expert Syst. Appl. (2010). https://doi.org/10.1016/j.eswa.2009.09.028
22. Yang, Z.R.: Mining SARS-CoV protease cleavage data using non-orthogonal decision trees: a novel method for decisive template selection. Bioinformatics (2005). https://doi.org/10.1093/bioinformatics/bti404
23. Lemonick, S.: Two groups use artificial intelligence to find compounds that could fight the novel coronavirus. Comput. Chem. (2020)
24. Rao, A.S.R.S., Vazquez, J.A.: Identification of COVID-19 can be quicker through artificial intelligence framework using a mobile phone-based survey in the populations when cities/towns are under quarantine. Infect. Control Hosp. Epidemiol. (2020). https://doi.org/10.1017/ice.2020.61
25. Hu, Z., Ge, Q., Jin, L., Xiong, M.: Artificial intelligence forecasting of covid-19 in china. arXiv Prepr. (2020)
26. Peng, M., et al.: Artificial intelligence application in COVID-19 diagnosis and prediction. Lancet. (2020)
27. Naudé, W.: Artificial intelligence vs COVID-19: limitations, constraints and pitfalls. Ai Soc. 1–5 (2020)

Computational IT Tool Application for Modeling COVID-19 Outbreak

Viroj Wiwanittkit and Suphatra Wayalun

Abstract Infection is an important problem in clinical medicine. Emerging infectious disease outbreak usually causes public health problem. Coronavirus disease 2019 (COVID-19) is a new emerging coronavirus infection that causes pandemic in 2020. The outbreak affects more than 200 countries around the world, causing illness in more than 11 million of world population. To understand the outbreak, in the depth epidemiological analysis is important. The use of novel information technology for assessment of the outbreak is interesting. The informatics technology is accepted for its advantage in public health. One of the important advantages is the clarification of the public health phenomenon. The IT tool and technique are applicable for study on outbreak of emerging infectious disease. The IT approach can help clarify the pattern of outbreak and can further give important data for disease containment. The IT tool can identify the descriptive details of the outbreak and it can also show the interrelationship between emerging infection to time and place. Additionally, the IT tool can also help further predict the trend or progression of the outbreak. This is a useful application for further public health policies planning for outbreak containment. In this chapter, the authors summarize and discuss on the usefulness and application of IT technology and tool for managing COVID-19. The examples are also given to help readers recognize the IT application for corresponding to COVID-19 outbreak. Using the standard tools, the clarification and prediction for the COVID-19 outbreak can be

V. Wiwanittkit (✉)
Dr DY Patil University, Pune, India
e-mail: wviroj@yahoo.com

V. Wiwanittkit
Hainan Medical University, Haikou, China

V. Wiwanittkit
Joseph Ayobabalola University, Ikeji Arakeji, Ilesa, Nigeria

V. Wiwanittkit
Faculty of Medicine, University of Nis, Nis, Serbia

S. Wayalun
Rajabhat University, Surin, Thailand
e-mail: bc2011com@gmail.com

© The Author(s), under exclusive license to Springer Nature Switzerland AG 2021
I. Arpaci et al. (eds.), *Emerging Technologies During the Era of COVID-19 Pandemic*, Studies in Systems, Decision and Control 348,
https://doi.org/10.1007/978-3-030-67716-9_6

successfully done. The IT tools are helpful for clinical epidemiological manipulation in public health management against COVID-19. Nevertheless, the use of the IT tool requires good primary data and there might be a problem if the collection technique of primary data from field work is not good.

Keywords COVID-19 · IT · Outbreak · Model

1 Introduction

Of the several groups of medical disorders, infection is an important group of disease. There are several kinds of infections such as bacterial, virus and parasitic infection. Some infectious diseases are endemic while the others sporadically occur. The occurrence of the new infectious that has never existed is an important consideration. The occurrence of a never existed infectious disease is known as "emerging infectious disease". A new emerging infectious disease is usually an important concern in public health. Since the new emerging infection is a new thing and there is usually no knowledge of it, therefore, it is difficult to diagnose, treat and prevent for the new emerging infection.

Emerging infectious disease outbreak usually causes public health problems and if there is no good control, a wide outbreak might occur [1]. Within the past decade, there are many new emerging infectious diseases. Many new infections become big problems globally. The good example is Zika virus infection, which is an arbovirus infection that is related to the congenital anomaly. However, a more problematic situation has just been caused by a new emerging coronavirus infection. The disease was firstly reported from Wuhan, China and it was firstly called Wuhan novel coronavirus infection. After its first occurrence in China, it was spread to many countries, starting from Thailand and others. The disease becomes the global public health issue and it is presently named as Coronavirus disease 2019 (COVID-19) [2].

COVID-19 is a new emerging coronavirus infection that already caused pandemic in 2020. At present (11 July 2020), the COVID-19 outbreak affects more than 200 countries around the world, causing illness in more than 11 million of world population. As a new emerging disease, scientist has to work hard to collect data on this new disease and seek the way to diagnose, treat and prevent the disease. Since the disease is a respiratory tract infection that can be easily transmitted via respiratory contact, the outbreak can easily occur and the rapid spread from a patent to the other/others is possible. The outbreak already occurs worldwide and there is an urgent need to contain it. In public health, the first step to contain any outbreak is the realization of its nature. To understand the outbreak, in the depth epidemiological analysis is necessary. Data collection, analysis and interpretation is the basis process in epidemiological assessment of any outbreak.

The classic approach for epidemiological assessment might be time consuming and requires a lot of the workforce. It is possible to use of novel information technology for assessment of the outbreak. Indeed, the informatics technology is accepted for its advantage in medicine and public health [3]. Theoretically, the combination between computational science and biomedicine is possible and can result to a new useful science. Biochemioinformatics is the good example of bridging science that conjoins between computational science and biomedicine. The medical informatics approach is proven for its usefulness in public health.ne of the important advantages is the clarification [4]. The medical informatics can help clarify the nature of public health problem. Additional, medical informatics can be used for prediction of the public health phenomenon. This is a useful application for future trend analysis and policy planning. The medical informatics can be applied for managing many medical problems, including to infectious disease.

In general, an emerging infectious disease usually caused the problem. As a new problem, there is usually a lack of knowledge on the new disease. It is usually not possible to give the correct diagnosis and treatment of the disease during the early outbreak period. Gathering of data of the disease is very important for public health manipulation. The data collection from classic public health surveillance is necessary. Applying new technologies for data management is important. The new technology can facilitate the process and help further analyze of the collected data. Without a good data analysis, it is difficult to appropriately management of the public health problem. The new tool might help clarify the nature of the new public health problem. Additionally, it can also help predict the future based on limited presently available data.

Regarding infectious disease, the advantage of medical informatics is approved. The medical informatics can help clarify and predict for problems on infectious disease [4]. This is also applied for the case of new emerging infectious disease. Since the emerging infectious disease is usually a big public problem. During the pandemic, such as COVID-19 pandemic, it requires a good and effective tool for data analysis for further clarification or prediction purpose. The classical techniques might be applied but the old techniques usually take time and it might not properly correspond to the rapid change during the outbreak crisis. To manage the influx of data and rapid change of the situation during am outbreak, a new computational technology might be helpful. The technology might shorten turnaround time for data manipulation and help provide useful data for containment of the outbreak.

In general, the IT tool and technique is applicable for assessment on outbreak of emerging infectious disease. The IT approach is helpful for clarifying the pattern of outbreak and it can further give important data for disease control and outbreak containment. The IT tool can identify the pattern of the outbreak and it can also show the time and place dimensions of the emerging infection. Additionally, the IT tool can also help further predict the trend of the outbreak. This means a scientific forecast of the outbreak is possible. The application of medical informatics for managing emerging infectious disease outbreak is a useful application for further public health policies planning for outbreak containment [5]. In this chapter, the authors summarize and discuss on the usefulness of applied medical informatics technique for managing COVID-19. The examples are also given to help reader better understand the IT application for corresponding to COVID-19 outbreak.

2 Application of IT Technology for Infectious Disease Outbreak Management

An application of IT technology for managing infectious disease is an interesting topic. It can be applied in case of infectious disease outbreak management (Table 1). There are several applications that are useful. The purpose of the application might be for clarification or prediction of the outbreak [4]. The classic IT technique can be well applied to achieve the aim on infectious disease outbreak containment.

The important concepts of medical IT for managing infectious disease outbreak include (a) database, (b) IT tool, (c) concept of clarification on infectious disease outbreak and (d) concept of prediction for infectious disease outbreak.

a. Database

Data is important for any IT manipulation. The data on the outbreak is an important source for further analysis and interpretation. The IT technology is useful for data collection, analysis and management. Several classic computer programs are useful for data management, such as Dbase, Excel, SPSS, etc. Additional, the IT database is also developed as data source for further mining.

In public health, there are many important databases on infectious disease. The example of a well-known public available database is PubMed (www.pubmed.com) [6]. The public database is the basic requirement for any further IT manipulation.

b. IT tool

IT tool is an important instrument for computational manipulation. At present, several IT tools are developed an available. Some IT tools are freely available and

Table 1 Application of IT technology for infectious disease outbreak management

Purpose	Examples of application
Clarification	• Natural history of infection
	• Spreading pattern
	• Host characteristic and pathogen characteristic
	• Background genetic and genomic pattern of infection
	• Endemic area and affected area of outbreak identification
	• Geographical distribution of outbreak
	• Chronological pattern of outbreak
	• Seasonal fluctuation of outbreak
Prediction	• Trend of disease transmission
	• Analysis of mutation possibility
	• Economic impact prediction
	• Cost effectiveness and cost utility analysis
	• Prediction of outbreak containment policies

helpful for medical informatics application for managing infectious disease outbreaks. The groups of IT tools that are useful for managing infectious disease outbreaks include database tool and manipulation tool. Database tool is the computation database. The database tool that is freely accessible via the internet is actually useful for medical informatics research. The good example is PubMed as already mentioned. This is an example of many IT manipulation tools that are freely available [6]. The tools in bioinformatics are good examples. The genomics tool can be helpful for comparative analysis and prediction. The molecular epidemiology of the pathogen that causes infectious disease outbreaks might be assessed by the biochemioinformatics tool. The phylogenetic research and gene ontology research are good examples [4].

c. Concept of clarification on infectious disease outbreak

The basic concept of clarification on infectious disease outbreak is to identify the characteristics and natural course of infectious disease outbreak. The first step is data collection by any technique, manual or electronic recording. The data collection is primarily performed and it has to be further manipulated by computational IT tool. The IT tool is useful for categorizing and grouping of data. It is also useful to identify interrelationships in time and place (space) dimension. A good example is the construction of GIS map showing infectious disease outbreak pattern using IT tool. This is an applicable for modelling geographical pathology of infectious disease outbreak. The IT tool is also useful for modeling the path of disease spreading or flow of outbreak which is a useful data for planning proper plan for outbreak containment.

d. Concept of prediction for infectious disease outbreak.

The basic concept of prediction for infectious disease outbreak is to present the possibility or trend of infectious disease outbreak. Similarly, the data collection is primarily performed and data must be further manipulated by computational IT tool. The IT tool is useful for finding relationship and construction of predictive model for prediction on the trend of infectious disease outbreak. A good example is the prediction of future size and area coverage of infectious disease outbreak.

3 Applied IT Technology for Modeling COVID-19 Outbreak

As already mentioned, applied IT tool is useful for understanding the emerging infectious disease outbreak. Modelling of outbreak is a useful application of IT tool. It is applicable for any outbreak including to new emerging disease outbreak. Regarding COVID-19, it is also useful. For modeling of the outbreak, the basic concept is to find a representative of the course and nature of infectious disease outbreak. Classically, the modelling of an outbreak is possible by using simple mathematical and statistical technique [6]. This is applicable for both clarification and prediction purposes. With advanced computational IT technology, the IT tool can play role in modeling of infectious disease outbreak. The application for the COVID-19 outbreak is also possible.

To create a mathematical model, the primary data from data collection on outbreak situation is necessary. The good data collection and verification for correctness is an important step. If there is error, or it is not reliable, further steps will not be acceptable. The important collected data will be used as primary parameters for further modeling process [7]. To create a mathematical model, finding for interrelationship among parameters has to be done. The mathematical equation will be written to represent the situation. The new IT tool will be useful for mathematical manipulation in modelling for infectious disease outbreak [7]. Regarding COVID-19 outbreak, there are some new reports on using medical informatics technology for modeling [8]. Those important reports are useful for public health policies planning against the COVID-19 pandemic. Due to the present uncontrollable COVID-19 pandemic, any available that might be applied for managing the problem should be used. The IT application might be a solution that helps about data manipulation regarding COVID-19 outbreak. How to apply a new computational technology for managing of the COVID-19 pandemic is an interesting research question. Here, the authors briefly discuss and present some examples of the IT application aiming at public health manipulation against COVID-19.

4 Examples of IT Application for Corresponding to COVID-19 Outbreak

As already mentioned, IT technology can be applied for corresponding to COVID-19 outbreak. This is for clarifying or predicting on COVID-19 outbreak. Here, the authors will give some brief examples of IT application for corresponding to COVID-19 outbreak that can help reader generate ideas for further application. The primary data for IT manipulation in all examples are public available data and the used dataset are the specific data of Thailand, a tropical country in Indochina, which is the standard official report recorded in referencing COVID-19 Time Series Data (https://data.world/shad/covid-19-time-series-data). Since all studied are in silico study without any human or animal subject involvement, the ethical approval by ethical committer is not required. The basic tools are used for these examples (Table 2).

a. IT tool for GIS modelling for COVID-19 outbreak

Here, the authors would like to present the example on GIS modeling by Map Check in for COVID-19 outbreak in Thailand. At first, the new virus was first discovered in an outbreak in Wuhan, Hubei Province, China, in late 2019 and quickly spreads to many countries. The first affected country outside China is

Table 2 Basic software tools used in the examples

No.	IT software tools	Descriptions
Map check in COVID-19		
1.	Google Earth Pro 7.3.3	Use Google Earth Pro Version 7.3.3
		– Map Check in
		– Report data Detection Screen
		– Report Confirmed COVID-19
		– Follow-Up Quarantine person
2.	Excel 2013	Database
Algorithm model		
1.	Raptor version 4.1.0.0001 15 November 2019	Rapid algorithmic prototyping tool for ordered reasoning: RAPTOR
		– Raptor is the software to use problem solving tool that enables the user to generate executable flowcharts and introduced to the computing discipline in order to develop problem solving skills and improve algorithmic thinking

Thailand [9]. Therefore, the Ministry of Public Health by the Department of Disease Control of Thailand. They prepared a situation report on the webpage Corona Virus Disease (COVID-19) domestically and globally, media knowledge, advice, including operational guidelines for agencies, related parties, and the general public to receive information thoroughly. Nowadays, People can use mobile phones to access to news information places including the status display and the coordinates that they live through the services of social media, whether Facebook, Instagram with check-in, and use the map service from Google Maps altogether.

Besides, the business community has adjusted marketing strategies to encourage customers to check-in at restaurants and various stores including product and service reviews. To be public relations, information and news that customers are interested in and want to receive services as well. Google Maps is a web mapping service developed by Google. It offers satellite imagery, aerial photography, street maps, 360° interactive panoramic Street View, real-time traffic conditions, and route planning for travel option consist of foot, car, bicycle, and air (in beta). In 2020, and in 2020, more than 1 million people are interested in using Google Maps each month [10], a User can display both browser and smartphone, both Android and IOS systems without any charge at all. Consequently, users prefer to use due to being convenient and easily accessible. In addition, Google has special projects such as improving public safety by helping users find convenient drug disposal locations, making invisible with project Air View, Get training and update with the Google News Lab, Discover your savings potential with Project Sunroof, Revealing the World's Fishing Fleet with Global Fishing Watch, Applying unique Google data and modeling capabilities to create new insights and action oriented tools to reduce GHG emissions, Earth Engine create a living map of forest loss, and Powering conservation science and storytelling using Google's mapping technologies, AI and the Cloud etc. [11]. The technique can help support GIS approach for geographical pathology application [12].

Here, the authors give an example on using the Google Maps tool developed by Google. Here, it focuses on how to apply the map check-in the area where Corona Virus Disease is found in Thailand by simulating check-in data from screening reports in Thailand at Suvarnabhumi Airport. Don Mueang Airport, Phuket Airport, Chiang Mai Airport and notifications from Erawan Medical Centre, University, Hotel, Guide Center and show the number of patients accumulated Number of new cases Number of deaths. The number of patients recovered and the Number of patients who can travel home from 1st February 2020 to 5th March 2020 as a guideline in applying technology to distribute information to the public Up to date on the situation of the novel coronavirus outbreak 2019. The Coronavirus 2019 detection map creation is from the Department of Disease Control of Thailand within 2 February 2020–5 March 2020, and there are steps for creating a check-in Covid-19 as followings.

The operation tools are Microsoft Excel 2010 software for database and Open-source Google Map application for applying medical informatics tool development. This paper Data used to create the maps check-in taken from Covid-19 infection status report by the Division of Disease Control of Thailand since 2 February, 2020 to 5 March 2020 (Department of Disease Control, 2020) and create a database with Microsoft Excel filename as "report.xlsx" In the first line, specify the desired field. In which we have defined 36 fields to collect data as Fig. 1. The coordinates can check by using a smartphone to notify the COVID-19 infection through the LINE application, as in Fig. 2. Longitude and latitude can send the location thought LINE mobile application or Line PC application, and the result will display the desired coordinates as shown in Fig. 3. The user can copy Latitude & Longitude to the file database in Excel. Then Google Sign in is done. The user should sing up the Google Account before start creates maps. Open Google Chrome browser and click browser sign in as Fig. 4 and click maps tools (Fig. 5) for further map construction process (Figs. 6, 7, 8, 9a, b, 10, 11, 12, 13 and 14).

Conclusively, the main idea of this example is using standard IT tool for construction of a GIS-based computational tool that is applicable for epidemiological

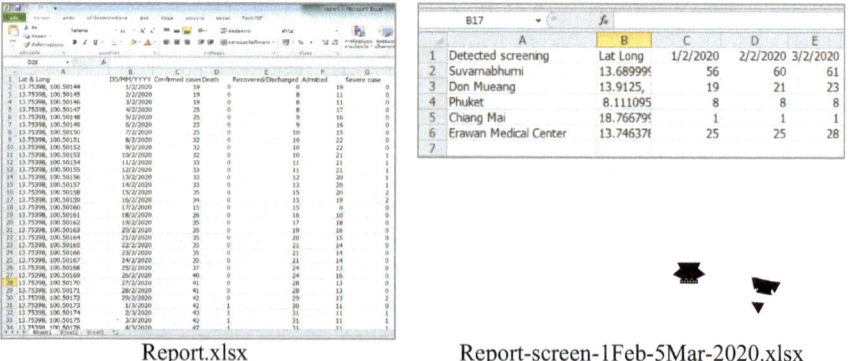

Report.xlsx   Report-screen-1Feb-5Mar-2020.xlsx

Fig. 1 Prepare database

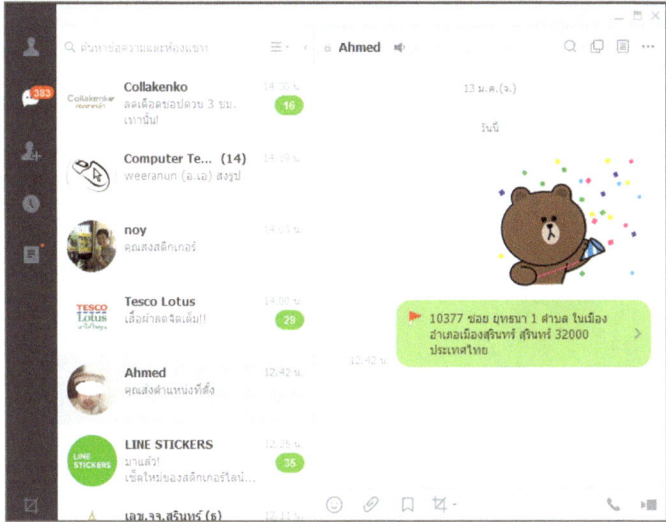

Fig. 2 Send location by line application

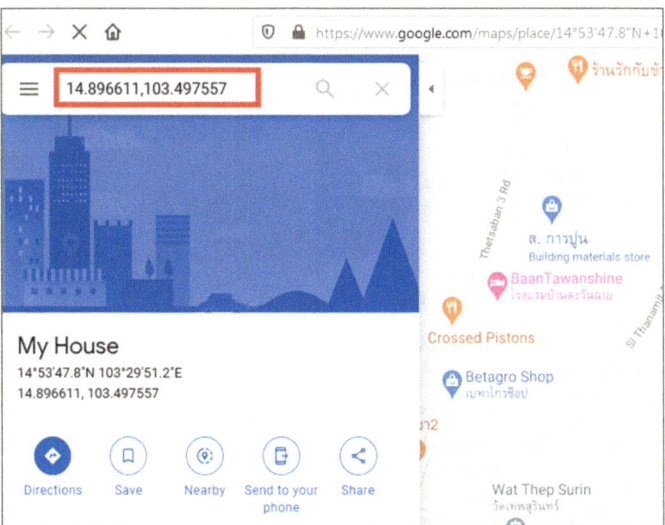

Fig. 3 Show latitude and longitude at Google Map

surveillance for COVID-19 outbreak. This work can confirm that IT-based GIS tool is useful for managing the outbreak crisis. This kind of IT tool might be designed and applied in any setting. The locally specific tool might be open to everyone in that setting for using as a referencing data for precautions on disease existence during the COVID-19 outbreak. This IT approach might be further adapted to cover

Fig. 4 Google sign in

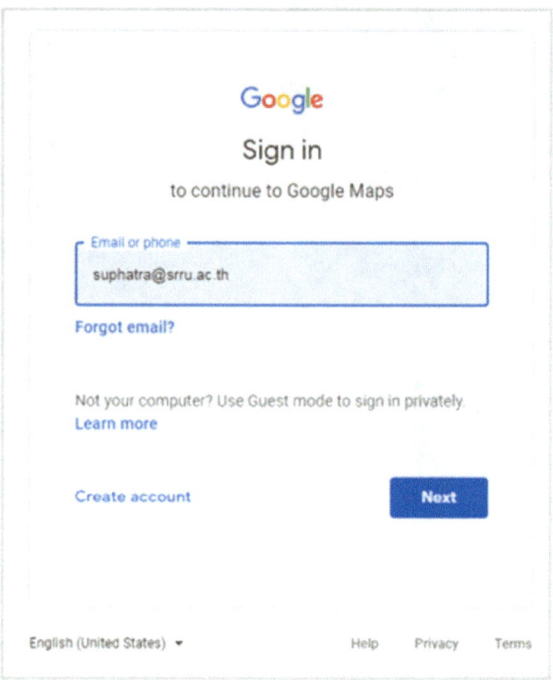

Fig. 5 Maps tools

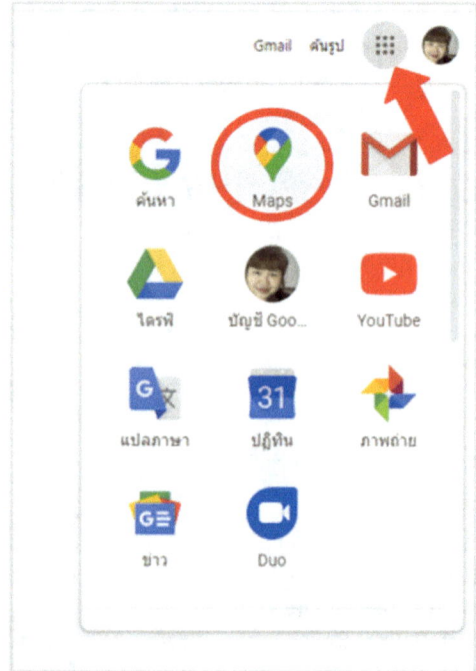

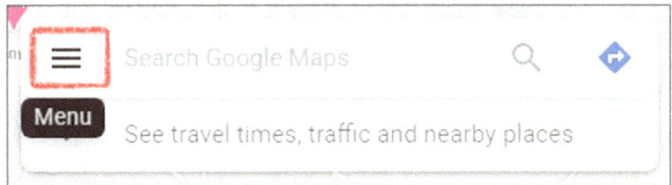

Fig. 6 Menu

Fig. 7 Your places

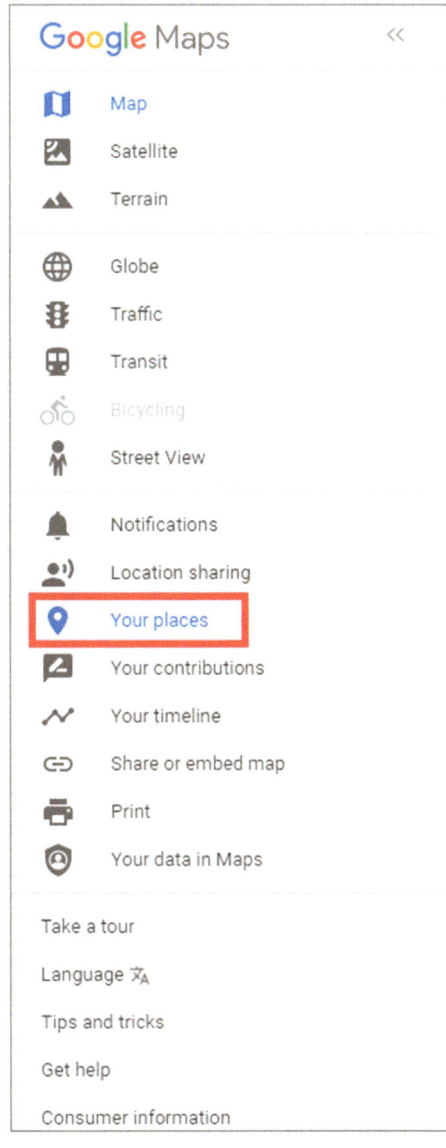

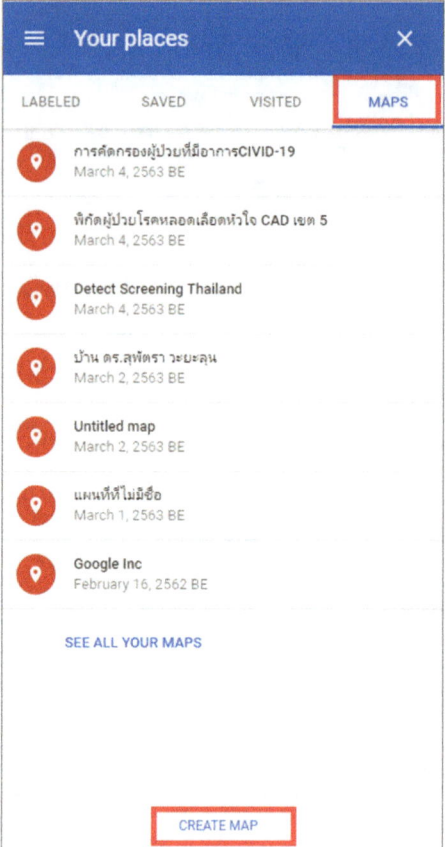

Fig. 8 Create map. In this step, change maps named as "Check-in Covid-19" and click "save

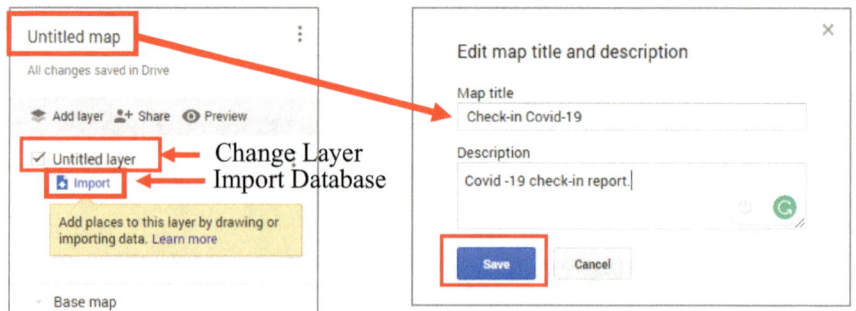

Fig. 9 **a** Change untitled map. **b** Edit map title and description

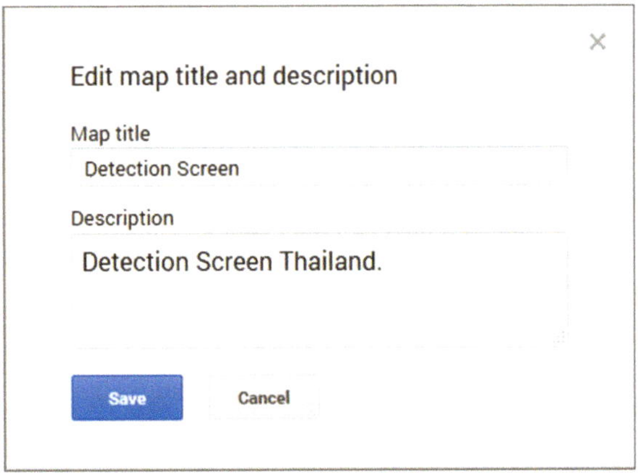

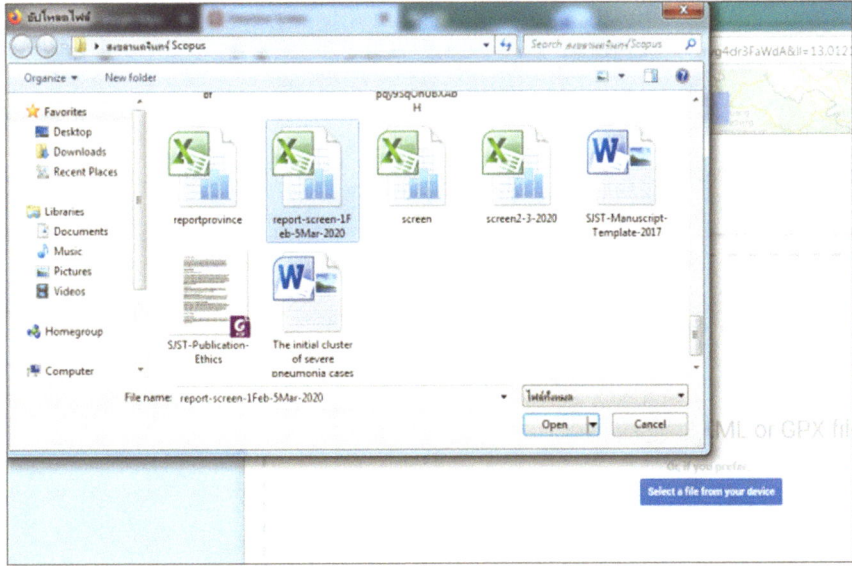

Fig. 10 Change layer name

Fig. 11 Select file report-screen-1 Feb–5 Mar-2020.xls

additional parameters that can help explain a more complex interrelationship between disease incidence and ecological background. For example, a GIS map might be constructed to represent interrelationship between incidence of COVID-19, mortality rate and background ecological factors in different settings [13].

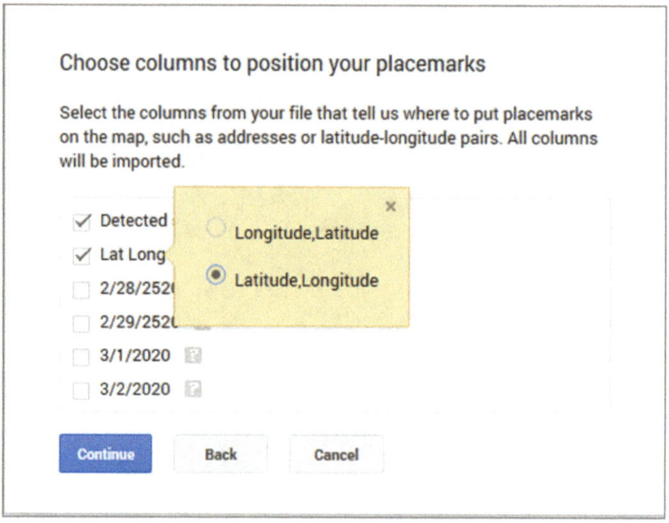

Fig. 12 Choose columns

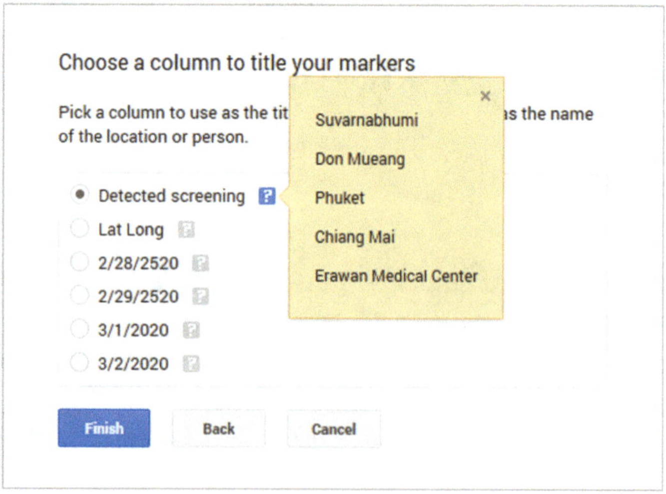

Fig. 13 Choose a column to title your makers

b. Distribution Tree Algorithm model in Southeast Asia's COVID-19 hotspot due to religious gathering in Malaysia

Malaysia became Southeast Asia's COVID-19 hotspot due to religious gatherings at the Petak Jamek Seri Petaling Mosque in Kuala Lumpur on 27 February 2020–1 March 2020 [14]. The situation of COVID-19 Infection in Malaysia on 23

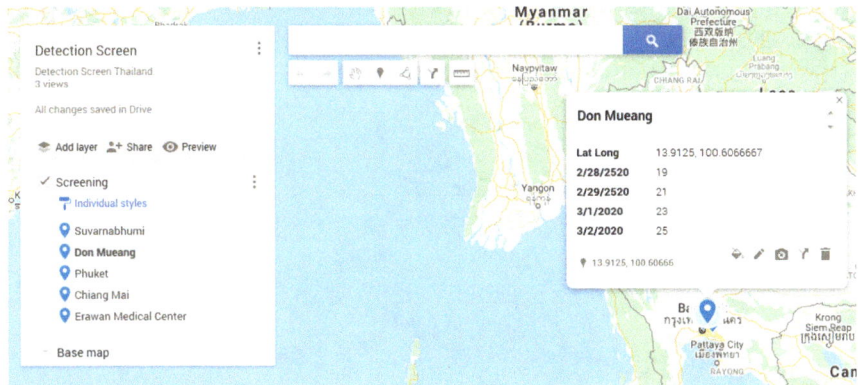

Fig. 14 Display an example marker in the position specified in the database

March 2020 reported 212 new coronavirus cases, bringing the national total to 1,518. The death toll has also increased to 14 of the new cases, 123 is from the cluster linked to the religious gathering in cluster account for 970. (62% of the total cases in the country), which is the highest number of infections in Southeast Asia. Moreover, it has led to infection in neighboring countries. The Muslim slept in paced tents outside the mosque, waking before down to knee on prayer mats rows lay out in this cavernous central hall. The COVID-19 coronavirus was passed unnoticed among them. Attendees Muslim held for 4 days worship at the Petak Jamek Seri Petaling mosque complex here has emerged as a source of hundreds of new coronavirus infections spanning Southeast Asia and unable to explain the distribution model. Therefore, this work, computational algorithm which is a good IT technique for modeling of outbreak is briefly mentioned [15, 16]. The method of Distribution Tree Algorithm model to explain the phenomenon of spread coronavirus has spread in all directions and uncontrolled. For modeling, a computational research and algorithm design is done. Open source software: RAPTOR software Version 5.129 for flowchart-based programming and output visualization is the instrument. Data source is from reported news since 17–23 March 2020 as in Table 3.

Analysis of the spread of the COVID-19 virus in religious gatherings at the Petak Jamek Seri Petaling mosque in Kuala Lumpur, Malaysia, where people gather for 4 days with people travelling from neighbor countries who do not know they are Infected with the Covid-19 virus. The integration of the infected person and normal people that has exposure to secretions of virus-infected people includes Coughing, sneezing, spit without careful caution by wearing a mask, using alcohol gels to

Table 3 The attendees at Petak Jamek Seri Petaling Mosque in Kuala Lumpur, Malaysia

Country	Attendees	Confirm case
1. Malaysia	14,500	970
2. Singapore	95	5
3. Brunei Darussalam	74	45
4. Cambodia	79	25
5. Philippines	215	77
6. Thailand	132	6
7. Vietnam	130	67
8. Indonesia	696	30
9. No country report	79	0
Total	16,000	1225

frequently clean hands, or spacing of at least 6 feet away, causing the virus to spread to many people and uncertain spread model, which can be explained from a tree in Fig. 15.

Distribution tree Analysis Unpredictable spread of the Model in Fig. 15 can explain as follows:

(1) The tree is a mosque that 16,000 religious gather at the ceremony on February 27, 2020–March 1, 1977. According to news reports, various agencies reported that the worshipers slept in the Mosque and tent outside the mosque in which

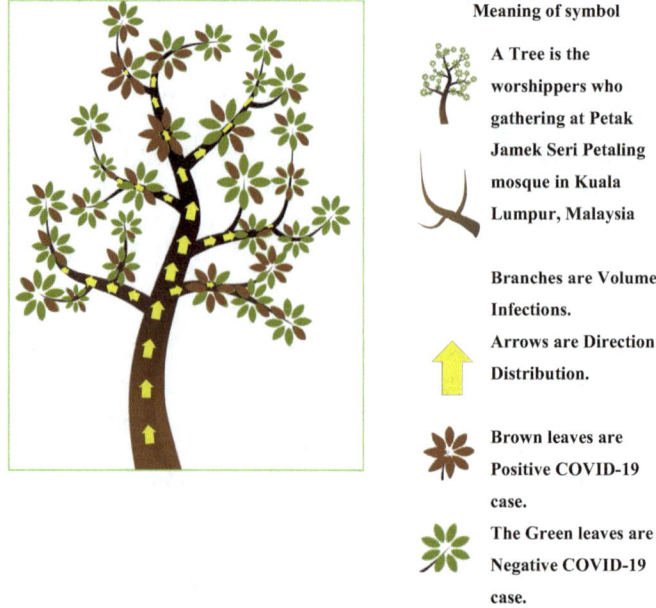

Meaning of symbol

A Tree is the worshippers who gathering at Petak Jamek Seri Petaling mosque in Kuala Lumpur, Malaysia

Branches are Volume Infections.
Arrows are Direction Distribution.

Brown leaves are Positive COVID-19 case.
The Green leaves are Negative COVID-19 case.

Fig. 15 Distribution tree analysis unpredictable spread of model

Table 4 Variable of distribution three algorithm model

No.	Field	Type of variable	Description
1	c_attendees	Number	Country attendee
2	n_attendees	Number	Number of attendees
3	sp	Number	Number of screening person
4	Result	Number	Result of screening type
5	c1	Number	Number of close up infected person
6	c2	Number	Number of the person who traveled to risk country
7	c3	Number	Number of the person who traveled to risk country
8	c4	Number	Number of Not wearing mask and social distance
9	confim_case	Number	Total of confirm case
10	N	Number	Number of positive case

those people have travelling in and out of the mosque and travelling round trip as well, therefore, the yellow arrow shows the direction of infinite spread and cannot predict in which direction it goes.

(2) The branches of the tree are volume Infections, that means a volume of infection from worshippers cohabitation behavior can expose to a coronavirus, including close up infected person, who travelled to risk country, leave outwear a mask and social distance, do not clean alcohol gel always, get other risk behavior, congenital disease, elderly, Obese etc.

(3) The arrows are a distribute direction. Coronavirus 2019 can move in any direction and beyond the control. Therefore, the spread of the COVID-19 virus is unpredicted and became Southeast Asia's COVID-19 hotspot.

(4) Brow leaves are positive COVID-19 cases, after the attendees leave out Malaysia to theirs country and they take screening. The result confirmed the case link to the Mosque Malaysia.

(5) The Green leaves are Negative COVID-19 cases after leaving out Malaysia and taking the screening. This satiation depends on their health and protection behavior or other reasons.

There are 10 variables in an algorithm model (Table 4).

The concepts mentioned from Distribution tree Analysis Unpredictable the spread of the Model in Fig. 1 and create an algorithm as follows.

1) Start.

2) Display 16,000 Attendees Religious assembly at at Petak Jamek Seri Petaling Mosque in Kuala Lumpur Malaysia" +"---"

3) Display 1 = Malaysia, 2 = Singapore, 3 = Brunei Darussalam, 4 = Cambodia, 5 = Philippines, 6 = Thailand, 7 = Vietnam, 8 = Indonesia, 9 = Not report"+"---------------------
--------------------------"

4) Input country attendees

5) Display Country attendees

7) Selection country attendees

Set Choice attendees value 1 = Malaysia, value 2 = Singapore, value 3 = Brunei Darussalam, value 4 = Cambodia, value 5 = Philippines, value 6 = Thailand, value 7 = Vietnam, value 8 = Indonesia, 9 = Not report

8) If country attendees =1 then

8.1 Input Number of attendees and display.

8.2 Input number of screen person and display.

8.3 Input Result of screening type

8.4 selection result value 1 = positive covid-19 case, Value 0 or other number =negative covid-19 case.

8.4.1 If result = 1 then (Mean Yes)

8.4.2 Input number of congenital disease and display

8.4.3 Input number of close up infected person and display

8.4.4 Input number of the person who traveled to risk country and display.

8.4.5 Input number of not wearing mask and social distance and display.

8.4.6 Process total of confirmed cases = $c_1+c_2+c_3+c_4$

8.4.7 Display Total of confirmed case.

8.4.8 If result = 0 or other number (Mean No)

8.4.9 Set variable confirm_case = 0

8.4.10 Display Number of negative case.

9) If country attendees =2 then follow as steps 8.1 to 8.4.10 then go 17)

10) If country attendees =3 then follow as steps 8.1 to 8.4.10 then go 17)

11) If country attendees =4 then follow as steps 8.1 to 8.4.10 then go 17)

12) If country attendees =5 then follow as steps 8.1 to 8.4.10 then go 17)

13) If country attendees =6 then follow as steps 8.1 to 8.4.10 then go 17)

14) If country attendees =7 then follow as steps 8.1 to 8.4.10 then go 17)

15) If country attendees =8 then follow as steps 8.1 to 8.4.10 then go 17)

16) If country attendees =9 then follow as steps 8.1 to 8.4.10 then go 17)

17) Display Attendees Country

18) Set N = screening person-confirm_case

19) Display Negative number and country attendees

20) End.

This work use Raptor software version 5.129, a flowchart-based programming, designed specifically to visualize the algorithm model and trace the execution through the flowchart (Figs. 16 and 17).

The programs, consist of 4 components as (1) work place (2) Display variables value (3) Output and symbols.

The result of Flowchart after complete generate algorithm. This algorithm uses sequential structure and branch structure to build Distribution Tree Algorithm model reference infection data source 8 countries and the other one no report. Represent the COVID-19 outbreaks in Malaysia that explain at entrance in one way and distribute in multiple exit as Fig. 18.

An algorithm process top to down and working separate modules 1–9 depending on user input attendees country. After the complete test, all modules can show the symbols evaluated and algorithm performance as Table 5.

Conclusively, this example shows an application of IT algorithm development to help explaining the pattern of COVID-19 outbreak. The work can confirm that the IT algorithm can fasten the simple, classic mathematical model technique for

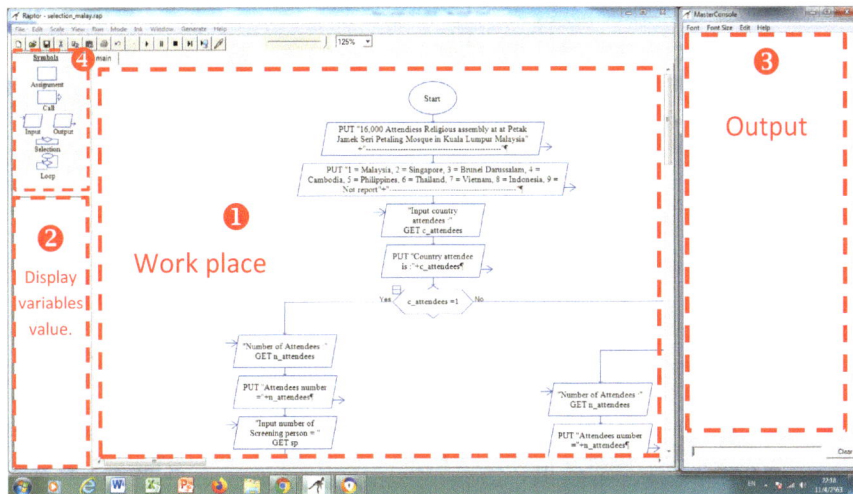

Fig. 16 Component program

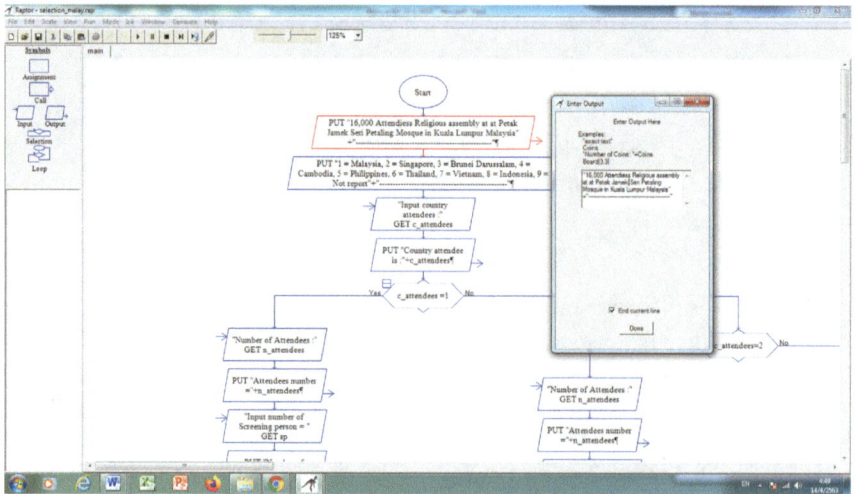

Fig. 17 Test algorithm step by step

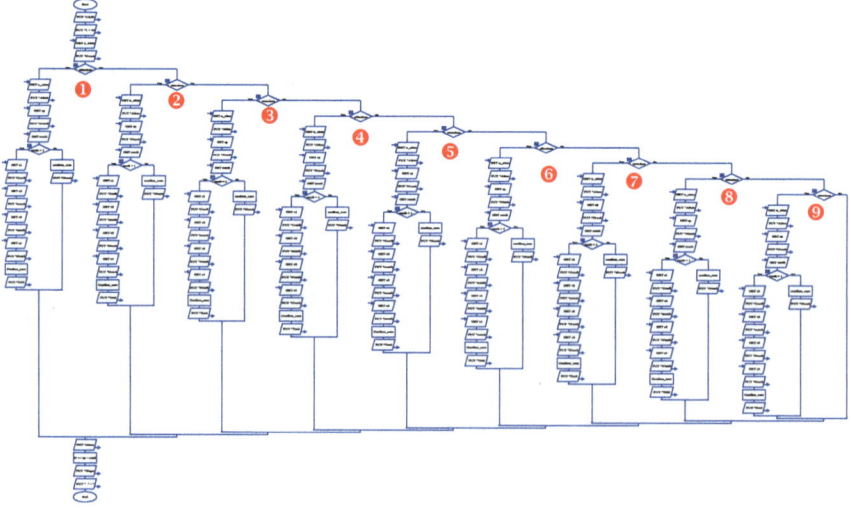

Fig. 18 Flowchart of distribution three algorithm model

modeling of the outbreak situation. This technique is a system based approach that focuses on the macro-scale data on the study setting. In fact, there might be some previous reports on info epidemiological study. For example, the analysis of data trends on Google or Twitter might be done and further used for referring to the clinical epidemiological data or development of a new clinical tool [17, 18]. For example, Google Trend data might be used for predicting COVID-19 incidence

Table 5 Result algorithm performance

No	Module	Type of screened	Number of symbol evaluated	Errors	Run complete	Accuracy algorithm
1	❶ Malaysia	1 = confirmed case	27	No	Complete	✓
		0 = Normal	19	No	Complete	✓
2	❷ Singapore	1 = confirmed case	28	No	Complete	✓
		0 = Normal	20	No	Complete	✓
3	❸ Brunei Darussalam	1 = confirmed case	29	No	Complete	✓
		0 = Normal	21	No	Complete	✓
4	❹ Cambodia	1 = confirmed case	30	No	Complete	✓
		0 = Normal	22	No	Complete	✓
5	❺ Philippines	1 = confirmed case	31	No	Complete	✓
		0 = Normal	23	No	Complete	✓
6.	❻ Thailand	1 = confirmed case	32	No	Complete	✓
		0 = Normal	24	No	Complete	✓
7	❼ Vietnam	1 = confirmed case	33	No	Complete	✓
		0 = Normal	25	No	Complete	✓
8	❽Indonesia	1 = confirmed case	34	No	Complete	✓
		0 = Normal	26	No	Complete	✓
9	❾No country report	1 = confirmed case	35	No	Complete	✓
		0 = Normal	27	No	Complete	✓

[19]. Nevertheless, the social media based might have problem with reliability. The algorithm development based on the official public health report might be more reliable.

The present technique can be applied in any setting and it can provide the useful data representing the nature of disease spreading in different settings. For this specific example, it can show that the disease existence in the studied setting is directly due to the importation of the COVID-19 patient. This can give an insight that an effective disease control should be based on strict control of any immigration.

5 Conclusion

Emerging infectious disease is an important problem in medicine. If there is a pandemic, its can widely affect global population. The novel IT technology can be useful for corresponding to the new emerging infection. In this chapter, the author briefly discuss on COVID-19, which is an important global public health problem. Computational IT tool can be applied for modeling COVID-19 outbreak. The application might be for clarification or prediction. The basic IT tools can be selected and applied depending on the different scenarios. The good examples of medical informatics application for managing of COVID-19 outbreak are GIS modeling and algorithm for explaining the outbreak. Also, the IT application can be the fundamental for more advance medical informatics approach for COVID-19 containment such as artificial Intelligence and machine learning tools development.

Conflict of interest None

References

1. Nii-Trebi, N.I.: Emerging and neglected infectious diseases: insights, advances, and challenges. Biomed. Res. Int. **2017**, 5245021 (2017). https://doi.org/10.1155/2017/5245021
2. Singhal, T.: A review of coronavirus disease-2019 (COVID-19). Indian J. Pediatr. **87**, 281–286 (2020). https://doi.org/10.1007/s12098-020-03263-6
3. Saeb, A.T.M.: Current bioinformatics resources in combating infectious diseases. Bioinformation **14**, 31–35 (2018). https://doi.org/10.6026/97320630014031
4. Wiwanitkit, V.: Utilization of multiple "omics" studies in microbial pathogeny for microbiology insights. Asian. Pac. J. Trop. Med. **3**, 330–333 (2013). https://doi.org/10.1016/S2221-1691(13)60073-8
5. Luna, D., et al.: Health informatics in developing countries: going beyond pilot practices to sustainable implementations: a review of the current challenges. Healthc. Inform. Res. **20**, 3–10 (2014). https://doi.org/10.4258/hir.2014.20.1.3
6. Fiorini, N., et al.: Best match: new relevance search for PubMed. PLoS. Biol. **16**, e2005343 (2018). https://doi.org/10.1371/journal.pbio.2005343
7. Wiwanitkit, V.: Applied medical mathematical modelling technique for epidemiology approach for web merging infection. Infect. Dis. Epidemiol. **3**, 043e (2017). https://doi.org/10.23937/2474-3658/1510043
8. Wynants, L., et al.: Prediction models for diagnosis and prognosis of COVID-19 infection: systematic review and critical appraisal. Version 2. BMJ. **369**, m1328 (2020). https://doi.org/10.1136/bmj.m1328
9. Yasri, S., Wiwanitkit, V.: Editorial: wuhan coronavirus outbreak and imported case. Adv. Trop. Med. Pub. Health. Int. **9**, 1–2 (2019)
10. Jianu, et al.: What Google Maps can do for biomedical data dissemination: examples and a design study. BMC. Res. Notes. **46**, 179 (2013). https://doi.org/10.1186/1756-0500-6-179
11. Shaw, N.: Understanding the use of geographical information systems (GIS) in health informatics research: a review. J. Innov. Health. Inform. **23**, 940 (2017). https://doi.org/10.14236/jhi.v24i2.940

12. Burney, A., et al.: Google Maps security concerns. J. Computer. Commun. **6**, 275–583 (2020). https://doi.org/10.4236/jcc.2018.61027
13. Zhang, G.H., Schwartz, G.G.: Spatial disparities in coronavirus incidence and mortality in the United States: an ecological analysis as of May 2020. J. Rural. Health. **36**, 433–445 (2020). https://doi.org/10.1111/jrh.12476
14. Elengoe, A.: COVID-19 outbreak in Malaysia. Osong. Public. Health. Res. Perspect. **11**, 93–100 (2020). https://doi.org/10.24171/j.phrp.2020.11.3.08
15. Wang, X., et al.: Comparing early outbreak detection algorithms based on their optimized parameter values. J. Biomed. Informatics. **43**, 97 (2010). https://doi.org/10.1016/j.jbi.2019.103181
16. Faverjon, C., Berezowski, J.: Choosing the best algorithm for event detection based on the intended application: a conceptual framework for syndromic surveillance. J. Biomed. Informatics. **85**, 126–135 (2018). https://doi.org/10.1016/j.jbi.2018.08.001
17. Arpaci, I., et al.: Analysis of Twitter data using evolutionary clustering during the COVID-19 pandemic. Compts. Mater. Continua. **65**, 193–204 (2020). https://doi.org/10.32604/cmc.2020.011489
18. Arpaci, I., et al.: The development and initial tests for the psychometric properties of the COVID-19 Phobia Scale (C19P-S). Pers. Individ. Differ. **164**, 110108 (2020). https://doi.org/10.1016/j.paid.2020.110108
19. Ayyoubzadeh, S.M., et al.: Predicting COVID-19 incidence through analysis of Google Trends data in Iran: data mining and deep learning pilot Study. JMIR. Public. Health. Surveill. **6**, e18828 (2020). https://doi.org/10.2196/18828

Efficient Twitter Data Cleansing Model for Data Analysis of the Pandemic Tweets

Belal Abdullah Hezam Murshed, Suresha Mallappa, Osamah A. M. Ghaleb, and Hasib Daowd Esmail Al-ariki

Abstract Twitter data generally tends to be unstructured and often very noisy, cluttered/disorganized, and clothed in informal language. In this paper, we propose an intelligent Twitter data cleansing model that can solve data quality problems associated with twitter text. This model can correct a wide variety of anomalies from slangs, typos, Elongated (repeated Characters), transposition, Concatenated words, complex spelling mistakes as unorthodox use of acronyms, manifold forms of abbreviations of same words, and word boundary errors. The effects of whole range of tasks of Twitter Data Cleansing Model (TDCM) on the performance of sentiment classification utilizing feature models and three common classifiers have been investigated and evaluated. We conducted our experiments on two sets of pandemics twitter datasets: COVID-19 and Dengue datasets. The primary objective of this paper is to both increase the accuracy and the quality of twitter data and to purify and cleanse twitter data for further analysis. The experiment results seem to indicate that the accuracy of sentiment classification increases once the data quality problems associated with the Twitter text are solved. In COVID-19 twitter dataset, the best performance obtained using Random forest classifier after cleansing the data in terms of accuracy, recall, and f1-score are found to be at 84.7%, 88.5%, and 86.3% respectively. However, the best performance in terms of precision at 84.5% was observed using SVM classifier when compared to that obtained with other

B. A. H. Murshed (✉) · S. Mallappa
Department of Studies in Computer Science, University of Mysore, Mysore, Karnataka, India
e-mail: belal.a.hezam@gmail.com

S. Mallappa
e-mail: sureshasuvi@gmail.com

O. A. M. Ghaleb
College of Engineering and Computer Science, Almustaqbal University, Qassim, Saudi Arabia
e-mail: oaghaleb-t@uom.edu.sa

H. D. E. Al-ariki
Al Saeed Faculty for Engineering and Information Technology,
Taiz University - Taiz-Technical Community College, Taiz, Yemen
e-mail: hasibalariki@gmail.com

© The Author(s), under exclusive license to Springer Nature Switzerland AG 2021
I. Arpaci et al. (eds.), *Emerging Technologies During the Era of COVID-19 Pandemic*, Studies in Systems, Decision and Control 348,
https://doi.org/10.1007/978-3-030-67716-9_7

classifiers. Further, in the Dengue twitter dataset, the best performance for cleansing data in terms of accuracy, precision and f1-score are observed to be 81.7%, 83.7% and 88.6% respectively using Random forest classifier. The best performance in terms of recall, however, is 94.9% and was obtained using SVM classifier when compared with those obtained with other classifiers.

Keywords Data cleansing · Data quality problem · Twitter · Natural language processing · Slang · Elongation · Data transformation · Informal data · COVID-19

1 Introduction

The increasing popularity of social media network and user-created web content has been generating massive quantities of data that are advantageous for a variety of applications such as sentiment analysis, information sharing, topic modeling, knowledge about the current affair. Most applications that use twitter data anticipate the tweets to be high-quality texts which are accurate, readable, and effective. For instance, high-quality data is desirable for training purposes, especially for topic modeling, and classification primarily because since high-quality dataset (training) enables training the classifier model to obtain a higher level of classification performance [1].

Training high-quality tweet classification requires needs high quality of data. Unfortunately, Twitter data tends to be noisy, extremely unstructured, and presents several challenges to traditional natural language processing technologies [2]. Twitter content is characterized by heterogeneity, which means that each tweet contains different kinds of entities, e.g., text, location, user, link, and hash-tag. Besides, twitter is replete with informal language with many words appearing in their abbreviated forms, e.g. "plz" short for "please". Since conventional text analysis technologies primarily concentrate on structured and formal texts, it is unrealistic to expect the same performance in tweet data. Thus, from the perspective of data analytics, Twitter data poses serious challenges and requires data cleansing tasks prior to building quality Twitter data set [3, 4]. In many domain-specific applications, text documents are typically unstructured and are casually worded with a lot of spelling grammar and mistakes [1]. In such cases, automatic detection of typographical errors correction of these error continues to be a challenging issue, especially in cases of unstructured text documents that include abbreviations, symbols, and acronyms related to application domains [1].

Prior to using Twitter data, it is necessary to cleanse and convert the data into a more formal and a more usable structured dataset to ensure that the subsequent stages of the process are easy, efficient, smooth, and effective. As a modern type of social media network, Twitter has many special characteristics, which renders it difficult to be dealt with using conventional text process technologies. In data analytics such as sentiment analysis, semantic analysis, tweet classification, topic discovery on tweets, feature extraction of the efficiency and efficacy are important to ensure both the performance and quality of the model. Research shows that Twitter data preprocessing can substantially impact model output [5, 6]. For instance, in order to analyses public discourse trends and patterns, a particular event or news story related to an individual can be defined such that she/he can determine and assess the manner in which it is being publicly debated. Nonetheless, owning to the dynamic nature of user-generated content, the user should consider the abbreviations, hashtags, and slang that could be related to the subject of interest [3].

Data of low quality could be dangerous because it can lead to incorrect or missing decisions, operations and strategies. Further, it can also slow down, the innovation processes. The losses for organizations caused by low data quality are estimated to be over billions of dollars per year [7]. The issue of bad data is a huge problem troubling close to 60% of enterprises suffering from it [8, 9]. Even though the process of data cleansing and methods have been research topics of interest for quite some time, the features of Twitter tend to reduce it to a less cumbersome it a non-trivial work. The key to leveraging the potential value presented by effective data analytics is the very process and methodology of data cleansing based on which analysis can be carried out. This, good data quality is fundamental to obtain good insights. The chief contribution of this work is the development of an intelligent Twitter data cleansing technique which can solve data quality problems associated with twitter data. As a means of justification for a need for a novel Twitter data purification model, those challenges typical of Twitter datasets but not very common in traditional datasets will be demonstrated.

The problems of Twitter Text Quality

Recent research has identified the quality of data to be a serious challenge commonly encountered in sentiment analysis and topic discovery [10]. Twitter posts tend to be inherently poor in terms of data quality which poses serious challenges to dealing and processing those data prominently featuring in big Twitter dataset. The statement of the problem can be described as follows:

How can we solve the data quality problems associated with Twitter text?

Twitter texts fail to adhere to standard rules of vocabulary, spelling and syntax. The twitter tweets are very short and usually written hurriedly in very short period of time, often in seconds or minutes, by people from varying education background and interests. As tweet writers fail to pursue the rules of grammar, they tend to include misspelt words, abbreviations, slangs, and domain specific terminologies that are not found in Standard English dictionaries Such, and their texts are barely comprehensible to people outside their fields of application. Moreover, Twitter data is very rich with a wide range of linguistic innovations including abbreviations, lengthening, concatenated words, and emoticons.

To address this problem, this study proposes an intelligent Twitter data cleansing model that can solve the data quality problems associated with twitter text. This model can also correct a wide variety of anomalies such as slangs, emoticons, typos, Elongated (repeated Characters), transposition, Concatenated Words, complex spelling mistakes, unorthodox use of acronyms, and manifold forms of abbreviations of the same words. It makes use of the generated knowledge to recognize unorthodox acronyms, and to string together similar words (correctly spelled and misspelled). This study also investigates and evaluates the effects of all the tasks associated with the Twitter Data Cleansing Model (TDCM) on the performance of sentiment classification utilizing three common classifiers. The experiment results seem to indicate that the performance of sentiment classification tends to increases after solving the data quality problems associated with the Twitter text.

The rest of this article is structured as follows: In Sect. 2, the literature survey related to preprocessing area is provided. Section 3 describes the proposed model (TDCM) which increases the data quality and cleanses twitter dataset. The details of the data collection, performance evaluation, and the experimental results are presented in Sect. 4. In Sect. 5, a concise conclusion is given.

2 Related Work

Some of the earlier works in literature have proposed different methods of pre-processing texts. Hemalatha et al. [11] proposed a method to perform certain common pre-processing tasks such as removal of special characters and URL, tokenization, and stemming in twitter to achieve sentiment analysis. In [12], Sun et al. suggested a preprocessing method on online financial text which can perform removal of numbers, URL, and punctuation, tokenization, extensions of contractions, removal of stop word, and lemmatization. Several pre-processing tasks such as feature correlation, n-gram models, and stemming in Arabic texts were suggested by Duwairi and El-Orfali [13]. All these tasks were carried out to reconnoiter the effects of the performance of sentiment analysis in Arabic. In [14], Rushdi-Saleh et al. proposed various Pre-processing steps such as stop word removal, n-gram generation, and stemming on movie reviews collected from various Arabic blogs and web pages. Jianqiang [15] debated the impact of various tasks of Preprocessing such as URL, negation, stop word, and repeated letters on the performance of sentiment classification conducted on twitter dataset. Several steps were presented by Lizaet et al. [16]. These tweets, after various stages of preprocessing, were transformed into a vector set of features utilizing Bag of Words (BOW). Several semantic similarity measures proposed by Murshed [17]. A pre-processing approach based on gathering of words of slang with coexisting words to determine the importance and the strength of sentiment of slang words used in tweets was suggested by Tajinder Singh [18]. Itisha et al. [19] presented a two-phased preprocessing approach, where in the first stage involves noise removals (URLs, punctuation marks, usernames, and elimination of stop words). The second stage involves normalization which includes the transformation of Non-standard words into formal words. Hussein et al. [20] proposed an approach to clean twitter data which includes several tasks such as segmentation of English tweets, elimination of stop words, cleaning, and stemming. The aim of [21] is to find out the advantages of various, twitter-specific, preprocessing techniques to preprocess the tweets efficiently rather than utilizing baseline text preprocessing methods. A preprocessing method, which consists of data collection, data cleaning, tokenization, stemming, and removal of stop words, was proposed by Naresh [22] for unstructured healthcare text data. Arpaci et al. [23] proposed a method utilizing an evolutionary clustering analysis aimed to analyze the COVID-19 Twitter dataset collected in a period of time and to describe the trend of public attention given to topics related to the COVID-19 pandemic. Further, Arpaci et al. [24] suggested an new type of the specific phobia diagnosis and initial tests for the psychometric properties of the COVID-19 Phobia Scale (C19P-S).

3   Proposed Model

3.1   Twitter Data Cleansing Model

The proposed twitter data Cleansing model starts with data collection obtained from Sect. 4.2. Our Twitter data cleansing model consists of four Phases (as shown in Fig. 1), namely, (1) Extraction and Filtering Tweets, (2) Removal of Noise, (3) Out of Vocabulary Cleansing (cleansing irrelevant vocabulary), (4) Transformation of tweets. These phases facilitate the process of saving storage space and analyzing time, (especially in cases of an enormous dataset), and also reducing data. Apart from enhancing the accuracy of the results of data analysis, these stages also ensure developing topic coherence. In addition of these advantages, these four phases transform the informal parts of tweets to formal ones. The complete details of each of these phases are described in the following sections.

3.1.1   Extraction and Filtering Tweets

Though the collected streaming tweets could be in various languages, our model, takes into account only the English tweets for analysis. Tweepy package permits specific language tweets to be fetched. The language parameter value such as 'en' is then assigned for those tweets with full texts collected in English language. We also observed the occurrence of many duplicate tweets when the tweeters tend to re-tweets the same text of tweets for other tweeters (called Re-tweet). Hence, to eliminate duplication, we used regular expression techniques based on NLP to seek hyperlinks in the text of tweets and to eliminate the same. Following this, the duplication tweets were dropped since they seemed to add no new knowledge to the dataset, thus proving to be inefficient in computational terms. Excluding these

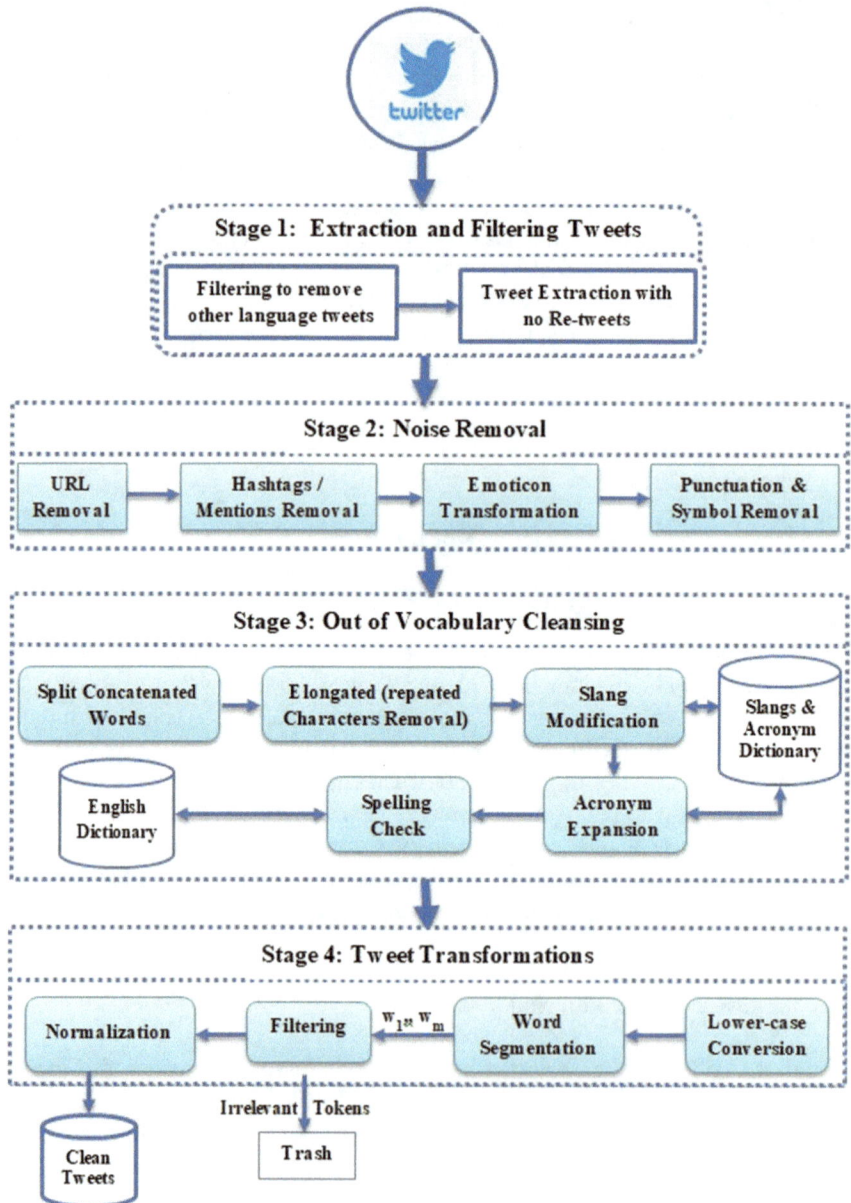

Fig. 1 Twitter data cleansing and transformation model

duplicate tweets tend to renders the twitter dataset to become all the more meaningful.

3.1.2 Noise Removal

- **URL removal**

Twitter tweets usually contain a lot of noise and unavailable data such as Uniform Resource Locator (URLs) for analysis. The embedded URL is commonly utilized to provide the source with the elaborated depiction of the content stated in the tweet— for instance, "Five lectures the Spanish flu can educate us about COVID-19 https://t.co/ANuMMUlPy7". However, the model in this study fails to take the URL for analyzing and extracting topics from tweets into account. Hence all these unnecessary data are deleted from tweets by applying preprocessing steps using a regular expression based on NLP. We only focused on meaningful English words containing in the tweets.

- **Hashtags/mentions removal**

Analysis tweets (system) and extracting topics from tweets need meaningful words that refer to a topic. Therefore, it is necessary to delete all unnecessary terms and signs from tweets. Twitter contains some of these terms/words which start with the signs '@' and '#'. User mentions are generally used in tweets to refer to someone else and these user mention entities start with a '@' sign, followed by the user's name e.g.: "Stay at home @Belal". Likewise, a hashtag is another entity associated with a tweet and refers to a specific tag or topic and usually begins with a '#' sign. This need not be a complete, meaningful word. At this stage of hashtags removal, these signs are eliminated using a regular expression in python language. However, our model restricts itself to deleting the hashtag sign, thus leaving the term intact. This is because the hashtag sign is typically followed by phrases or words describing the topic being discussed.

- **Emoticon and Emoji Transformation**

The text of tweet can include emojis which are encoded in Unicode and contains emoticons. These noises such as emoticons and emojis can affect the process of data purification and feature extraction. In the following example of a tweet, "Plz, #stayAtHome don't go outside. COVID-19 is verrrry dangerous u'\U0001f61f' contains Emoticon and ends with a Unicode encoded "worried face" icon. Both these emoticons and the emojis will be transformed into their respective word format. We have used two wonderful emoji and Emoticons data dictionaries developed by NeelShah[1] in order to convert emoticons and emoji to word format and to utilize a regular expression based on the NLTK package in python platform.

- **Punctuation and symbol removal**

Generally Twitter tweets also include numbers, punctuations marks and symbols. Utilizing regular expression will eliminate all such inclusions and supply only the important data. This step further reduces the storage of the data set and holds only the effective data used for topic modeling and classification methods.

3.1.3   Out of Vocabulary Cleansing

The stage of anomaly detection determines the Out-of-vocabulary words/terms (i.e., those terms/words do not exist in the English dictionary) such as elongated words (e.g., stayyyyyyyyyy), slangs (e.g., plz, luv), concatenated words (e.g., StayAtHome). This phase involved many tasks, and each task is described in a concise manner.

a. **Concatenated Words**

In Twitter data, the tweet size has a character limit of 140 and tweets tend to contain very limited information typically needed for reliable extraction. Often times, two or more words (such as 'StayAtHome') are commonly concatenated to reduce the size of tweet. Such words/terms should be split into their components forms before utilizing them in data analytics. At this stage, we use regular expression techniques that split those words into their components for other analysis

b. **Contraction Replacement (Apostrophes)**

Text normalization is a much required step in data cleansing in order to correct errors in words or texts, Apostrophes give rise to significant word sense ambiguity since it may exemplify contraction for a word. For example, contractions such as "hasn't", "did't", and "he'll" commonly occur in tweets. Therefore, the Contractions in Twitter data should be transformed into regular lexicons before processing and this data is also very important for sentiment analysis. In this step, public lexicon that contains all contractions is used. Therefore contractions should be substituted with the corresponding extended expressions of the language. The first step to be considered is identifying the pattern, followed by substitution. Each instance of the pattern is substituted by an equivalent replacement pattern.

c. **Elongated Words**

Most users write their tweets and status updates informally resulting in the occurrences of repeated characters and Elongation in Tweets. At this step, those

---

[1]https://github.com/NeelShah18/emot/blob/master/emot/emo_unicode.py.

terms that are lengthened by an unneeded repetition of characters are shortened. These elongated terms are used to articulate emotions such as 'loooooove' and so on. Eliminating these unnecessary letters to emerge with a standard meaningful word is an important stage in preprocessing of tweets. Some of these methods have reduced the repeated characters to 2 by eliminating surplus letters. This step introduces an approach to convert these irritating Elongated (repeating) characters to a meaningful English word using a regular expression module known as back-references. A backreference is a popular method to refer to a previously matched group in a regular expression. This approach matches and excludes repeated characters. Before eliminating repeated characters recursively, a WordNet lookup is which guarantees the creation of removal of unnecessary letters. A class consisting of a replace () method which takes an individual word and returns a more proper and accurate version of that word is constructed, thus eliminating all the suspicious, recurring characters.

d. Slangs Modifications

The tweets text has certain anomalies that are not typically found in traditional text cleaning issues. Slang is one such anomaly, which is quite prevalent in Twitter posts and includes the use of abbreviations, acronyms or internet slang. These slangs usually tend to reduce the number of characters during writing tweets. For instance, incorrect/nonstandard words such as 'Plz', 'Gr8', and 'luv' are not found in English language and it is necessary to convert such slang words into correct lexicons in English, before rendering them beneficial for data analysis. To achieve this, we built a dictionary consists of 2864 slang which furnishes a set of all possible slangs as lookup dictionaries for transformation objectives. The slang lookup is carried out by contrasting it with the internet slang.

e. Spelling Correction

Yet another extremely significant step in the cleansing of twitter data is spelling correction. Typos are commonly prevalent in twitter posts and correcting these spelling errors is important before starting analysis. Pyspellchecker, a package in python, is used at this stage to correct spelling.

3.1.4 Tweet Transformations

a. Lower-case conversion

Lower case conversion is the process of converting words of the tweets to the same form. At this stage, all the tweets are transformed into lowercase in order to supply a constant format for all tweets utilizing python lower string function.

b. Word segmentation(Tokenization)

The second step of the Tweet transformations phase is the word segmentation process (Tokenization), which means dividing a tweets, sentences, and phrases into a collection of lexical units (Words) known as tokens that are both methodologically advantageous and linguistically significant. Tokenization is typically the most significant and primary task in the plurality of text processing applications. We have used NLTK library [25] to tokenize the tweets in python language. For instance, consider the tweet "COVID-19 assaults immune system as HIV doctors afraid" When this tweet passes through tokenization, the outcome of this task would be the following tokens ['COVID-19', 'assaults', 'immune', 'system, 'as, 'HIV', 'doctors', 'afraid'].

c. **Filtering irrelevant words**

Tweets can include common and frequent words that might not contribute much to the meaning and importance of tweets. Such words/terms are known as stop words (some of these words, like 'is', 'he', 'an', 'she', 'a', etc.), which add more noise to NLP. Hence, it is important filtering these stop words from tweets during the Twitter data cleansing model. The removal of stop words is carried out by python, and all the stop words are eliminated from tweets utilizing the corpus of NLTK stop words, except for those that refers to positive or negative feeling.

d. **Normalization**

The normalization phase aims to decrease the variation in tweets data by transforming words/terms to their base form stem. In addition to that, the purpose of this stage is to reduce redundancy, eliminate disparate suffixes from tweets to economize time and storage space. For example, the words 'converter', 'converts', 'converted', 'converting', 'conversion' can be converted and stemmed to the word 'Convert'. In the tweets mining analysis, the stemming phase is the most popular phase because it assists in focusing the analysis on the basic form of the terms/ words, instead of distinguishing between various terms/words which can introduce ambiguity in text/tweets mining methods. The Porter Stemmer algorithm [26], the most widely used approach in English, has been utilized in the proposed work.

3.2 Feature Extraction of the Model

The feature extraction model, which indicates a method that determines how certain features have been used to classify new data into a particular class, plays a significant role in proposing classification. Throughout the classification of the texts or tweets, various feature selection models were presented for sentiment analysis of the twitter data. Most investigators present state-of-the-art results on twitter data for sentiment analysis utilizing a unigram feature model [27, 28]. In this research, we have utilized word N-gram features models and Sklearn library to extract the features.

3.3   Twitter Sentiment Classification Process

To evaluate the quality and performance of proposed Twitter Data Cleansing model, we have utilized three common supervised techniques namely Multinomial Naïve Bayes (MNB), Random Forest (RF) Classifier, and Support Vector Machine (SVM) for performing sentiment analysis. Further, the performance of each of these classifiers has been assessed based on the following three metrics: Accuracy, precision, Recall, and F1-score. All classifiers have been implemented using a machine learning sklearn package in Python. For the process of classification, each twitter dataset has been split into two portions - one portion for training (80%) and 20% the other for testing. We have used the Sklearn library for classification purposes.

4   Experiments and Analysis

4.1   Experimental Setup

The experiments have been performed on an Intel Core i7-3210 M CPU @ 2.5 GHz machine with 16 GB RAM and Windows 10. These experiments have been executed using Python3.7 and Pycharm IDE. All graphics have been generated using Origin pro 8. We have performed our experiments on the Pandemics twitter datasets (as described in Sect. 4.2). The proposed data cleansing model has been incorporated with several tools such as a slang and acronym dictionary which supply a set of all versions of slangs, abbreviations as lookup dictionaries for transformation purposes and an English dictionary. Moreover, tools such as Twitter streaming API and "Tweepy" provide access to twitter to extract tweets. While Natural Language Toolki (NLTK) was used for common preprocessing, we have used packages such as SpellChecker that offer support for methods such as a 'spell', and 'correction' methods to check and correct the spelling in the words of tweets. Likewise, while the Sklearn library was used for feature extraction and classification, we have utilized sklearn.feature_extraction module to extract features form our pandemics twitter datasets.

4.2   Data Collection

To display the differences and effects of the proposed twitter data cleansing model, the following datasets were collected to verify the model. The Twitter Streaming Application Programming Interface (API) has been used to extract tweets from twitter depending on trending events for a period of three months starting from 13th Jan, 2020 to 30th April, 2020 with the use of language filter 'en' set to English using the relevant specific filter words provided as keywords. Tweepy is an

Table 1 Details of pandemic datasets

Description of datasets	Dataset name	No. of tweets
Corona Virus	COVID-19	501,231
Cholera	Cholera	26,068
Swine flu	Swine Flu	34,901
Dengue fever	Dengue fever	1,967
Malaria	Malaria	117,386
Ebola virus disease (EVD)	EVD	444,3
Chikungunya	Chikungunya	685
Total number of tweets		**686,681**

open-sourced Python library that enables python to access Twitter and uses its API. We just placed filter terms specifically for pandemic-related tweets such as COVID-19, Cholera, swine flu, dengue fever, malaria, Ebola Virus Disease (EVD), and Chikungunya. The complete details of the tweets collected in Pandemics dataset are given in Table 1.

4.3 *Performance Evaluation*

It is necessary to evaluate the Twitter data cleansing model of the Pandemic tweets to measure both the performance and efficiency of the proposed model. Appropriate performance metrics must be selected to evaluate the accuracy and quality of the data both before and after data cleansing. Therefore, the key metrics taken into account in this research are as follows: Accuracy, precision (P), recall (R), and also F1-score. These measures are calculated on the basis of the values of True Negative (TN), True Positive (TP), False Negative (FN), and False Positive (FP) assigned classes. Accuracy can be formulated, as given in Eq. 1.

$$\text{Accuracy} = \frac{TP + TN}{TP + TN + FP + FN} \tag{1}$$

Precision metric quantifies as the number of true positive out of all positively assigned Tweets and this metric is expressed as in Eq. 2.

$$P = \frac{TP}{TP + FP} \tag{2}$$

Recall metric quantifies as the number of true positive out of the actual positive tweets. This metric can be defined as given in Eq. 3.

$$R = \frac{TP}{TP + FN} \tag{3}$$

Finally, f1-score provides a single score that combine both precision and recall into a single metric. F1-score is calculated as expressed in Eq. 3.

$$F1 - \text{score} = \frac{2 * P * R}{P + R} \tag{4}$$

where the values of its range are between 0 and 1 and the closer it is to 1, which means better results.

4.4 Experimental Results

The experimental results obtained have been compared in this section, and presented in two parts. The first part presents details of the empirical case study of the proposed TDCM and the second part presents details related to the evaluation performance and effectiveness of the proposed twitter data cleansing model on sentiment classification. Firstly, in our model, only English tweets are taken into account for analysis. We filtered all the tweets with language filter 'en' in the First Phase of our model to remove all non-English tweets from the Corpus Twitter dataset, as shown in Fig. 3 (1.1). Further, we have also removed all Re-tweets and duplication of tweets as shown in the sample of duplication in Fig. 3 (1.2). Thus, the number of tweets after removing duplication stand at 166,482, 7,047, 11,341, 1,078, 35,989, 1,676, and 261 tweets for each of these datasets COVID-19, Cholera, Swine Flu, Dengue fever, and Malaria, Ebola Virus Disease (EVD), and Chikungunya respectively, as shown in Table 2. Figure 2 shows the comparison of the number of tweets both before and after removal of duplication.

The second phase of our model is noise removal, which removes URLs, Hashtags "#", user mentions "@", emoticon transformation, punctuation marks

Table 2 Details of datasets after removing duplicate

Dataset name	No. of tweets	No. of tweets without duplicate
COVID-19	501,231	166,482
Cholera	26,068	7,047
Swine Flu	34,901	11,341
Dengue fever	1,967	1,078
Malaria	117,386	35,989
EVD	444,3	1,676
Chikungunya	685	261
Total # of Tweets	**686,681**	**223,874**

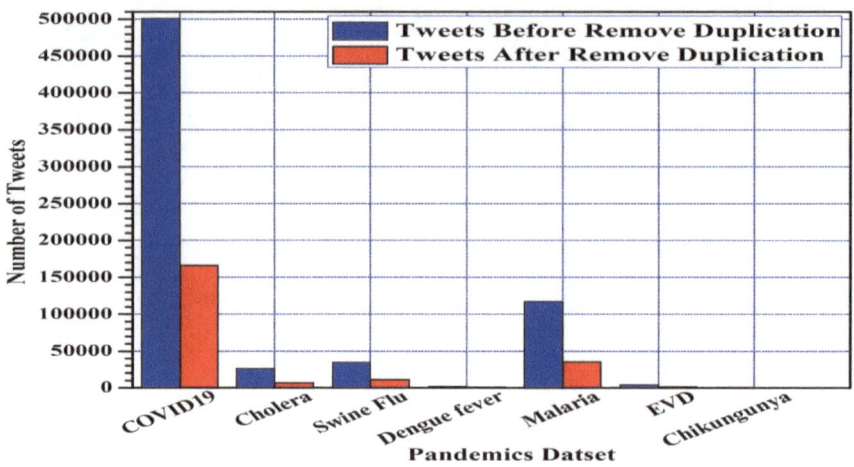

Fig. 2 Number of tweets before and after removing duplicates and re-tweets in pandemic datasets

and special characters as shown in Fig. 3 (2.1), (2.2), (2.3), (2.3), (2.4) respectively. The third phase of our model is cleansing of Out of Vocabulary words such as concatenated words. All the concatenated words have been split into the their component words, as shown in Fig. 3 (3.1) For example, the word "StayAtHooooooooommmme" is divided into three words "Stay", "At", and "Hooooooooommmme". Besides, all the characters that are repeated in elongated words have also been removed as shown in Fig. 3 (3.2), the word "Hooooooooommmme" converted to "home". The next step in this phase is modification of slang words during which informal words are converted to their formal equivalents words. For example abbreviations "ASAP" and "COVID" are converted to their respective formal equivalents namely "As soon as possible" and "coronavirus", as shown in Fig. 3 (3.3). The last step in this phase is correction of wrongly spelt words—all the errors in words such as "tronsmited" and "petient" are corrected to "Transmitted" and "patient", as shown in Fig. 3 (3.4). The final last phase of our model is Tweet Transformation which consists of Lower case conversion, word segmentation, and filtering irrelevant words and normalization, as shown in Fig. 3 (4).

Table 3 shows the number of tokens before cleansing tweets and the detailed results of the number of tokens in each phase of the twitter data cleansing framework for all the gathered datasets. Besides, Fig. 4 compares the token number in each task of the data cleansing model for all the datasets collected.

In the second part of experimental result, we have utilized three most commonly used supervised techniques namely MNB, RF and SVM classifiers to evaluate the performance and effectiveness of the proposed twitter data cleansing model on sentiment classification. We have also assessed the performance of each classifier based on the following metrics: Accuracy, precision, Recall, and f1-score. The

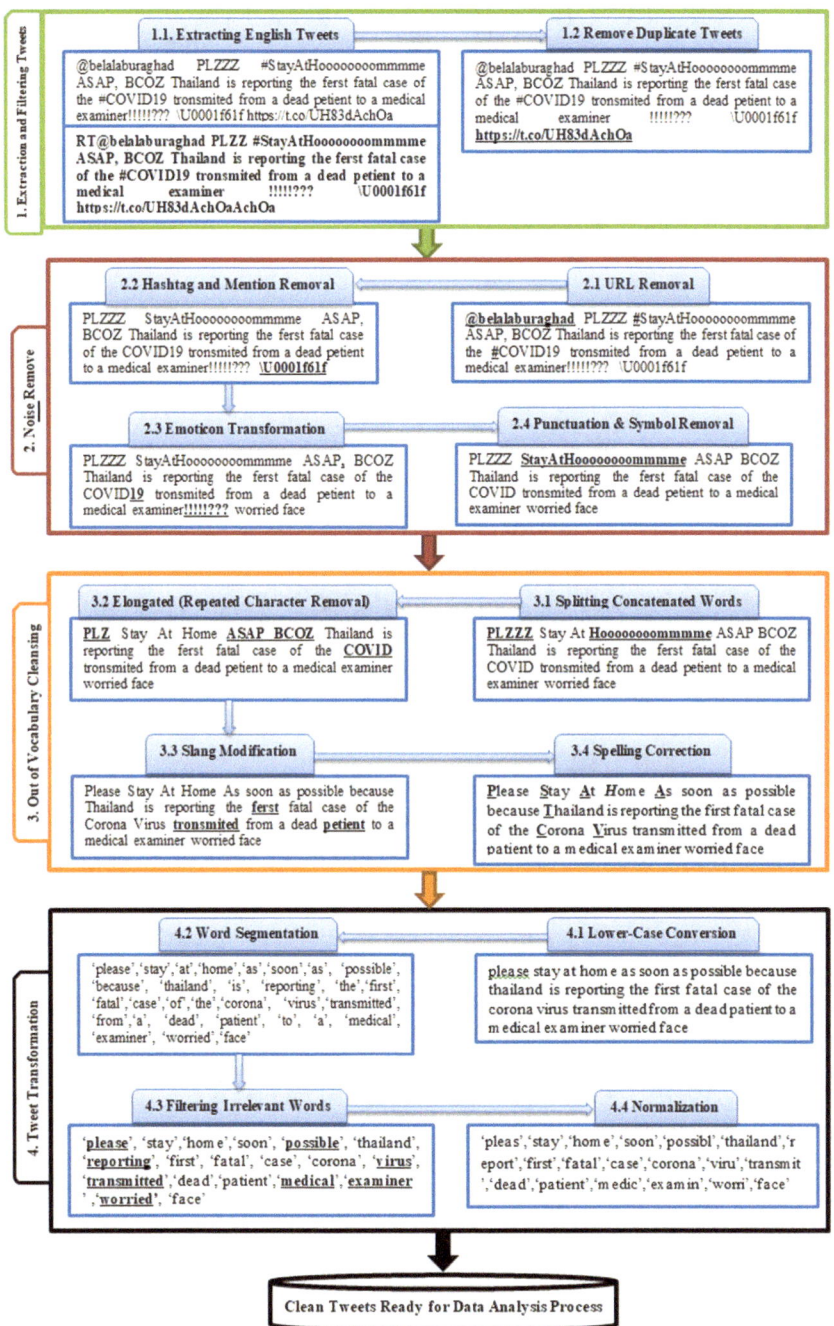

Fig. 3 Tasks of the twitter data cleansing model for a tweet of COVID-19 with tweet ID 125004982361449000

Table 3 Number of tokens before and after each task of Twitter data cleansing model

Dataset name	COVID-19	Cholera	Swine Flu	Dengue	Malaria	EVD	Chikungunya
# of Tweets without duplicate	**166,482**	**7,047**	**11,341**	**1,078**	**35,989**	**1,676**	**261**
# of Token after remove duplicate tweets	4,281,084	202,690	345,741	31,685	973,411	53,538	7,995
# of Token after remove URL	4,053,410	191,689	325,392	30,508	916,545	51,366	7,642
# of Token after replace apostrophe	4,306,820	202,891	348,457	32,004	981,074	53,899	8,036
# of Token after splitting concatenate	4,069,551	188,870	312,690	30,271	930,839	50,676	7,627
# of Token after remove repeated letters	4,069,551	188,870	312,690	30,271	930,839	50,676	7,627
# of Token after slang modification	4,419,308	205,072	342,093	32,367	1,001,803	55,010	8,086
# of Token after spelling correction	4,419,396	205,080	342,104	32,369	1,001,855	55,013	8,078
# of Token after remove stop words	2,748,922	119,524	192,441	19,661	603,885	31,659	5,156

experiments have been conducted on 29,062 tweets chosen randomly from COVID-19 twitter dataset, and 1,078 tweets, chosen from the dengue twitter dataset.

All the tasks of the twitter data cleansing model on the three selected classifiers namely: MNB, SVM, and RF classifiers have been tested to show the performance of the proposed model. Table 4 shows a significant increase in the performance and quality of the classification after cleansing the data of COVID-19 twitter dataset in n-gram feature model, with the highest accuracy of 84.7% performed in the RF Classifier after cleansing the data, followed by 81.1% in SVM classifier and 74.5% in MNB classifier. Similarly, in the case of Dengue twitter dataset, it can be observed from Table 5 that the best levels of accuracy of this dataset after cleansing the data are at 81.7% performed in the RF Classifier, followed by 81.2% in SVM classifier and 79.7% in MNB classifier.

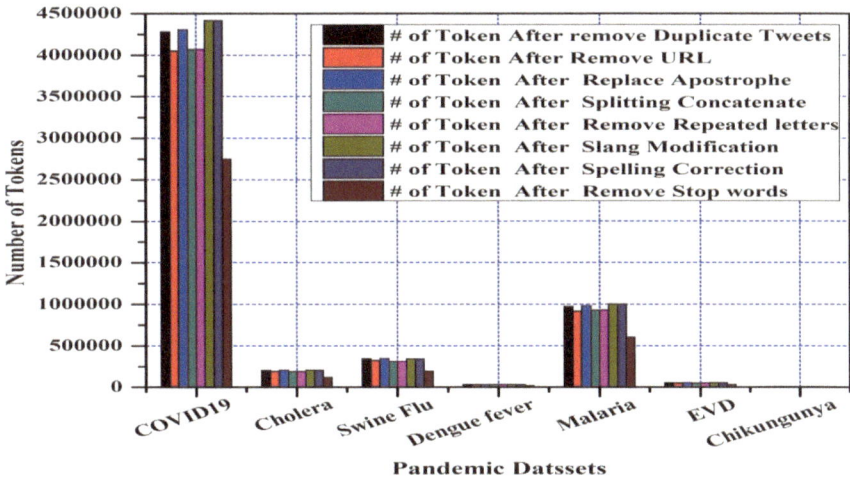

Fig. 4 Comparison of the tokens number in each task of the data cleansing model for each pandemic

Table 4 Classification results of the twitter data cleansing model for COVID-19 Twitter dataset using MNB, SVM, and RF classifiers

Methods metrics	MNB classifier		SVM classifier		RF classifier	
	Before date cleansing	After date cleansing	Before date cleansing	After date cleansing	Before date cleansing	After date cleansing
Accuracy	0.723	0.745	0.771	0.811	0.816	**0.847**
Precision (P)	0.733	0.748	0.801	**0.845**	0.817	0.842
Recall (R)	0.775	0.812	0.779	0.828	0.855	**0.885**
F1-score	0.754	0.778	0.790	0.836	0.836	**0.863**

Table 5 Classification results of the data cleansing model for dengue Twitter dataset using MNB, SVM, and RF classifiers

Methods metrics	MNB classifier		SVM classifier		RF classifier	
	Before date cleansing	After date cleansing	Before date cleansing	After date cleansing	Before date cleansing	After date cleansing
Accuracy	0.717	0.797	0.651	0.812	0.667	**0.817**
Precision (P)	0.738	0.829	0.649	0.822	0.673	**0.837**
Recall (R)	0.725	0.795	0.925	**0.949**	0.904	0.942
F1-score	0.731	0.815	0.763	0.881	0.772	**0.886**

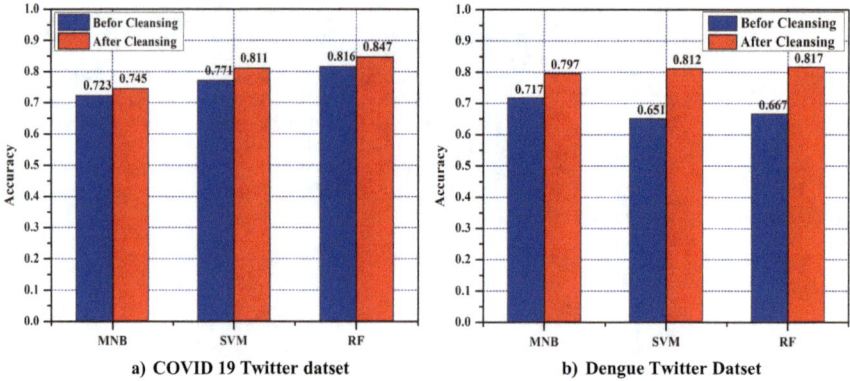

Fig. 5 Classification accuracies results using NBM, SVM, and random forest (RF) classifiers, **a** COVID-19 twitter datset, **b** dengue twitter datset

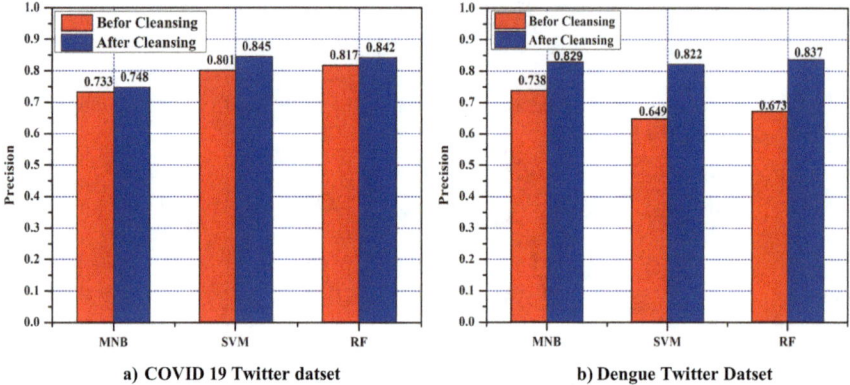

Fig. 6 Classification precision results using NBM, SVM, and random forest (RF) classifiers, **a** COVID-19 twitter datset, **b** dengue twitter datset

Figure 5 (a) and (b) show a significant increase in accuracy in both datasets. It can also be seen that quality in terms of accuracy in COVID-19 is higher when compared to the dengue twitter dataset. Likewise, the Recall performance of Dengue dataset using SVM classifiers is close to 1 around 94.9% that indicate a high performance of the classification achieved using SVM classifier, while the best recall result of the COVID-19 is 88.5% obtained using the RF classifier.

Similarly, Fig. 6 (a) and (b) show that COVID-19 and Dengue datasets has the highest precision after cleansing the data when compared to the methods before cleansing. The precision values obtained using the SVM at 0.845 seems to indicate that this classifier is the best classifier on COVID-19. Precision on the dengue dataset using the RF classifier is at 0.837, which seems to indicate that the RF classifier is the best classifier on the Dengue dataset.

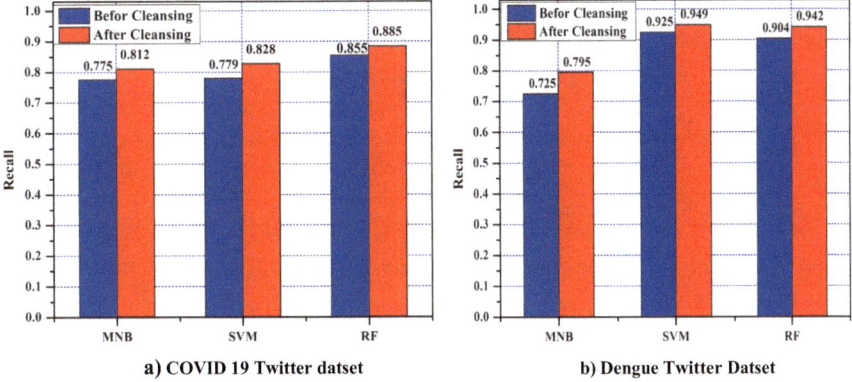

Fig. 7 Classification recall results using NBM, SVM, and random forest (RF) classifiers, **a** COVID-19 twitter datset, **b** dengue twitter datset

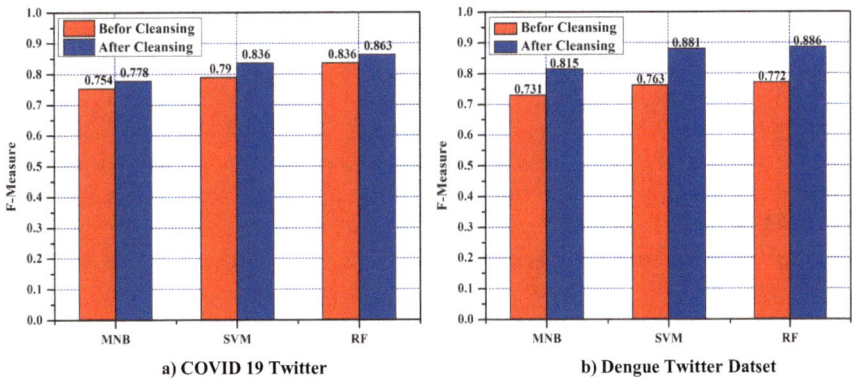

Fig. 8 Classification F1-score results using NBM, SVM, and random forest (RF) classifiers, **a** COVID-19 twitter, **b** dengue twitter datset

Table 6 The percentage of increasing accuracy, precision, recall and f1-score after cleansing to before cleaning dataset for classification

Datasets	COVID-19 Twitter dataset			Dengue Twitter dataset		
Classifiers metrics	MNB	SVM	RF	MNB	SVM	RF
Accuracy	2.2	**4.0**	3.1	8.0	**16.1**	15.0
Precision (P)	1.5	**4.4**	2.5	9.1	**17.3**	16.4
Recall (R)	3.7	**4.9**	3.0	7.0	2.40	**3.80**
F1-score	2.4	**4.6**	2.7	8.4	**11.8**	11.4

From Fig. 7 (a), it is evident that after cleansing the data, RF classifier has higher values of recall on COVID-19 Twitter dataset, while SVM classifier has higher values of recall on Dengue dataset after cleansing when compared to others classifiers, as shown in Fig. 7 (b). With respect to the parameter of recall, the recall of Dengue dataset shows higher values than that of COVID-19 dataset.

Finally, the performance of F1-score of both datasets COVID-19 and Dengue are shown in Fig. 8 (a) and (b). Hence, the use of Twitter Data Cleansing Model followed by filtering the texts data reduces noise and solves the problem of data quality problem associated with twitter texts data and also improves the performance of classification.

Table 6 shows that the percentage of increasing performance in terms of accuracy, precision, recall, and F1-score of all classifiers after the use of Twitter Data Cleansing Model in the n-grams feature model on COVID-19 and Dengue twitter datasets. At the end of all tasks in the proposed model on COVID-19 twitter dataset, the maximum performance with respect to accuracy was found to be at 4.0%, obtained using SVM classifier, when compared to those obtained with MNB and RF classifiers. In dengue dataset, the maximum improvement of accuracy using SVM was at 16.1%, when compared to those obtained with other classifiers. Further, cleansing the data on the COVID-19 dataset seems to increase Precision, Recall, and F1-score of MNB, SVM, and RF and the desired levels of improvements in terms of Precision, Recall, and F1-score using the SVM were found to be at 4.4%, 4.9%, and 4.6%. In addition, the performance of MNB, SVM, and RF in terms of Accuracy, precision, recall, and F1-score appear to increase on the Dengue dataset. However, using SVM classifier on the dengue twitter data set seems to increase Accuracy, precision and F1-score increase at 16.1%, 17.3% and 11.8% respectively. In terms of Recall, the best performance using RF classifier can be seen to be at 3.8%.

5   Conclusion

This study presented an efficient Twitter Data Cleansing Model (TDCM) which can solve the data quality problems associated with twitter data and extract both qualitative and quantitative data from the Twitter social media. The data was in the form of tweets chosen on trending topics such as COVID-19, cholera, etc. The presence of slang, acronyms, bad grammar, typos, duplications, concatenated words, repeated characters in a word, compounded with the undesirable content such as URLs, expressions, stop words, etc. present in these datasets make it incredibly difficult to obtain meaningful insight from these data sets. The challenges associated with twitter data were solved using the proposed TDCM Model, which consists of four phases namely Extraction and Filtering Tweets, noise removal, Out of vocabulary cleansing and tweets transformation. Each part of a tweet was converted into an enhanced form before being utilized as an input to the data analytics and mining systems. Comprehensive experimental results that show that the appropriate TDCM can significantly improve the quality and efficiency of data have also been presented. In COVID-19 twitter dataset, the best performance in terms of accuracy, recall, and f1-score using Random forest classifier was observed to be at 84.7%, 88.5%, and 86.3% respectively. However, the best performance in terms of precision using SVM classifier was found to be at 84.5%. In the case of Dengue twitter dataset, the best performance in terms of accuracy, precision and f1-score using Random forest classifier were found to be at 81.7%, 83.7% and 88.6% respectively. However, the best performance in terms of recall using SVM classifier was found to be at 94.9% when compared to those obtained using other classifiers. Future studies could focus on analyzing the effect of Twitter data cleansing model on the performance of topic modeling. The proposed method has not considered data quality in terms of data streaming. In the future, we could address this problem.

References

1. Huang, Y., Murphey, Y.L., Ge, Y.: Intelligent typo correction for text mining through machine learning. Int. J. Knowl. Eng. Data Min. 3(2), 115 (2015)
2. Kireyev, K., Palen, L., Anderson, K.M.: Applications of topics models to analysis of disaster-related twitter data. NIPS Work. Appl. Top. Model. Text Beyond, Canada, Whistler 1 (2009)
3. Kim, A.E., Hansen, H.M., Murphy, J., Richards, A.K., Duke, J., Allen, J.A.: Methodological considerations in analyzing twitter data. J. Natl. Cancer Inst. Monogr. 2013(47), 140–146 (2013)
4. Torunoglu, D., Cakirman, E., Ganiz, M.C., Akyokus, S., Gurbuz, M.Z.: Analysis of preprocessing methods on classification of Turkish texts. Int. Symp. Innovations Intell. Syst. Appl. IEEE 2011, 112–117 (2011)
5. Denny, M.J., Spirling, A.: Assessing the consequences of text preprocessing decisions. Available SSRN 2849145 (2016)
6. Boyd-Graber, J., Mimno, D., Newman, D.: Care and feeding of topic models: problems, diagnostics, and improvements. In: Airoldi, E.M., Blei, D., Erosheva, E.A., Fienberg, S.E., (eds.) Handbook of Mixed Membership Models and Their Applications, pp. 225–254. Chapman and Hall/CRC (2014)
7. Dey, D., Kumar, S.: Reassessing data quality for information products. Manage. Sci. 56(12), 2316–2322 (2010). https://doi.org/10.1287/mnsc.1100.1261
8. Han, J., Chen, K., Wang, J.: Web article quality ranking based on web community knowledge. Computing 97(5), 509–537 (2015)
9. Nurse, J.R., Rahman, S.S., Creese, S., Goldsmith, M., Lamberts, K.: Information quality and trustworthiness: a topical state-of-the-art review. Int. Conf. Comput. Appl. Netw. Secur. (ICCANS 2011) (2011)
10. Chinnov, A., Kerschke, P., Meske, C., Stieglitz, S., Trautmann, H.: An overview of topic discovery in twitter communication through social media analytics. Twenty-first Am. Conf. Inf. Syst, Puerto Rico (2015)
11. Hemalatha, I., Varma, G.P.S., Govardhan, A.: Preprocessing the informal text for efficient sentiment analysis. Int. J. Emerg. Trends Technol. Comput. Sci. 1(2), 58–61 (2012)
12. Sun, F., Belatreche, A., Coleman, S., McGinnity, T.M., Li, Y.: Pre-processing online financial text for sentiment classification: a natural language processing approach. In: IEEE Conference on Computational Intelligence for Financial Engineering and Economics (CIFEr), London, IEEE, pp. 122–129 (2014)

13. Duwairi, R., El-Orfali, M.: A study of the effects of preprocessing strategies on sentiment analysis for Arabic text. J. Inf. Sci. **40**(4), 501–513 (2014). https://doi.org/10.1177/0165551514534143
14. Rushdi-Saleh, M., Martín-Valdivia, M.T., Ureña-López, L.A., Perea-Ortega, J.M.: OCA: opinion corpus for Arabic. J. Am. Soc. Inf. Sci. Technol. **62**(10), 2045–2054 (2011)
15. Jianqiang, Z.: Pre-processing boosting twitter sentiment analysis? In: IEEE International Conference on Smart City/SocialCom/SustainCom (SmartCity), IEEE, pp. 748–753 (2015)
16. Indra, S.T., Wikarsa, L., Turang, R.: Using logistic regression method to classify tweets into the selected topics. In: 2016 International Conference on Advanced Computer Science and Information Systems (ICACSIS), IEEE, pp. 385–390 (2016)
17. Murshed, B.A.H., Mallappa, S., Ahmed, F.A.M., Al-Ariki, H.D.E.: Semantic analysis on big twitter dataset for automatic topic modeling. Test Eng. Manag. **83**, pp. 14657–14684 (2020)
18. Singh, T., Kumari, M.: Role of text pre-processing in twitter sentiment analysis. Procedia Comput. Sci. **89**, 549–554 (2016)
19. Gupta, I., Joshi, N.: Tweet normalization: a knowledge based approach. In: International Conference on Infocom Technologies and Unmanned Systems (Trends and Future Directions) (ICTUS), IEEE, pp. 157–162. Dubai, United Arab Emirates (2017)
20. Al-Khafaji, D.H.K., Habeeb, A.T.: Efficient algorithms for preprocessing and stemming of tweets in a sentiment analysis system. IOSR J. Comput. Eng. **19**(3), 44–50 (2017)
21. Ramachandran, D., Parvathi, R.: Analysis of twitter specific preprocessing technique for tweets. Procedia Comput. Sci. **165**, 245–251 (2019)
22. N. P. K M., K. P, Preprocessing methods for unstructured healthcare text data. Int. J. Innov. Technol. Explor. Eng. **9**(2), 715–719 (2019)
23. Arpaci, I., et al.: Analysis of twitter data using evolutionary clustering during the COVID-19 pandemic. Comput. Mater. Contin. **65**(1), 193–204 (2020)
24. Arpaci, I., Karataş, K., Baloğlu, M.: The development and initial tests for the psychometric properties of the COVID-19 Phobia Scale (C19P-S). Pers. Individ. Dif. **164**, 110108 (2020). https://doi.org/10.1016/j.paid.2020.110108
25. Joakim, C.: Explore python, machine learning, and the NLTK library. IBM Dev. Work. (2012)
26. Porter, M.F.: An algorithm for suffix stripping. Program **40**(3), 211–218 (2006)
27. Go, A., Bhayani, R., Huang, L.: Twitter sentiment classification using distant supervision. In: Processing, CS224N Project Report, pp. 1–6 (2009)
28. Pak, A., Paroubek, P.: Twitter as a corpus for sentiment analysis and opinion mining. Proc. 7th Int. Conf. Lang. Resour. Eval. Lr. pp. 1320–1326 (2010)

Feature Based Automated Detection of COVID-19 from Chest X-Ray Images

Shawli Bardhan and Sukanta Roga

Abstract Nowadays the biggest challenge for health care is controlling the pandemic of Coronavirus disease 2019 (COVID-19). Radiological investigation combining with machine learning can serve as a standardized methodology for detecting COVID-19. Chest X-ray imaging is the most feasible radiological test for COVID-19. Machine learning-based automated classification of COVID-19 from chest X-ray images can act as an assistive method to the medical experts for accurate diagnosis of disease. Aiming at this, the study focused on developing a simplified method of X-ray image based computerized COVID-19 detection through conventional feature extraction and classification approach. The method of X-ray image based COVID-19 detection consists of only two main steps: feature extraction, and classification. In feature extraction, a total of 55 X-ray image texture features is extracted from seven different groups. Classification of COVID-19 has been performed using those extracted features through four different popularly used classifiers. The overall analysis of the study has been performed over two datasets. The Random Forest classifier generates the best accuracy of 98.6% and 98.9% for dataset 1 and 2 with the area under the curve (AUC) values 0.99 and 1 respectively. The outcome of our study provides optimal accuracy of COVID-19 classification using X-ray images compare to existing popular studies in this domain. The reliable and less-complex feature of the proposed method may serve it as a computerized X-ray image based COVID-19 detection mechanism, especially in rural areas where medical experts are not available.

Keywords COVID-19 · X-ray image · Feature extraction · Classification · Detection

S. Bardhan (✉)
Department of Computer Science, Indian Institute of Information Technology Una, Una, Himachal Pradesh, India
e-mail: shawli.cse@gmail.com

S. Roga
Department of Mechanical Engineering, Visvesvaraya National Institute of Technology, Nagpur, Maharashtra, India
e-mail: rogasukanta@gmail.com

© The Author(s), under exclusive license to Springer Nature Switzerland AG 2021
I. Arpaci et al. (eds.), *Emerging Technologies During the Era of COVID-19 Pandemic*, Studies in Systems, Decision and Control 348,
https://doi.org/10.1007/978-3-030-67716-9_8

1 Introduction

The Coronavirus disease 2019 (COVID-19) comes under the communicable category and spreading all over the world with rapid growth. At the ending of 2019, the COVID-19 was first found in Wuhan city of China as pneumonia. The perilous disease syndromes in the human body are similar to Severe Acute Respiratory Syndrome (SARS). Due to COVID, patients go through syndromes like cough, fever, sore throat, fatigue, headache, shortness of breathing, muscle pain, etc. [1]. In the present scenario, COVID-19 is wide spreader around the world. According to the report of the World Health Organization (WHO), the outbreak of the disease infected a vast number of people, and death and affected cases are rapidly increasing each day. The total number of death due to COVID-19 reaches 3,49,095 and total infected cases are 54,88, 825 [2]. A total of 217 countries are under the epidemic of COVID-19. The gold standard for the diagnosis of COVID-19 is the reverse-transcription polymerase chain reaction (RT-PCR) test [3, 4]. But the sensitivity of the RT-PCR test is not satisfactory. The negative test outcome through RT-PCR does not interpret the complete absence of the COVID virus in the body [4–6]. So, medical imaging is popularly used as a complementary modality for the diagnosis of COVID-19. X-ray, computed tomography (CT), lung ultrasound, magnetic resonance imaging (MRI) [7, 8] plays a vital role in the area of COVID-19 diagnosis. Among the above-mentioned imaging techniques, CT and X-rays are most popular for COVID-19. However, CT imaging contains negative issues related to a shortage of CT facilities in a rural area, the cost of CT imaging, etc. Chest X-ray is a low-cost and available imaging methodology, and will likely be the frequently utilized imaging modality for detection and monitoring COVID-19 related abnormality. Generally image features like lung consolidation, lesions with ground-glass opacities, pulmonary fibrosis, multiple lesions, bilateral lower lobe consolidations, peripheral air space opacities, diffuse air space, etc. are indicating characteristics for COVID-19 positive [9–12]. The use of X-ray imaging for identification of the above mentioned features can act as a key diagnosis method for COVID-19 and also for monitoring the progress of treatment. A combination of artificial intelligence (AI) and imaging technology can further strengthen the capacity of X-ray image analysis for the identification of COVID-19 characteristics and also help in the computerized image-based diagnosis of the disease.

Seeing the present necessity of computerized COVID-19 detection, the study aims to classify normal and COVID-19 X-ray images by extracting textures features of images. The advantage of the study includes methodological simplicity with higher accuracy compared to existing popular researches (Table 1).

The rest of the paper is organized as Sect. 2 describes the existing image analysis-oriented research on COVID-19. Section 3 details the methodology of the study. Sections 4 and 5 contains the result and discussion respectively. Finally, Sect. 6 concludes the overall study directing the future scope.

Table 1 Literature review on X-ray image analysis based COVID-19 classification

Literature	Aim	Method	Subjects	Accuracy (%)/Result
Ozturk et al. [8]	Classification in between COVID+, COVID−, and pneumonia X-ray	DarkCovidNet	COVID-19(+): 125 Pneumonia: 500 No-Findings: 500	2 class: 98.08% 3 class:87.02%
Apostolopoulos, Ioannis D., and Tzani A. Mpesiana [13]	Classification in between COVID, pneumonia, and normal X-ray	Convolutional neural network (CNN)	Dataset 1: COVID-19(+):224 Pneumonia: 700 No-Findings: 504 Dataset 2: COVID-19(+):224 Pneumonia: 714 No-Findings: 504	2 class: 96.78% 3 class: 94.72%
Narin, Ali, Ceren Kaya, and Ziynet Pamuk. [14]	Classification in between COVID-19 (+), and COVID-19 (−)	Deep CNN ResNet-50	COVID-19(+): 50 COVID-19 (−): 50	98%
Sethy, Prabira Kumar, and Santi Kumari Behera [15]	Classification in between COVID-19 (+), and COVID-19 (−)	ResNet50 + SVM	COVID-19(+): 25 COVID-19 (−): 25	95.38%
Abbas, Asmaa, Mohammed M. Abdelsamea, and Mohamed Medhat Gaber [16]	Classification in between COVID-19 (+), and COVID-19 (−)	DeTraC (Decompose, Transfer, and Compose) CNN model	COVID-19(+):105 SARS: 11 Normal: 80	95.12%
Zhang, Jianpeng, et al. [17]	Classification in between COVID-19 (+), and others	ResNet	COVID-19(+): 70 Others: 1008	Sensitivity: 0.96, Specificity: 0.71 AUC: 0.952
Wang, Linda, and Alexander Wong [18]	COVID and non-COVID X-ray classification	COVID-Net	COVID-19(+): 53 COVID-19 (−): 5526 Healthy: 8066	92.6%
Hemdan, Ezz El-Din, Marwa A. Shouman, and Mohamed Esmail Karar [19]	Classification in between COVID-19 (+), and normal X-ray	COVIDX-Net	COVID-19(+):25 Normal: 25	90%

2 Literature Review and Contribution of the Study

Accurate analysis of COVID-19 related X-ray images of the chest may serve as an initial detection scheme of abnormality in the medical domain. Existing studies detailed some methodologies for computerized automated COVID-19 detection using X-ray images. Ozturk et al. [8] developed a COVID-19 detection methodology using raw X-ray images of the chest. They performed classification in between COVID and no-findings and achieved 98.08% accuracy. In multiclass classification, they categorized COVID, no findings, and pneumonia and produced 87.02% accuracy. The DarkNet model used in their study for classification. Apostolopoulos, Ioannis D., and Tzani A. Mpesiana [13] applied the existing convolutional neural network (CNN) oriented transfer learning methodology for COVID, pneumonia, and normal X-ray classification. For 2 class classification, they got 96.78% accuracy using Deep-learning, and for three-class, maximum achieved accuracy is accuracy 94.72%. Narin, Ali, Ceren Kaya, and Ziynet Pamuk. [14] used deep transfer learning for the detection of COVID using X-ray images. Their study shows 98% classification accuracy using pre-trained ResNet50 architecture. The study performed by Sethy, Prabira Kumar, and Santi Kumari Behera [15] detected COVID affected X-ray images by combining the ResNet50 CNN model and support vector machine (SVM) classifier. The feature outcome of CNN is fed to SVM classifier for the categorization and obtained 95.38% accuracy. Abbas, Asmaa, Mohammed M. Abdelsamea, and Mohamed Medhat Gaber [16] performed a similar classification using deep CNN named as DeTraC (Decompose, Transfer, and Compose) model. Their obtained accuracy was 95.12% for classification. Zhang, Jianpeng, et al. [17] also did the same classification using deep learning methodology and obtained an accuracy of 96% for COVID-19 cases and 70.65% for non-COVID-19 cases. Wang, Linda, and Alexander Wong [18] achieved a classification accuracy of 92.6% for COVID and non-COVID X-ray categorization using COVID-Net deep CNN architecture. For classification in between COVID positive and COVID negative, Hemdan, Ezz El-Din, Marwa A. Shouman, and Mohamed Esmail Karar [19] used a deep learning network named as Covidx-net. Their maximum obtained accuracy is 90%.

Observation of the above review work indicates that existing work on X-ray image based COVID detection is performed by using deep learning-based CNN architecture. The maximum achieved accuracy is 98.08%. But, the existing works contain complex data models. Also, the use of deep learning requires high topological knowledge, deep parametric knowledge, and large dataset. Therefore it is difficult to adopt CNN architecture for the zone with less skilled manpower. Focusing on the scenario, in our study we focused on developing a simplified computerized methodology by using feature-based conventional classification strategy with optimal accuracy for X-ray image categorization aiming at COVID-19 detection.

Fig. 1 Flow diagram for X-ray image based COVID-19 detection

3   Methodologies of COVID-19 Detection

The chest X-ray image consists of salient features related to COVID-19. Computerized extraction of those features will help in the detection of the disease. In this study, we aim to develop a simplified methodology for chest X-ray image based abnormality detection. The methodology consists of only four important steps including data collection as shown in Fig. 1 and described below.

3.1   Data Collection and System Requirement

The analysis has been performed over two chest X-ray image datasets. The image data are collected from Kaggle repository (link: https://www.kaggle.com/ tawsifurrahman/covid19-radiography-database) and database developed by Cohen JP (link: https://github.com/ieee8023/COVID-chestxraydataset). The dataset 1 contains both the COVID positive and negative X-ray images. In our study, we used all the 219 COVID positive X-ray images and 250 normal X-ray images. All the images are in Portable Network Graphics (PNG) file format. The size of each X-ray is 1024 × 1024 pixels. Most of the images are in a grayscale format where some are in color format. The color (RGB) images are converted into grayscale before processing. The dataset 2 also contains 125 COVID positive X-ray images, 500 pneumonia-related X-ray images, and 500 normal chest X-ray images. The images are of variable size and stored in png/jpg/jpeg format. We used the COVID-19 and normal X-ray images for this study. For both the dataset, COVID-19 oriented X-ray images are labeled as 1, and normal X-ray images are labeled as 0. The datasets are dedicated to research purposes and for the development of useful and impactful methodologies for COVID detection.

The overall is performed on MATLAB R2013a software. The system specification is Windows 10 with 64-bit Operating system, 4 GB RAM.

3.2   Feature Extraction

In this study, classification in between COVID-19 and normal raw chest X-ray images is performed by analyzing texture level features. The extracted features are used as the input of the classifier as shown in Fig. 1. The details of the extracted features are given below.

a. **First-order statistical features** [20]:

Image pixel value-oriented simple statistical summary is extracted through first-order statistical features. The texture information related to the neighborhood intensity distribution of each pixel is not considered here. The histogram shows the rate of occurrence of each pixel intensity in an image. Based on each pixel intensity occurrence rate in the image the extracted first-order statistical features are mean, variance, kurtosis, skewness, standard deviation, entropy, and energy. A total of seven features are extracted from histogram characteristics.

b. **Spatial gray level dependence Matrices features** [21]:

Spatial gray level dependence Matrices features are calculated by extracting pixel intensity from $0°$, $45°$, $90°$, and $135°$ angle with distance value 1 from center pixel (C). The horizontal pixels with center pixels are at $0°$ and the vertical pixels with C are at $90°$. The top left and bottom right pixels for the 3×3 image are at $135°$ with the center. Finally, the bottom left and top right pixels from C are at $45°$ with the center. The Fig. 2 shows each pixel degree position clearly. 13 different features are extracted from special gray level dependence Matrices namely angular second moment, variance, contrast, correlation, local homogeneity, entropy, sum variance, difference variance, sum average, sum entropy, difference entropy, and information measure of correlation from x and y-axis (individually). All the mentioned features are extracted into two groups, mean group and difference group. In mean group average of feature value for a particular feature extracted from all the four angles are calculated. Similarly, in the difference group, the difference between the minimum and maximum values for a particular feature extracted from all four angles is calculated. Therefore a total of 26 features are extracted.

c. **Gray level difference statistics features** [22]:

The gray level difference based statistical features are calculated from X-ray images depending on the absolute difference present in between pairs of gray levels considering displacement ∂. For a given displacement ∂, four features are calculated. Those features are mean, contrast, entropy, and energy. The mentioned features are calculated considering $\partial = (0, 1), (1, 1), (1, 0), (1, -1)$. Here $\partial \equiv (\Delta x, \Delta y)$.,

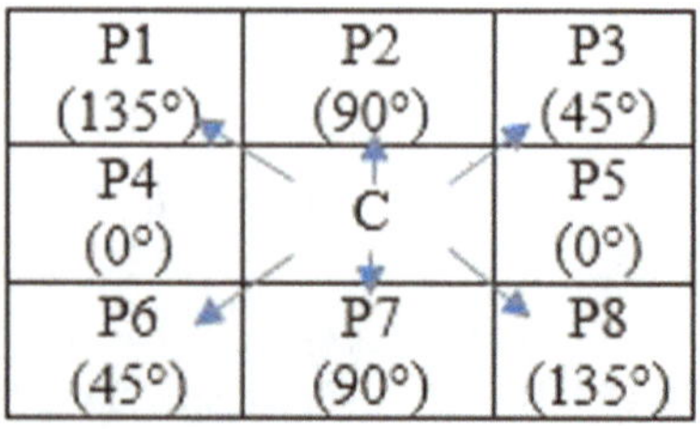

Fig. 2 Angular position of each pixel (P) from center pixel (C) for a 3×3 image

P1 (135°)	P2 (90°)	P3 (45°)
P4 (0°)	C	P5 (0°)
P6 (45°)	P7 (90°)	P8 (135°)

Therefore, ∂ indicates a vector in (x, y) plane for an image $f(x, y)$. The mean value ∂ is taken for each feature.

d. **Neighborhood gray-tone difference matrix** [23]:

In image processing, information related to spatial changes in intensity distribution can be found by analyzing the gray tone of a pixel comparing with its neighborhood pixels. The neighborhood gray-tone difference matrix provides the same for spatial feature extraction. For the generation of the matrix, let $I(x, y)$ be the gray tone of a pixel at the position (x, y) having value i. Then the average gray tone considering 8-neighborhood pixels is,

$$P_i = P(x, y) = \frac{1}{W - 1} \left[\sum_{m=-s}^{s} \sum_{n=-s}^{s} I(x + m, y + n) \right], (m, n) \neq (0, 0)$$

Here, s represents the size of the neighborhood and $W = (2s + 1)^2$.

Following the above, ith entry in the Neighborhood gray-tone difference matrix is,

$$N(i) = \begin{matrix} \sum |i - \overline{P_i}| & \text{for } i \in A_i \text{ if } A_i \neq 0 \\ 0, & \text{otherwise} \end{matrix}$$

Here, $\{A_i\}$ is the set of all pixels with gray tone i.

Using the Neighborhood gray-tone difference matrix, five features are extracted in this study: Coarseness, contrast, complexity, busyness, and strength.

Coarseness indicates the uniformity in intensity distribution in an image. Higher the coarseness indicates increased uniformity in an image. In a coarse-textured image, the basic patterns related textures of the image are large. Therefore the intensity change is less in a coarse texture. The contrast of an image indicates the clarity in area based intensity change. High contrast indicates intensity difference in between neighborhood region is large. The feature Complexity of image depends on the presence of primitives or patches in a texture. A complex texture contains more patches that indicate content information is high in the image. The complexity of an image is also related to contrast and busyness features. The busyness of an image indicates a rapid change in intensity distribution compares to neighborhood intensity. Hence higher the frequency of spatial changes in pixel intensity values indicates greater busyness in an image.

The feature texture strength is complex to define. It is correlated with coarseness and contrast. In general, in a high texture strength-based image, the primitives that comprise it are clearly visible and definable. Such an image contains a high degree of visual clarity.

e. **Statistical feature matrix** [24]:

The statistical feature matrix performs a texture analysis of the image by measuring the statistical features of pixel pairs with multiple distances. In this statistical feature matrix, the size of the matrix is based on the maximum distance considered, not on the gray level. The extracted properties in this study are coarseness, roughness, periodicity, and contrast. The study also considered the maximum inter-sample spacing distance as 4, i.e., $L_r = L_c = 4$.

f. **Texture energy measure** [25]:

LL, SS, EE, LE, LS, ES are six different kernels and generated from three simple vectors of length 3 for measuring texture energy. These are L3 $\equiv$ (1, 2, 1), S3 $\equiv$ (−1, 2, −1), and E3 $\equiv$ (−1, 0, 1). The vectors represent the one-dimensional operations of center-weighted local averaging, symmetric second differencing for spot detection, and symmetric first differencing for edge detection. If these vectors are convolved with themselves or with each other, we obtain five vectors of length 5. Those derived vectors are L5 = (1, 4, 6, 4, 1), S5 = (−1, 0, 2, 0, −1), E5 = (−1, −2, 0, 2, 1) where L5 again performs local averaging, S5 is spot detector, and E5 is the edge detector. Multiplying the column vectors of length 5 by row vectors of the same length, we obtain 5 × 5 masks. Using the masks, the texture energy is extracted by convolving them with the X-ray images. The final masks used in our experiments are as shown in Fig. 3. Finally, from this feature group, a total of six energy features are extracted from six different kernel/mask.

g. **Fourier power spectrum** [26]:

The Fourier transform of an image $I(x, y)$ is expressed as,

$$F(u, v) = \int\limits_{-\infty}^{\infty} \int\limits_{-\infty}^{\infty} e^{-2r\sqrt{-1}(ux + vy)} I(x, y) dx dy$$

The Fourier power spectrum is represented as $|F|^2 = FF^*$. Here * indicates the complex conjugate. In the case of a coarse-textured image, $|F|^2$ is high near the

LL:

1	4	6	4	1
4	16	24	16	4
6	24	36	24	6
4	16	24	16	4
1	4	6	4	1

SS:

1	0	-2	0	1
0	0	0	0	0
-2	0	4	0	-2
0	0	0	0	0
1	0	-2	0	1

EE:

1	2	0	-2	-1
2	4	0	-4	-2
0	0	0	0	0
-2	-4	0	4	2
-1	-2	0	2	1

LE:

-1	-2	0	2	1
-4	-8	0	8	4
-6	-12	0	12	6
-4	-8	0	8	4
-1	-2	0	2	1

LS:

-1	0	2	0	1
-4	0	8	0	4
-6	0	12	0	6
-4	0	8	0	4
-1	0	2	0	1

ES:

1	0	-2	0	-1
2	0	-4	0	-2
0	0	0	0	0
-2	0	4	0	2
-1	0	2	0	1

Fig. 3 Final kernels/masks for textural energy measure

origin. For fine texture, $|F|^2$ will be spread out more. To analyze the texture coarseness using the Fourier power spectrum, here in the study the annular-ring and wedge sampling features of X-ray images are extracted using the following equations.

Annular ring: $\phi_{x_1, x_2} \equiv \sum_{x_1^2 \le u^2 + v^2 < x_2^2} |F(u, v)|^2$

Wedge: $\phi_{\theta_1, \theta_2} = \sum_{\theta_1 < \tan^{-1}(v/u) < \theta_2} |F(u, v)|^2$

Here, x_1, x_2 our inner and outer ring radius. θ is wedge slope. In this study, the value of x_1, x_2 is 2, 4 and θ_1, θ_2 is $0°, 45°$ respectively. Also, $0 < u, v < P - 1$ for a given $P \times P$ image.

3.3 Classification

The extracted features from chest X-ray images are fed to the classifier for COVID-19 abnormality detection. For that, four states of the art classifiers are used namely Support Vector Machine (SVM), Random Forest (RF), K Nearest Neighbour (KNN), and Linear Discriminant Analysis (LDA). The ten-fold cross-validation is performed for classifier accuracy analysis and 30% of the overall dataset is used for testing purposes. The parameter description of each classifier is mentioned here for a clear understanding.

SVM [27]: SVM classifier is a popularly used method for medical image classification. The required parameters for SVM are regularization parameter(C), kernel type, degree, and kernel coefficient. Here we used C = 1.0, linear kernel, and kernel coefficient as 0.1. The degree is not required here as we are using linear kernel for our analysis.

RF [20, 27]: RF classifier is designed based on combining multiple decision trees to obtain enhanced accuracy of classification. The required parameters for RF with values used for this study are number of tree: 100, quality of each split: Gini, the maximum number of feature for best split: auto, maximum tree depth: None, the minimum number of sample for the split: 20, the minimum size of end node/leaf:1, minimum weight fraction leaf: 0.0, maximum leaf node size: None, minimum impurity decrease: 0.0, bootstrap: True, cross-validation method (oob_score): False, processor number: None, random state: None, verbosity:0, warm_start: False, balanced subsample weight: None.

KNN [20, 27]: KNN is a simple, versatile, and easy to understand method of classification. The parameters tuning performed for KNN in this study are number of neighbors: 5, weight function: uniform, the algorithm used: auto, leaf size: 20, power parameter: 2, distance metric: Minkowski, arguments for metric function: None, number of parallel job for neighborhood search: None.

LDA [20, 27]: Through fitting class conditional density of the data and by using Bayes rule, the classifier generates a linear decision boundary for classification. In this study, the Singular value decomposition is used as a solver with rank estimation threshold value 0.0001. The other parameters for the LDA classifier are set as the default value, i.e., None in this study.

3.4 Validation

The validation of the study outcome has been performed by comparing the result of the classification with clinical findings. Here seven validation parameters are used: true positive, true negative, false positive, false negative, sensitivity, specificity, and accuracy. True negative indicates that clinically COVID-19 marked X-ray images are accurately detected by the classifier. True positive is clinically normal detected X-ray images are also classified as normal through the classifier. False-negative indicates that clinically normal identified X-ray images are detected as COVID-19 through the classifier and false-positive is reverse of a false negative. The sensitivity, specificity, and accuracy are calculated by following the equations below.

$$\text{Sensitivity:} \frac{\text{True Positive}}{\text{True Positive} + \text{False Negative}}$$

$$\text{Specificity:} \frac{\text{True Negative}}{\text{True Negative} + \text{False Positive}}$$

$$\text{Accuracy:} \frac{\text{True Positive} + \text{True Negative}}{\text{True Positive} + \text{True Negative} + \text{False Positive} + \text{False Negative}} \times 100\%$$

Higher the value of all the measures indicates better performance of the classifier.

4 Results

In this study, the classification of the chest X-ray images for COVID-19 detection has been performed using the extracted textured features. A total of 55 features are extracted from seven different groups. The extracted features are fed to the four different classifiers with parameter tuning for optimal classification accuracy. The experiment has been performed over two datasets of COVID-19 X-ray images as mentioned in Sect. 3.1. Table 2 shows the outcome of classification using validation parameters. Observation of Table 2 shows that the Random Forest (RF) classifier generates maximum accuracy, i.e., 98.6% and 98.9% for both the dataset (dataset 1 and dataset 2) respectively. The sensitivity of the classifier is also maximum, i.e., 1 for both the datasets with specificity 0.97 and 0.95 respectively. Higher the sensitivity and specificity of classification indicates that most of the COVID-19 oriented X-ray images are accurately classified. The LDA classifier is also able to provide a high degree of classification with value 97.9% and 98.4% for datase1 and dataset 2 respectively. The classification outcome of each classifier is determined by taking the mean of ten-fold cross-validation. Figure 4 shows the

Table 2 Classification outcome

Classifier used	True positive	False negative	False positive	True negative	Sensitivity	Specificity	Accuracy (%)
Dataset 1							
SVM	71	2	15	53	0.97	0.78	87.9
RF	83	0	2	56	1	0.97	98.6
KNN	74	2	10	55	0.97	0.85	91.2
LDA	75	1	2	63	0.99	0.97	97.9
Dataset 2							
SVM	145	1	19	23	0.99	0.55	89.4
RF	145	0	2	41	1	0.95	98.9
KNN	144	3	14	27	0.98	0.66	90.9
LDA	146	1	2	39	0.99	0.95	98.4

Fig. 4 Classification accuracy of each fold in ten-fold cross validation for **a** dataset 1, **b** dataset 2

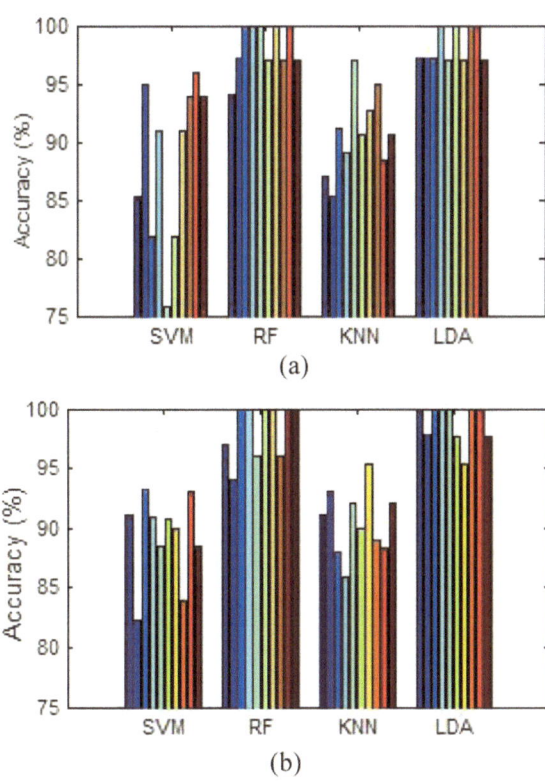

accuracy of each fold for all the classifiers of both the datasets. Observation of Fig. 4 indicates that fold wise accuracies of SVM and KNN is highly variable and low compare to RF and LDA. For RF, fold base intra accuracies are very much similar to each other. As accuracy wise RF generates an optimal result of

Fig. 5 ROC curve of RF
classifier **a** dataset 1, **b** dataset
2

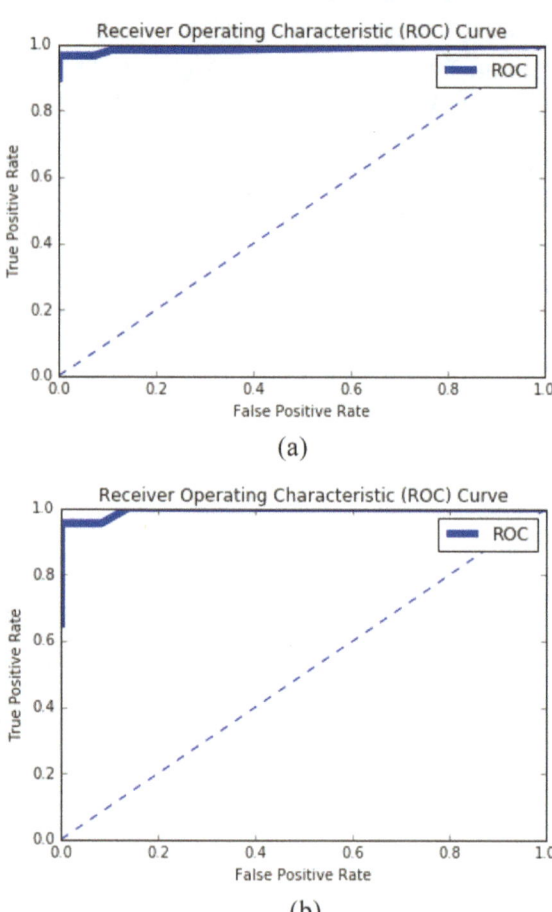

(a)

(b)

classification, so the Receiver Operating Characteristic (ROC) curve of RF outcome
for both the dataset is given in Fig. 5. The area under the curve (AUC) for dataset 1
is 0.99 and dataset 2 is 1. The ROC curve also shows that RF classifier provides
significant performance for X-ray image feature-based COVID-19 classification.

5 Discussion

The performance of the overall methodology is dependent on feature extraction and
classifier selection. For optimal outcome, multiple classifiers are used as detailed in
Sect. 3.3 and the effect of classifier in the accuracy of classification is shown in
Table 2. In the case of feature extraction, a total of 55 texture features are extracted
from seven different groups. Among them, some groups of features contain a higher

Fig. 6 Feature group base classication accuracy for **a** dataset 1, **b** dataset 2

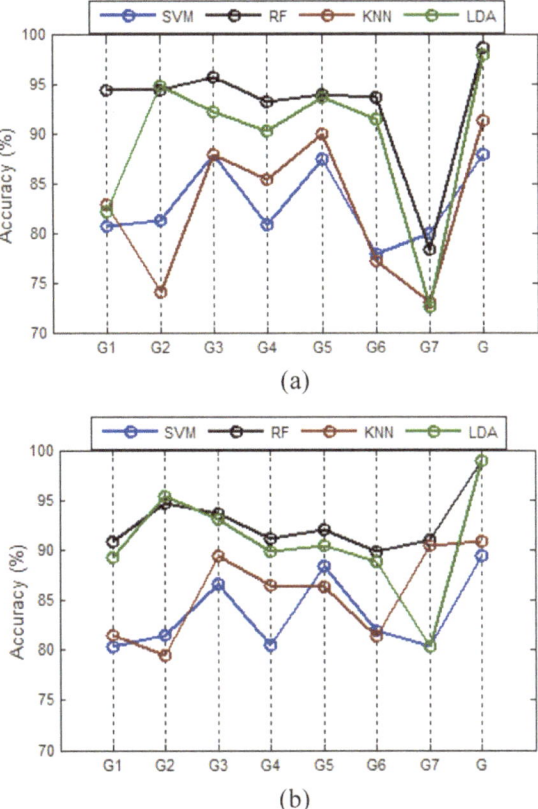

degree of differentiable features for COVID oriented X-ray and normal X-ray classification. Figure 6 shows the classification accuracy of each group of features using four different classifiers. Here G1 to G7 represents seven different feature groups, i.e., G1: First order statistical features, G2: Spatial gray level dependence Matrices features, G3: Gray level difference statistics features, G4: Neighborhood gray-tone difference matrix features, G5: Statistical feature matrix features, G6: Texture energy measure features, G7: Fourier power spectrum features and G indicates the combination of all extracted features. The X-axis of Fig. 6 represents each group individually (G1 to G7) and G. Observation of Fig. 6 shows that in case of feature group- wise classification also, RF classifier generates optimal result for most of the groups. The G3 for dataset 1 and G2 for dataset 2 individually generates maximum accuracy of classification ($\leq 96\%$). But the combination of all feature groups (G) increases the accuracy of classification. The increase of accuracy through G is noticeable in the case of all four classifiers.

The maximum classification accuracy achieved through this study is 98.6% and 98.9% respectively for dataset 1 and dataset 2. Existing studies on X-ray image based COVID-19 classification also shows higher accuracy using CNN

Table 3 Comparative study of X-ray image based COVID-19(+) and COVID-19(−) classification

Study	Dataset details	Method used	Accuracy (%)
[8]	COVID-19(+): 125, No-Findings: 500	CNN (DarkCovidNet)	98.08
[13]	Dataset 1: COVID-19(+): 224, No-Findings: 504 Dataset 2: COVID-19(+): 224, No-Findings: 504	CNN	96.78
[14]	COVID-19(+): 50, COVID-19 (−): 50	CNN (Deep CNN, ResNet-50)	98
[15]	COVID-19(+): 25, COVID-19 (−): 25	CNN + SVM (ResNet50+)	95.38
[16]	COVID-19(+) and SARS: 116, Normal: 80	CNN(**DeTraC**))	95.12
[17]	COVID-19(+): 70, Others: 1008	CNN(ResNet)	Sensitivity: 0.96, Specificity: 0.71 AUC: 0.952
[18]	COVID-19(+): 53, COVID-19 (−) and healthy: 13,592	CNN (COVID-Net)	92.6
[19]	COVID-19(+): 25, Normal: 25	CNN (COVIDX-Net)	90
Our Study	Dataset 1: COVID-19(+): 219, Normal: 250 Dataset 2: COVID-19(+): 125, No-Findings: 500 (same dataset used in [8])	Texture Feature with Random Forest (RF) classifier.	Dataset 1: Sensitivity: 1, Specificity: 0.97, Accuracy: 98.6, AUC: 0.99 Dataset 2: Sensitivity: 1, Specificity: 0.95, Accuracy: 98.9, AUC: 1

architecture. Table 3 shows the comparative study of presented methodology and existing popular works on X-ray image base COVID detection. Observation shows that all the existing methodologies used the CNN architecture for classification and achieved a maximum 98.08% accuracy. In our study, we used the conventional classification strategy by extracting texture features of X-ray images and using those features as the input of the Random Forest (RF) classifier. The proposed conventional strategy of classification provides the highest accuracy among all existing studies. The used dataset 2 is as same as the dataset used by Ozturk et al. [8] in their study.

6 Conclusion

COVID-19 is the infectious disease caused by the most recently discovered coronavirus. This new virus and disease were unknown before the outbreak began in Wuhan, China, in December 2019. COVID-19 is now a pandemic affecting many countries globally [28–32]. This study proposes a conventional feature extraction-based method for COVID-19 detection using chest X-ray image datasets. The method is applied over two datasets and generates 98.6% and 98.9% accuracy of classification in between COVID affected X-ray images and normal X-ray images using RF classifier with AUC as 0.99 and 1 respectively. A total of 55 texture features are extracted for this purpose from seven different groups. Group-wise analysis of classification shows that individually Gray level difference statistics features (G3) and Spatial gray level dependence Matrices features (G2) for dataset 1 and dataset 2 respectively generate maximum accuracy of 95.7% (using RF classifier) and 95.4% (using LDA classifier). A combination of all the feature groups improves the accuracy by 2.9% and 3.5% for dataset 1 and dataset 2 respectively. The existing popular studies of the same classification generates a maximum accuracy of 98.08% using CNN architecture. Also, the use of feature extraction dependent conventional classification methodology is not present in existing studies. Though CNN architectures are used for X-ray image base COVID classification but contain the difficulty related to a limited number of image data, the requirement of data augmentation, architecture complexity, hyper-parameter tuning, slow processing, and lack of expert knowledge. Our suggested method generates better accuracy compared to CNN based methods for COVID-19 classification and also less complex, faster, and applicable for small datasets. The future direction of the method will aim to identify the region of abnormality in the chest using CT and X-ray images.

Acknowledgements Authors would also like to thank Dr. M.K. Bhowmik, Assistant Professor, Department of Computer Science and Engineering, Tripura University, Suryamaninagar 799022, Tripura, India for his support during knowledge development in the area of image feature extraction and classification.

References

1. Singhal, T.: A review of coronavirus disease-2019 (COVID-19). Indian J. Pediatr. 1–6 (2020)
2. Coronavirus.: https://www.who.int/emergencies/diseases/novel-coronavirus-2019 (2020). Retrieved 27 May 2020
3. W. H. Organization.: Clinical management of severe acute respiratory infection when Novel coronavirus (2019-nCoV) infection is suspected: interim Guidance
4. National Health Commission of the People's Republic of China, Diagnosis and treatment protocol for COVID-19 (trial version 7). http://en.nhc.gov.cn/2020-03/29/c_78469.htm
5. Ai, T., et al.: Correlation of chest CT and RT-PCR testing in coronavirus disease 2019 (COVID-19) in China: a report of 1014 cases. Radiology. 200642 (2020)

6. Fu, H., et al.: Association between clinical, laboratory and CT characteristics and RT-PCR Results in the follow-up of COVID-19 patients. medRxiv (2020)

7. Dong, D., Tang, Z., Wang, S., et al.: The role of imaging in the detection and management of COVID-19: a review [published online ahead of print, 2020 Apr 27]. IEEE Rev. Biomed. Eng. (2020). https://doi.org/10.1109/rbme.2020.2990959. https://doi.org/10.1109/rbme.2020.2990959

8. Ozturk, T., et al.: Automated detection of COVID-19 cases using deep neural networks with X-ray images. Comput. Biol. Med. 103792 (2020)

9. Xie, X., Zhong, Z., Zhao, W., Zheng, C., Wang, F., Liu, J.: Chest CT for typical 2019-nCoV pneumonia: relationship to negative RT-PCR testing. Radiology (2020)

10. Shi, H., et al.: Radiological findings from 81 patients with COVID-19 pneumonia in Wuhan, China: a descriptive study. Lancet. Infect. Diseas. 242020 (2020). https://doi.org/10.1016/s1473-3099(20)30086-4

11. Kanne, J.P.: Chest CT findings in 2019 novel coronavirus (2019-nCoV) infections from Wuhan, China: key points for the radiologist. Radiology 295(1), 16–17 (2020). https://doi.org/10.1148/radiol.2020200241

12. Jacobi, A., et al.: Portable chest X-ray in coronavirus disease-19 (COVID-19): a pictorial review. Clin. Imaging. 64, 35–42 (2020). https://doi.org/10.1016/j.clinimag.2020.04.001apostolopoulos

13. Apostolopoulos, I.D., Mpesiana, T.A.: Covid-19: automatic detection from x-ray images utilizing transfer learning with convolutional neural networks. Phys. Eng. Sci. Med. 1 (2020)

14. Narin, A., Kaya, C., Pamuk, Z.: Automatic detection of coronavirus disease (COVID-19) using x-ray images and deep convolutional neural networks. Preprint at arXiv:2003.10849 (2020)

15. Sethy, P.K., Behera, S.K.: Detection of coronavirus disease (COVID-19) based on deep features. Preprints 2020030300 (2020)

16. Abbas, A., Abdelsamea, M.M., Gaber, M.M.: Classification of COVID-19 in chest X-ray images using DeTraC deep convolutional neural network. Preprint at arXiv:2003.13815 (2020)

17. Zhang, J., et al.: Covid-19 screening on chest x-ray images using deep learning based anomaly detection. Preprint at arXiv:2003.12338 (2020)

18. Wang, L., Wong, A.: COVID-Net: a tailored deep convolutional neural network design for detection of COVID-19 cases from chest X-Ray images. Preprint at arXiv:2003.09871 (2020)

19. Hemdan, E.E.D., Shouman, M.A., Karar, M.E.: Covidx-net: a framework of deep learning classifiers to diagnose covid-19 in x-ray images. Preprint at arXiv:2003.11055 (2020)

20. Bhowmik, M.K., et al (2017) Designing of ground truth annotated DBT-TU-JU breast thermogram database towards early abnormality prediction. IEEE J. Biomed. Health Inf. https://doi.org/10.1109/jbhi.2017.27405.00

21. Haralick, R.M., et al.: Textural features for image classification. IEEE Trans. Syst. Man. Cybern. SMC. 3(6), 610–621 (1973). https://doi.org/10.1109/tsmc.1973.43093.14

22. Weszka, J.S., et al.: A comparative study of texture measures for terrain classification. IEEE Trans. Syst. Man. Cybern. SMC. 64, 269–285 (1976). https://doi.org/10.1109/tsmc.1976.54087.77

23. Amadasun, M., King, R.: Textural features corresponding to textural properties. IEEE Trans. Syst. Man. Cybern. 19(5), 1264–1274 (1989). https://doi.org/10.1109/21.44046

24. Wu, C.M., Chen, Y.C.: Statistical feature matrix for texture analysis. CVGIP Graph Models Image Process. 54(5), 407–419 (1992). https://doi.org/10.1016/1049-9652(92)90025-s

25. Laws, K.I.: Texture energy measures. DARPA Image Understanding Workshop, pp. 47–51. DARPA, Los Altos, CA (1979)

26. Wu, C.M., et al.: Texture features for classification of ultrasonic liver images. IEEE Trans. Med. Imaging. 11(2), 141–152 (1992). https://doi.org/10.1109/42.141636

27. Guyon, I., Weston, J., Barnhill, S., Bapnik, V.: Gene selection for cancer classification using support vector machines. Mach. Learn. 46(1–3), 389–422 (2002)

28. Arpaci, I., Alshehabi, S., Al-Emran, M., Khasawneh, M., Mahariq, I., Abdeljawad, T., Hassanien, A.E.: Analysis of Twitter data using evolutionary clustering during the COVID-19 pandemic. Comput. Mater. Continua. **65**(1), 193–204 (2020). https://doi.org/10.32604/cmc. 2020.011489

29. Arpaci, I., Karataş, K., Baloğlu, M.: The development and initial tests for the psychometric properties of the COVID-19 Phobia Scale (C19P-S). Personality Individ. Differ. **164**, 110108 (2020). https://doi.org/10.1016/j.paid.2020.110108

30. Arpaci, I.: A hybrid modeling approach for predicting the educational use of mobile cloud computing services in higher education. Comput. Hum. Behav. **90**, 181–187 (2019). https:// doi.org/10.1016/j.chb.2018.09.005

31. Arpaci, I., Al-Emran, M., Al-Sharafi, M.A., Shaalan, K.: A novel approach for predicting the adoption of smartwatches using machine learning algorithms. In: Recent Advances in Intelligent Systems and Smart Applications, pp. 185–195. Springer, Cham (2021)

32. Arpaci, I.: What drives students' online self-disclosure behavior on social media? A hybrid SEM and artificial intelligence approach. Int. J. Mobile Commun. **18**(1) (2020). https://doi. org/10.1504/IJMC.2020.105847

Indoor Air Quality Monitoring Systems and COVID-19

Jagriti Saini⃝, Maitreyee Dutta⃝, and Gonçalo Marques⃝

Abstract Coronavirus pandemic is proven as wreaking havoc with the rising number of cases since its first identification in Wuhan, China. This invisible threat has taken approximately 1.06 M lives with more than 36.5 M cases worldwide as on October 9th, 2020. World Health Organization (WHO) estimated the mortality rate for COVID-19 to be around 3–4% with higher risk to the people that have underlying medical conditions such as respiratory disease, diabetes, cancer, heart disease, asthma, and kidney disease. Numerous public health experts have defined a clear link between coronavirus death rates and long-term exposure to air pollution, especially $PM_{2.5}$ and NO_2 levels. Medical health experts are working hard to find a reliable treatment for COVID-19. The need for real-time monitoring systems to promote indoor air quality is another critical concern that demands the attention of the research community. The main contribution of this study is to present the connection between COVID-19 pandemic, public health and indoor air quality while addressing the importance of real-time monitoring systems for public health and wellness. This chapter presents the necessity of developing indoor air quality monitoring systems for hospitals, schools, offices and homes for enhanced health and well-being.

Keywords COVID-19 · Coronavirus · Public health · Indoor air quality · Monitoring systems

J. Saini · M. Dutta
National Institute of Technical Teacher's Training and Research, Chandigarh 160019, India
e-mail: jagritis1327@gmail.com

M. Dutta
e-mail: d_maitreyee@yahoo.co.in

G. Marques (✉)
Polytechnic of Coimbra, ESTGOH, Rua General Santos Costa, Oliveira do Hospital
3400-124, Portugal
e-mail: goncalosantosmarques@gmail.com

© The Author(s), under exclusive license to Springer Nature Switzerland AG 2021
I. Arpaci et al. (eds.), *Emerging Technologies During the Era of COVID-19 Pandemic*, Studies in Systems, Decision and Control 348,
https://doi.org/10.1007/978-3-030-67716-9_9

1 Introduction

The novel coronavirus disease 2019 (COVID-19) outbreak in Wuhan, Hubei Province, China has now become a worldwide panic with a rapid spread in 216 countries and territories [1, 2]. The total number of confirmed cases has now increased above 36.5 M with the death toll of more than 1.06 M [1]. The first case of coronavirus was reported in China in December 2019, and it was declared as a pandemic by the World Health Organization (WHO) on March 11th, 2020 [3]. As compared to MERS and SARS, this disease has spread more rapidly around the globe due to increased adaptation of the virus in different environments [4]. As this virus can spread from an infected person, contaminated surfaces and objects, several countries have installed a lockdown [5]. Considering the impact, leading economists have already predicted a shrink of the global economy due to novel coronavirus, leaving more burden on developing countries with limited resources [6, 7]. In this scenario, slowing down the impact of COVID-19 is critical to promote public health and well-being. Medical health experts are making hard efforts to find a reliable treatment for COVID-19 patients. Moreover, it has become equally relevant to understand the inherent causes of the rising number of deaths.

1.1 COVID-19 and Underlying Illness

The data obtained from Italy and China state that most of the deaths due to COVID-19 were reported among persons with some severe underlying health conditions such as chronic lung disease, cardiovascular disease, diabetes mellitus, obesity, hypertension, asthma and severe respiratory health issues [8–12]. Figure 1 shows the association between COVID-19 cases and underlying health conditions. A study presented by [13] report that coronavirus patients belonging to the cities with a higher level of air pollution before the outbreak of this pandemic are more likely to die from this new infection as compared to the patients in the cleaner parts of the world. The study was conducted over 3087 counties in the United States while covering 98% of the population. Furthermore, the results show a higher mortality rate for people with long-term exposure to $PM_{2.5}$ levels. The repeated exposure to dust and harmful gases presented in the environment adversely affects respiratory and cardiovascular systems [14–17]. Consequently, it is further linked to the severity of COVID-19 cases with worse symptoms of infection among people affected by air pollution [13, 18, 19]. The recent study [13] reveals that an increase of only 1 $\mu g/m^3$ in $PM_{2.5}$ is responsible for 8% increase in the COVID-19 death rate (95% confidence interval [CI]: 2%, 15%). The above stats report that it is equally important to address the inherent causes of the rising mortality rate due to this virus [20]. Moreover, the impact of this virus is linked to the immunity levels of the patients [21]. Therefore, it is crucial to find measures to improve the overall

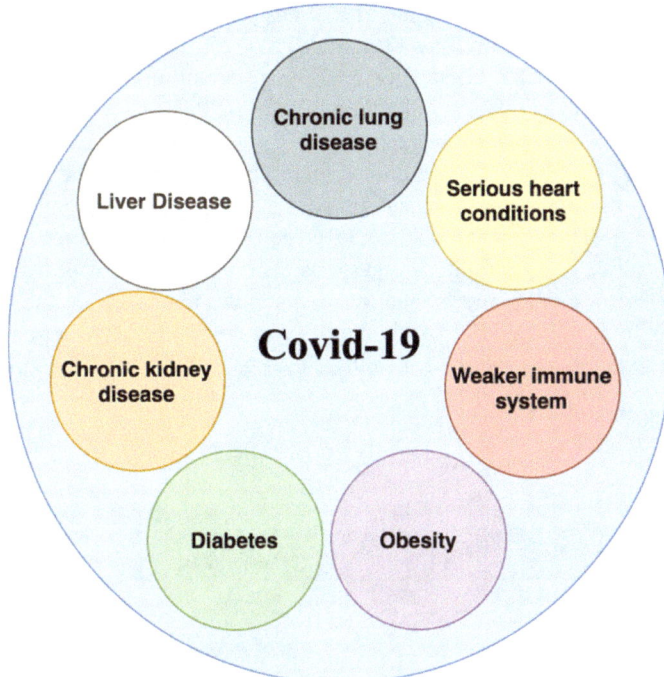

Fig. 1 Association between COVID-19 and underlying health conditions

community lifestyle. These efforts will improve the chances of survival of patients affected by a pandemic and open new opportunity to save the world from such invisible attacks in the future.

1.2 Indoor Air Pollution

The impact of indoor air pollution (IAP) on human beings is almost 2–5 times higher as compared to outdoor air pollution [22]. As people spend nearly 90% of their time indoors (homes, offices, schools and hospitals), they are more affected by the pollutants present in the building environment [23]. Even in the present scenario, when most of the countries have announced complete lockdown to implement social distancing for reducing the spread of COVID-19, people are spending most of their time indoors. As a result, the risk of IAP has increased several times, contributing to a considerable fall in the immunity due to rising chances of respiratory health problems [24, 25]. The impact is even worsening the underlying symptoms of other chronic health problems such as cancer, heart disease, kidney disease, diabetes and asthma [26]. Consequently, making people more vulnerable to the spread of COVID-19 disease [27].

Table 1 shows data about coronavirus deaths from the United States (Released by New York City Health on May 13th, 2020) [28]. These stats reveal a 75% share of total deaths in the USA because of underlying conditions.

Figure 2 presents a relationship between COVID-19 deaths cases and underlying conditions in the UK on 19th March 2020 [29].

In hospitals, the number of patients is increasing every day. The regular monitoring of air pollution levels is critical to avoid the chances of worsening of disease symptoms [30]. The beforementioned facts demand the installation of real-time indoor air quality (IAQ) monitoring systems. These systems can provide relevant real-time data to support the evaluation of air pollution levels in the buildings. Furthermore, the application of IAQ monitoring systems ensures better efficiency of ventilation systems to promote a healthy lifestyle [31, 32]. IAQ monitoring has

Table 1 Coronavirus death stats from all states of the USA

Age	Number of deaths	With underlying conditions	Without underlying conditions	Unknown (if any underlying cond.)
75 + years old	7,419	5,236	2	2,181
65–74 years old	3,788	2,801	5	982
45–64 years old	3,413	2,851	72	490
18–44 years old	601	476	17	108
0–17 years old	9	6	3	0
Total	15,230	11,370 (75%)	99 (0.7%)	1,551 (24.7%)

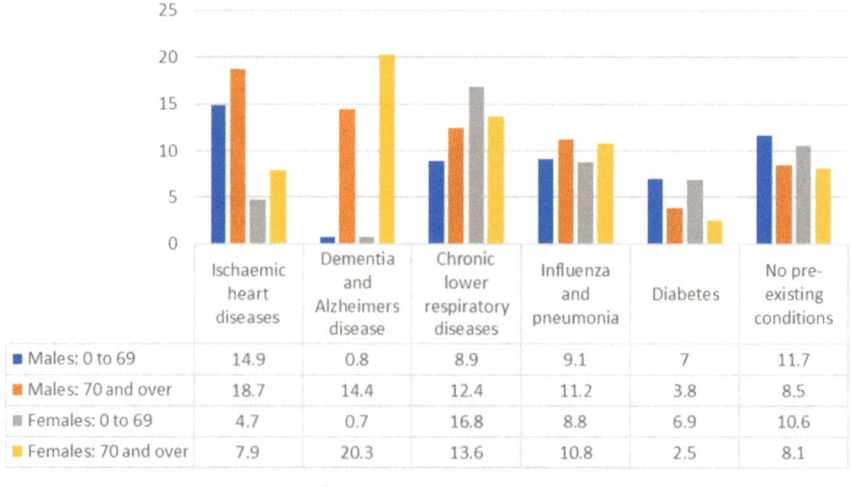

Fig. 2 Stats about COVID-19 deaths and underlying health conditions in the UK

been a crucial field of concern for a broad community of researchers around the world. Moreover, it is critical to perform an in-depth analysis of the topic to find out reliable answers to prevent exposure risk. Therefore, IAQ monitoring systems are crucial in the current pandemic scenario as an enhanced instrument to ensure the quality of the building ventilation systems.

The objective of this study is to bridge the gap between COVID-19, public health and indoor air quality while addressing the need for real-time monitoring systems. Furthermore, this work can be useful for public health professionals, researchers, students, industry experts and policymakers around the world.

2   Indoor Air Quality and COVID-19

Air quality in hospitals is considered a significant risk factor for a wide range of health consequences among patients and working staff who visit the premises. Several studies link the spread of infections, not just COVID-19, but also other existing diseases with IAQ [33–36]. Numerous researchers have also investigated the sources, levels, characteristics, and measures of bioaerosol in hospitals [37, 38]. Furthermore, medical facilities, COVID-19 patient quarantine centres, and nursing homes are equally affected by IAP. The use of chemical-rich detergents, cleaning solutions and medical treatments can also worsen the risk of infections [39]. Li et al. investigated the relationship between bioaerosol and ventilation arrangements. They recommended that adequate flow of fresh air into the hospital premises could regulate the bioaerosol level to a considerable extent [40]. Several hospitals have traditional ventilation systems that drive air from patient areas to the remaining circulation areas.

Consequently, airborne pathogens can easily spread the virus to the entire area leading to nosocomial outbreaks. The use of adequate ventilation systems with a reliable prediction of future airflow patterns is required to address this problem. Airflow patterns prediction is crucial to ensure enhanced ventilation and control in the buildings. It not only reduces the travel of pathogens in the operating environments but also minimizes the energy consumptions in the hospital areas.

2.1   COVID-19: Association to Biomass Usage

The critical concern to COVID-19 outbreak and IAQ levels is associated with the lifestyle of people in the middle and low-income countries. Globally, 2.8 billion people, 41% of the households, rely on solid fuels, including biomass and coal for their routine heating and cooking needs [41]. There is strong evidence between household air pollution and potential health problems such as stillbirth, low birth weight, lung cancer, cataract, chronic obstructive pulmonary disease, acute lower respiratory infections, asthma, cardiovascular disease, and chronic bronchitis [42–

45]. The underlying metabolic health is further linked to the rising number of COVID-19 cases [46, 47].

Migrant workers and refugees that are living in fragile conditions are more vulnerable to COVID-19. Almost 80% of refugees live in middle and low-income countries. These people spend most of their time in poorly ventilated houses while using solid fuels as continuous sources of cooking and heating. Moreover, these rooms do not have adequate chimneys that could take smoke outside [48]. Due to lack of evidence, the policymakers rarely prioritize actions to control an outbreak of COVID-19 on such groups. However, from a research point of view, it is essential to explore the association between COVID-19 and biomass smoke. Rigorous testing with real-time surveillance of IAQ must be performed in all such communities to take vigorous actions to control the situation.

2.2 COVID-19 and Ventilation Issues

IAQ measures are equally necessary for modern housing facilities due to excessive use of chemical-rich detergents, deodorants, and cleaning products [26]. Although modern residential complexes are equipped with advanced ventilation arrangements, and consequently, there are still lesser options for circulation of fresh air. The defective ventilated buildings tend to accommodate kitchen odours, residues of scents, perfumes and other chemical-rich consumer products for longer time [49]. Steinemann et al. reported 37 different consumer products, including cleaning agents, laundry products and air freshener that are directly linked to the development of volatile organic compounds (VOCs). Repeated exposure to VOCs is further related to decreased lung function and oxidative stress while causing severe airways inflammation [50]. Win-Shwe et al. report a close connection between indoor chemical sensitivity reactions and VOCs exposure to immune dysfunction [51]. In a recent study, Giamarellos-Bourboulis et al. reported that patients with immune dysregulation are more likely to experience severe respiratory failure by COVID-19 [52]. Tay et al. also studied the interaction between the immune system and COVID-19. This study reports a subsequent contribution of the dysfunctional immune response of the human body to the progression of viral disease [53]. Therefore, the developed nations are not protected from the impact of COVID-19 due to poor health quality of the residents.

Figure 3 shows the combined effect of COVID-19 and IAP on human beings. The real-time monitoring of IAQ, along with adequate measures for enhanced ventilation facilities promote the overall health and well-being in the community. Consequently, these systems promote public health and make people less vulnerable to infectious diseases such as COVID-19.

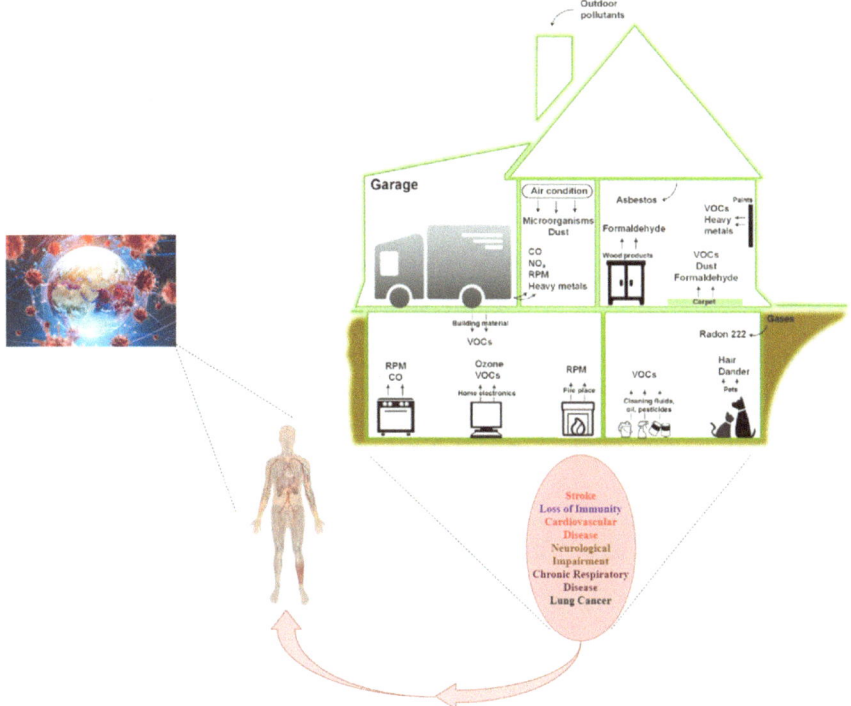

Fig. 3 The combined effect of COVID-19 and IAQ on human beings

3 IAQ Monitoring Systems: A Missed Opportunity

Public health experts advised people to stay indoors as much as possible to reduce the spread of COVID-19. However, the current home isolation requirements can further increase the exposure to IAP. Moreover, people that are already affected by chronic respiratory diseases and are spending more time in the poorly ventilated homes are, therefore, more exposed to the disease symptoms. The situation also promotes an adverse impact on their immunity levels. Consequently, these patients are more vulnerable to this novel coronavirus.

Furthermore, the patients that are already admitted in the hospitals may also experience a rise in disease symptoms due to inadequate ventilation arrangements. In particular, in low-income countries where medical care centres are equipped with limited facilities.

More hospital admissions, frequent emergency room visits, and increased movements of staff members also result in poor IAQ, leaving existing patients in a higher risk of mortality due to COVID-19 [54, 55]. These critical scenarios require a reliable solution for regular monitoring of IAQ in hospitals, apartments, offices, and other indoor public spaces. It can serve as a preventive tool to control the inherent cause (respiratory health problems, low immunity levels and other underlying symptoms) behind increased mortality rate due to COVID-19 [56]. The requirements are not restricted to monitoring systems. There is a critical need for real-time IAQ prediction systems that can provide advance notifications about expected changes in indoor pollutant levels. The latest technologies such as the Internet of Things (IoT) [57, 58] and Wireless Sensor Networks (WSN) [59, 60] provide a versatile platform for the development and implementation of IAQ monitoring systems. Furthermore, the potential of Artificial Intelligence (AI), Fuzzy Logic (FL) and Machine Learning (ML) can be utilized to make predictions for enhanced real-time control and management [61–63].

The real challenge in the development of real-time monitoring systems is in the selection of accurate sensors for measuring indoor pollution levels [64]. Industry experts have designed multiple sensor units for measuring numerous pollutant levels. However, real-time implementation requires critical analysis over power efficiency, cost-effectiveness and reliability of these sensing technologies [65]. Reports reveal a rising concern of IAQ levels and their health impacts on low income and middle-income countries [66]. Roberton et al. also analyzed the indirect effects of COVID-19 pandemic on low income and middle-income countries [67]. Consequently, it is essential to perform a critical analysis of the use of adequate sensing technologies.

3.1 Existing Solutions for IAQ Monitoring

Researchers around the world are already working in this direction. Numerous authors have presented several competitive solutions to deal with the rising health concerns due to reduced IAQ levels [23].

Marques et al. designed a cost-effective air quality supervision system using IoT for enhanced living conditions in indoor environments [68]. They used IoT architecture for designing a Wi-Fi module named as iAir. This module includes a multi-gas sensor (MiCS-6814) and an ESP8266 microcontroller. This system is capable enough to detect several harmful gases in indoor environments, including propane, ethanol, carbon monoxide, and nitrogen dioxide.

Idrees et al. also designed an edge computing-based IoT system for low-cost air pollution monitoring for measuring six different pollutant levels, including O_3, SO_2, NO_2, CO, $PM_{2.5}$ and PM_{10} [69]. It was efficient enough to reduce the computational requirements on sensing nodes by almost 70% while ensuring a considerable reduction of 23% in the overall power consumption.

Tiele et al. presented a low-cost and portable monitoring device for measuring IAQ levels. This system focused on the impact of total VOCs, PM_{10}, $PM_{2.5}$, CO, CO_2, temperature, and humidity as well [70]. The experimental results show successful monitoring of IAQ parameters, and the proposed system also ensures easy installation.

Yu et al. followed an advanced system for improving IAQ levels while providing in-depth analysis on monitoring, prediction and pre-action [71]. The authors used ARIMA model for improving the accuracy of the system while predicting IAQ trends. They also utilized fuzzy-logic based decision modelling for designing an energy-efficient feedback control mechanism.

These studies show the scope of real-time monitoring systems for management of IAP levels. Such systems can be installed in modern apartments, rural homes, hospitals, cafeterias, schools, and offices to monitor IAQ. Several researchers designed IAQ monitoring and prediction systems that could provide instant notifications via SMS or email for the rise of pollutant levels above desired threshold levels [23, 72]. Figure 4 provides a general architecture of IAQ monitoring system.

These IAQ monitoring systems can be further linked to monitoring the impact of IAP on COVID-19 patients in medical care centres. The caregivers and medical health experts are also vulnerable to COVID-19 spread. Since they visit patient care units multiple times; an automated monitoring system can provide real-time analysis remotely. These systems can also be installed in coronavirus quarantine centres and residential apartments to monitor the IAQ conditions. It is possible to install low-cost sensors to address PM, CO_2, and NO_2 levels while ensuring a cost-effective solution for low and middle-income countries. Low-cost sensors can reduce the financial burden on government and policymakers during this pandemic [73–75].

Moreover, these sensors have enough accuracy to provide qualitative analysis on IAQ for taking preventive steps. The IAQ monitoring using cost-effective sensors enable the correct evaluation of ventilation procedures on time. Consequently, the exposure risk is decreased, and unhealthy IAQ scenarios can be avoided. Smart sensing technologies provide a reliable solution to help medical health experts,

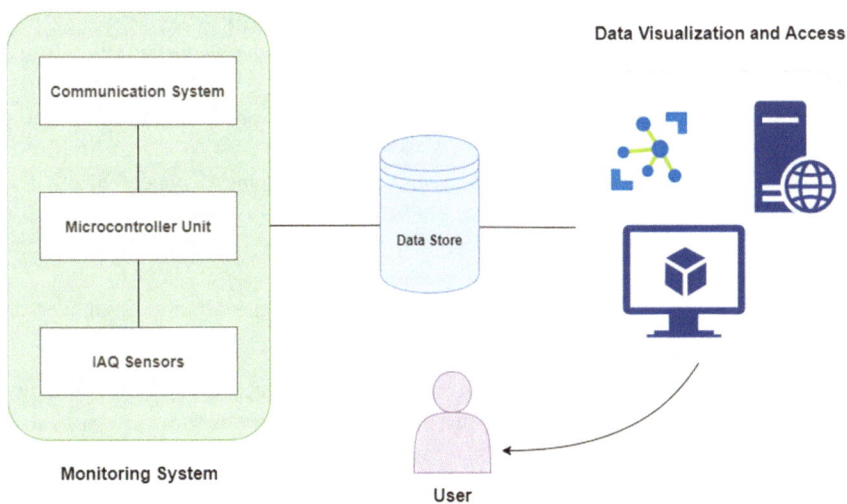

Fig. 4 The general architecture of IAQ monitoring systems

caregivers, policymakers, and government authorities to provide better control on inherent causes of rising mortality rate due to coronavirus pandemic.

Currently, without effective treatments or medicines to reduce the ongoing pandemic ventilation is suggested as a critical element to face the spread of the SARS-CoV-2 virus [76, 77]. Nevertheless, it is relevant to understand that outdoor air used for ventilation can also have low air quality which tends to happen especially for underserved communities facing environmental injustice scenarios. The use of cost-effective sensors is recommended for qualitative assessment of the ventilation conditions and can help in maintaining healthy indoor conditions for the occupants.

4 Conclusion

The lack of information, unfamiliarity with COVID-19, conflicting information and human perception are critical aspects for rising threats of this pandemic. As people these days are spending more time indoor, repeated exposure to low air quality can pose a severe threat to their respiratory health and immunity levels while making them more vulnerable to the attack of virus. Consequently, it is critical to address the concern ahead of time to reduce the rate of fatalities worldwide. In this study, the impact of COVID-19 on patients with underlying medical health conditions, especially respiratory disease, was analyzed. This study also established a link between IAQ and the rising threat of pandemic while proposing the need for real-time monitoring and predictions systems for enhanced living environments. The evidence presented in this study show potential of smart sensing technologies to ease the burden of global mortality rate due to COVID-19.

This chapter has the following outcomes:

- Provides valuable information about the association between COVID-19 and underlying illnesses.
- Discuss the impact of IAQ on public health while providing insights about relevant chronic health problems.
- Highlight the importance of IAQ monitoring for enhanced patient health and healthy building environment.
- Present details about existing research in the field of IAQ monitoring system development.
- Provide baseline information for the upcoming researchers and policymakers to deal with the pandemic and associated IAQ challenges.

The findings of this work are useful not just for research point of view, but also to develop new policies for establishing robust control measures for the spread of COVID-19. Furthermore, it proposes new guidelines for following adequate safety measures related to IAQ levels during this pandemic attack.

References

1. World Health Organization: Coronavirus. In: World Health Organization (2020). https://www.who.int/emergencies/diseases/novel-coronavirus-2019. Accessed 17 May 2020
2. World Health Organization: COVID-19 situation reports. In: World Health Organization (2020). https://www.who.int/emergencies/diseases/novel-coronavirus-2019/situation-reports. Accessed 17 May 2020
3. Cucinotta, D., Vanelli, M.: WHO declares COVID-19 a pandemic. Acta Biomed. **91**, 157–160 (2020). https://doi.org/10.23750/abm.v91i1.9397
4. Vellingiri, B., Jayaramayya, K., Iyer, M., et al.: COVID-19: a promising cure for the global panic. Sci. Total Environ. **725**, 138277 (2020). https://doi.org/10.1016/j.scitotenv.2020.138277
5. World Health Organization: Modes of transmission of virus causing COVID-19—mplications for IPC precaution recommendations (2020). https://www.who.int/news-room/commentaries/detail/modes-of-transmission-of-virus-causing-covid-19-implications-for-ipc-precaution-recommendations. Accessed 17 May 2020
6. Nicola, M., Alsafi, Z., Sohrabi, C., et al.: The socio-economic implications of the Coronavirus and COVID-19 pandemic: a review. Int. J. Surg. (2020). https://doi.org/10.1016/j.ijsu.2020.04.018
7. Ozili, P., Arun, T.: Spillover of COVID-19: impact on the Global Economy. SSRN Electron. J. (2020). https://doi.org/10.2139/ssrn.3562570
8. Balakrishnan, K., Dey, S., Gupta, T., et al.: The impact of air pollution on deaths, disease burden, and life expectancy across the states of India: the Global Burden of Disease Study 2017. Lancet Planet. Health **3**, e26–e39 (2019). https://doi.org/10.1016/S2542-5196(18)30261-4
9. Jiang, F., Deng, L., Zhang, L., et al.: Review of the clinical characteristics of coronavirus disease 2019 (COVID-19). J. Gen. Intern. Med. **35**, 1545–1549 (2020). https://doi.org/10.1007/s11606-020-05762-w
10. Mitter, S.S., Vedanthan, R., Islami, F., et al.: Household fuel use and cardiovascular disease mortality: golestan cohort study. Circulation **133**, 2360–2369 (2016). https://doi.org/10.1161/CIRCULATIONAHA.115.020288
11. Sun, Q., Yue, P., Ying, Z., et al.: Air pollution exposure potentiates hypertension through reactive oxygen species-mediated activation of Rho/ROCK. Arterioscler. Thromb. Vasc. Biol. **28**, 1760–1766 (2008). https://doi.org/10.1161/ATVBAHA.108.166967
12. Weisel, C.P.: Assessing exposure to air toxics relative to asthma. Environ. Health Perspect. **110**, 527–537 (2002)
13. Wu, X., Nethery, R.C., Sabath, B.M., et al.: Exposure to air pollution and COVID-19 mortality in the United States: a nationwide cross-sectional study. Epidemiology (2020)
14. Goldizen, F.C., Sly, P.D., Knibbs, L.D.: Respiratory effects of air pollution on children. Pediatr. Pulmonol. **51**, 94–108 (2016). https://doi.org/10.1002/ppul.23262
15. Graudenz, G.S., Oliveira, C.H., Tribess, A., et al.: Association of air-conditioning with respiratory symptoms in office workers in tropical climate. Indoor Air **15**, 62–66 (2005). https://doi.org/10.1111/j.1600-0668.2004.00324.x
16. Patel, V., Kantipudi, N., Jones, G., et al.: Air pollution and cardiovascular disease: a review. Crit. Rev. Biomed. Eng. **44**, 327–346 (2016). https://doi.org/10.1615/CritRevBiomedEng.2017019768
17. Samet, J.M., Bahrami, H., Berhane, K.: Indoor air pollution and cardiovascular disease: new evidence from Iran. Circulation **133**, 2342–2344 (2016). https://doi.org/10.1161/CIRCULATIONAHA.116.023477
18. Chen, K., Wang, M., Huang, C., et al.: Air pollution reduction and mortality benefit during the COVID-19 outbreak in China. Lancet Planet Health (2020). https://doi.org/10.1016/S2542-5196(20)30107-8

19. Guan, L., Zhou, L., Zhang, J., et al.: More awareness is needed for severe acute respiratory syndrome coronavirus 2019 transmission through exhaled air during non-invasive respiratory support: experience from China. Eur. Resp. J. **55**, 2000352 (2020). https://doi.org/10.1183/13993003.00352-2020

20. Ogen, Y.: Assessing nitrogen dioxide (NO2) levels as a contributing factor to coronavirus (COVID-19) fatality. Sci. Total Environ. **726**, 138605 (2020). https://doi.org/10.1016/j.scitotenv.2020.138605

21. Altmann, D.M., Douek, D.C., Boyton, R.J.: What policy makers need to know about COVID-19 protective immunity. The Lancet **395**, 1527–1529 (2020). https://doi.org/10.1016/S0140-6736(20)30985-5

22. US EPA O: Indoor Air Quality. In: US EPA (2017). https://www.epa.gov/report-environment/indoor-air-quality. Accessed 17 May 2020

23. Saini, J., Dutta, M., Marques, G.: A comprehensive review on indoor air quality monitoring systems for enhanced public health. Sustain. Environ. Res. **30**, 6 (2020). https://doi.org/10.1186/s42834-020-0047-y

24. Ezzati, M., Kammen, D.M.: The health impacts of exposure to indoor air pollution from solid fuels in developing countries: knowledge, gaps, and data needs. Environ. Health Perspect. **110**, 1057–1068 (2002). https://doi.org/10.1289/ehp.021101057

25. Kurmi, O.P., Lam, K.B.H., Ayres, J.G.: Indoor air pollution and the lung in low- and medium-income countries. Eur. Respir. J. **40**, 239–254 (2012). https://doi.org/10.1183/09031936.00190211

26. Apte, K., Salvi, S.: Household air pollution and its effects on health. F1000Research **5**, 2593 (2016). https://doi.org/10.12688/f1000research.7552.1

27. Zhu, Y., Xie, J., Huang, F., Cao, L.: Association between short-term exposure to air pollution and COVID-19 infection: evidence from China. Sci. Total Environ. **727**, 138704 (2020). https://doi.org/10.1016/j.scitotenv.2020.138704

28. NYC Health: COVID-19: Data—NYC Health (2020). https://www1.nyc.gov/site/doh/covid/covid-19-data.page. Accessed 18 May 2020

29. Office for National Statistics: Measuring pre-existing health conditions in death certification—deaths involving COVID-19: March 2020—Office for National Statistics (2020). https://www.ons.gov.uk/peoplepopulationandcommunity/birthsdeathsandmarriages/deaths/methodologies/measuringpreexistinghealthconditionsindeathcertificationdeathsinvolvingcovid19march2020. Accessed 18 May 2020

30. Barcelo, D.: An environmental and health perspective for COVID-19 outbreak: meteorology and air quality influence, sewage epidemiology indicator, hospitals disinfection, drug therapies and recommendations. J Environ Chem Eng. (2020). https://doi.org/10.1016/j.jece.2020.104006

31. Marques, G., Pitarma, R.: Indoor air quality monitoring for enhanced healthy buildings. In: Abdul Mujeebu, M. (ed.) Indoor Environmental Quality. IntechOpen (2019)

32. Satheesan, M.K., Mui, K.W., Wong, L.T.: A numerical study of ventilation strategies for infection risk mitigation in general inpatient wards. Build. Simul. (2020). https://doi.org/10.1007/s12273-020-0623-4

33. Gan, W.H., Lim, J.W., Koh, D.: Preventing intra-hospital infection and transmission of coronavirus disease 2019 in health-care workers. Safety Health Work **11**(2), S2093791120030161X (2020). https://doi.org/10.1016/j.shaw.2020.03.001

34. Maki, D.G., Crnich, C.J., Safdar, N.: Chapter 51—Nosocomial infection in the intensive care unit. In: Parrillo, J.E., Dellinger, R.P. (eds.) Critical Care Medicine, 3rd edn, pp. 1003–1069. Mosby, Philadelphia (2008)

35. Slack, R.C.B.: 69 - Hospital infection. In: Greenwood, D., Barer, M., Slack, R., Irving, W. (eds.) Medical Microbiology, Eighteenth edn, pp. 718–726. Churchill Livingstone, Edinburgh (2012)

36. Wong, S.C.Y., Kwong, R.T.-S., Wu, T.C., et al.: Risk of nosocomial transmission of coronavirus disease 2019: an experience in a general ward setting in Hong Kong. J. Hosp. Infect. **105**, 119–127 (2020). https://doi.org/10.1016/j.jhin.2020.03.036

37. Chamseddine, A., El-Fadel, M.: Exposure to air pollutants in hospitals: indoor–outdoor correlations. In: Brebbia, C.A. (ed.) WIT Transactions on the Built Environment, 1st ed.,, pp. 707–716. WIT Press (2015)

38. El-Sharkawy, M.F., Noweir, M.E.H.: Indoor air quality levels in a University Hospital in the Eastern Province of Saudi Arabia. J. Family Commun. Med. **21**, 39–47 (2014). https://doi.org/10.4103/2230-8229.128778

39. Gola, M., Settimo, G., Capolongo, S.: Chemical pollution in healing spaces: the decalogue of the best practices for adequate indoor air quality in inpatient rooms. Int. J. Environ. Res. Public Health **16** (2019). https://doi.org/10.3390/ijerph16224388

40. Li, Y., Leung, G.M., Tang, J.W., et al.: Role of ventilation in airborne transmission of infectious agents in the built environment? a multidisciplinary systematic review. Indoor Air **17**, 2–18 (2007). https://doi.org/10.1111/j.1600-0668.2006.00445.x

41. Amegah, A.K., Jaakkola, J.J.: Household air pollution and the sustainable development goals. Bull. World Health Organ. **94**, 215–221 (2016). https://doi.org/10.2471/BLT.15.155812

42. Chafe, Z.A., Brauer, M., Klimont, Z., et al.: Household cooking with solid fuels contributes to ambient PM2.5 air pollution and the burden of disease. Environ. Health Perspect. **122**, 1314–1320 (2014). https://doi.org/10.1289/ehp.1206340

43. Josyula, S., Lin, J., Xue, X., et al.: Household air pollution and cancers other than lung: a meta-analysis. Environ. Health **14**, 24 (2015). https://doi.org/10.1186/s12940-015-0001-3

44. McCracken, J.P., Wellenius, G.A., Bloomfield, G.S., et al.: Household air pollution from solid fuel use: evidence for links to CVD. Glob. Heart **7**, 223–234 (2012). https://doi.org/10.1016/j.gheart.2012.06.010

45. Pope, D.P., Mishra, V., Thompson, L., et al.: Risk of low birth weight and stillbirth associated with indoor air pollution from solid fuel use in developing countries. Epidemiol. Rev. **32**, 70–81 (2010). https://doi.org/10.1093/epirev/mxq005

46. Banerjee, A., Pasea, L., Harris, S., et al.: Estimating excess 1-year mortality associated with the COVID-19 pandemic according to underlying conditions and age: a population-based cohort study. Lancet (2020). https://doi.org/10.1016/S0140-6736(20)30854-0

47. Endocrinology TLD & (2020) COVID-19: underlying metabolic health in the spotlight. Lancet Diabetes Endocrinol. (2020). https://doi.org/10.1016/S2213-8587(20)30164-9

48. Ghergu, C., Sushama, P., Vermeulen, J., et al.: Dealing with indoor air pollution: an ethnographic tale from urban slums in Bangalore. Int. J. Health Sci. **6**, 348–361 (2016)

49. Steinemann, A.: Volatile emissions from common consumer products. Air Qual. Atmos. Health **8**, 273–281 (2015). https://doi.org/10.1007/s11869-015-0327-6

50. Kwon, J.-W., Park, H.-W., Kim, W.J., et al.: Exposure to volatile organic compounds and airway inflammation. Environ Health **17**(1) (2018). https://doi.org/10.1186/s12940-018-0410-1

51. Win-Shwe, T.-T., Fujimaki, H., Arashidani, K., Kunugita, N.: Indoor volatile organic compounds and chemical sensitivity reactions. Clin. Dev. Immunol. 2013. https://doi.org/10.1155/2013/623812

52. Giamarellos-Bourboulis, E.J., Netea, M.G., Rovina, N., et al.: Complex immune dysregulation in COVID-19 patients with severe respiratory failure. Cell Host Microbe S1931312820302365 (2020). https://doi.org/10.1016/j.chom.2020.04.009

53. Tay, M.Z., Poh, C.M., Rénia, L., et al.: The trinity of COVID-19: immunity, inflammation and intervention. Nat. Rev. Immunol. 1–12 (2020). https://doi.org/10.1038/s41577-020-0311-8

54. Liu, P., Wang, X., Fan, J., et al.: Effects of air pollution on hospital emergency room visits for respiratory diseases: urban-suburban differences in Eastern China. Int. J. Environ. Res. Public Health **13**(3), 41 (2016). https://doi.org/10.3390/ijerph13030341

55. Zhang, H., Niu, Y., Yao, Y., et al.: The impact of ambient air pollution on daily hospital visits for various respiratory diseases and the relevant medical expenditures in Shanghai, China. Int. J. Environ. Res. Public Health **15**(3), 425 (2018). https://doi.org/10.3390/ijerph15030425

56. Zakaria Abouleish, M.Y.: Indoor air quality and coronavirus disease (COVID-19). Public Health (2020). https://doi.org/10.1016/j.puhe.2020.04.047
57. Hapsari, A.A., Hajamydeen, A.I., Abdullah, M.I.I.: A review on indoor air quality monitoring using IoT at campus environment. Int. J. Eng. Technol. **7**, 55–60 (2018). https://doi.org/10.14419/ijet.v7i4.22.22191
58. Sun, S., Zheng, X., Villalba-Díez, J., Ordieres-Meré, J.: Indoor air-quality data-monitoring system: long-term monitoring benefits. Sensors **19**, 4157 (2019). https://doi.org/10.3390/s19194157
59. Lozano, J., Suárez, J.I., Arroyo, P., et al.: Wireless sensor network for indoor air quality monitoring. Chem. Eng. Trans. **30**, 319–324 (2012). https://doi.org/10.3303/CET1230054
60. Yi, W., Lo, K., Mak, T., et al.: A survey of wireless sensor network based air pollution monitoring systems. Sensors **15**, 31392–31427 (2015). https://doi.org/10.3390/s151229859
61. Elbayoumi, M., Ramli, N.A., Fitri Md Yusof, N.F.: Development and comparison of regression models and feedforward backpropagation neural network models to predict seasonal indoor $PM_{2.5-10}$ and $PM_{2.5}$ concentrations in naturally ventilated schools. Atmos. Pollut. Res. **6**, 1013–1023 (2015). https://doi.org/10.1016/j.apr.2015.09.001
62. Kolokotsa, D.: Artificial intelligence in buildings: a review of the application of fuzzy logic. Adv. Build. Energy Res. **1**, 29–54 (2007). https://doi.org/10.1080/17512549.2007.9687268
63. Masih, A.: Machine learning algorithms in air quality modeling. Glob. J. Environ. Sci. Manag. **5**, 515–534 (2019). https://doi.org/10.22034/GJESM.2019.04.10
64. Kumar, P., Skouloudis, A.N., Bell, M., et al.: Real-time sensors for indoor air monitoring and challenges ahead in deploying them to urban buildings. Sci. Total Environ. **560–561**, 150–159 (2016). https://doi.org/10.1016/j.scitotenv.2016.04.032
65. Revel, G., Arnesano, M., Pietroni, F., et al.: Cost-effective technologies to control indoor air quality and comfort in energy efficient building retrofitting. Environ. Eng. Manag. J. **14**, 1487–1494 (2015). https://doi.org/10.30638/eemj.2015.160
66. Bruce, N., Perez-Padilla, R., Albalak, R.: Indoor air pollution in developing countries: a major environmental and public health challenge. Bull. World Health Organ. **78**, 1078–1092 (2000)
67. Roberton, T., Carter, E.D., Chou, V.B., et al.: Early estimates of the indirect effects of the COVID-19 pandemic on maternal and child mortality in low-income and middle-income countries: a modelling study. Lancet Glob. Health (2020). https://doi.org/10.1016/S2214-109X(20)30229-1
68. Marques, G., Pitarma, R.: A cost-effective air quality supervision solution for enhanced living environments through the internet of things. Electronics **8**, 170 (2019). https://doi.org/10.3390/electronics8020170
69. Idrees, Z., Zou, Z., Zheng, L.: Edge computing based IoT architecture for low cost air pollution monitoring systems: a comprehensive system analysis. Design Considerations Dev. Sens. **18**, 3021 (2018). https://doi.org/10.3390/s18093021
70. Tiele, A., Esfahani, S., Covington, J.: Design and development of a low-cost, portable monitoring device for indoor environment quality. J. Sens. **2018**, 1–14 (2018). https://doi.org/10.1155/2018/5353816
71. Yu, T.-C., Lin, C.-C.: An intelligent wireless sensing and control system to improve indoor air quality: monitoring, prediction, and preaction. Int. J. Distrib. Sens. Netw. **11**, 140978 (2015). https://doi.org/10.1155/2015/140978
72. Okokpujie, K., Noma-Osaghae, E., Odusami, M., et al.: a smart air pollution monitoring system. Int. J. Civil Eng. Technol. **9**, 799–809 (2018)
73. Bulot, F.M.J., Russell, H.S., Rezaei, M., et al.: Laboratory comparison of low-cost particulate matter sensors to measure transient events of pollution. Sensors **20**, 2219 (2020). https://doi.org/10.3390/s20082219
74. Li, J., Mattewal, S.K., Patel, S., Biswas, P.: Evaluation of nine low-cost-sensor-based particulate matter monitors. Aerosol Air Qual. Res. **20**, 254–270 (2019). https://doi.org/10.4209/aaqr.2018.12.0485

75. Liu, M., Lin, J., Boersma, K.F., et al.: Improved aerosol correction for OMI tropospheric NO_2 retrieval over East Asia: constraint from CALIOP aerosol vertical profile. Atmos. Meas. Tech. **12**, 1–21 (2019). https://doi.org/10.5194/amt-12-1-2019

76. Morawska, L., Tang, J.W., Bahnfleth, W., et al.: How can airborne transmission of COVID-19 indoors be minimised? Environ. Int. **142**, 105832 (2020). https://doi.org/10.1016/j.envint.2020.105832

77. Sun, C., Zhai, Z.: The efficacy of social distance and ventilation effectiveness in preventing COVID-19 transmission. Sustain. Cities Soc. **62**, 102390 (2020). https://doi.org/10.1016/j.scs.2020.102390

Leveraging Digital Transformation Technologies to Tackle COVID-19: Proposing a Privacy-First Holistic Framework

Ebru Gökalp, Kerem Kayabay, and Mert Onuralp Gökalp

Abstract The COVID-19 outbreak has caused unprecedented shocks to public health and economies around the world. The emergent technologies creating digital transformation across all industries have started to transform businesses in the last years. Although utilized by the healthcare industry before COVID-19, some of these technologies gained significant attention since they can alleviate the pandemic crisis. A holistic framework can still help (1) understand the relationship between fragmented technology solutions, (2) figure out how to combine fragmented solutions, (3) integrate data sources, and (4) introduce integrated digital solutions. Correspondingly, we explore the use of emergent technologies by proposing a privacy-first holistic framework by integrating digital transformation technologies to tackle the COVID-19 pandemic. The framework consists of data sources, technologies, users, applications for diagnosis, treatment, and prevention, and users. We then discuss the benefits and challenges of the framework. The proposed framework aims to provide a more efficient and dynamic healthcare system to reduce the death toll, the risk of virus transmission, and healthcare givers' workload and stress levels.

Keywords COVID-19 · Digital transformation · Emerging technologies · Data science · Blockchain

E. Gökalp (✉)
Department of Technology and Knowledge Management, Baskent University,
Ankara, Turkey
e-mail: ebrug@baskent.edu.tr; eg590@cam.ac.uk

Institute for Manufacturing, University of Cambridge, Cambridge, United Kingdom

K. Kayabay · M. O. Gökalp
Department of Information Systems, Graduate School of Informatics,
METU, Ankara, Turkey
e-mail: kayabay@metu.edu.tr

M. O. Gökalp
e-mail: gmert@metu.edu.tr

© The Author(s), under exclusive license to Springer Nature Switzerland AG 2021
I. Arpaci et al. (eds.), *Emerging Technologies During the Era of COVID-19
Pandemic*, Studies in Systems, Decision and Control 348,
https://doi.org/10.1007/978-3-030-67716-9_10

149

1 Introduction

The COVID-19 is a respiratory disease caused by the recently discovered coronavirus attacking the respiratory system and causing fever, cough, breathlessness, and fatigue. The first case was reported in Wuhan City, China, in December 2019. After then, it has spread across the world in a short amount of time. On March 11, the World Health Organization (WHO) declared COVID-19 as a pandemic disease after reporting 118,000 cases and 4,300 deaths across 114 countries. According to the latest COVID-19 statistics published by WHO [1], the total number of cases has exceeded 20 million people, and the total number of COVID-19 deaths has reached 700,000, spread across 213 countries, with the US having the highest number cases.

The healthcare system would be overwhelmed by the pandemic in a short amount of time when the number of severe coronavirus cases spikes sharply, as seen in the case of Italy. The high rise of the novel coronavirus cases can cause a significant scarcity of resources, hospital beds, and healthcare professionals. To prevent these kinds of problems by reducing the spread of the virus and delaying the peak of hospitalization, there are some possible solutions, as lockdown and quarantine, which have some national and international drawbacks, such as the economic recession and mental problems of the population.

Since all organizations providing non-essential products and services are forced to shut down during lockdown, billions of people are at risk of losing their jobs. The estimated loss on the global economy caused by the COVID-19 is forecasted to exceed 5.5 trillion dollars in the next 18–24 months. This is the same as losing Japan.[1] There is still no specific treatment or vaccine found to alleviate this pandemic when this paper was written. The increase in the number of laboratory-confirmed COVID-19 cases at an alarming rate and shortcomings of lockdown has caused the need for urgent countermeasures to reduce the disastrous effects of this pandemic.

The emergent technologies creating digital transformation across all industries are the digital twin, the Internet of Things (IoT) and connected devices, artificial intelligence, cyber-physical systems, integration, social media and platforms, blockchain, everything-as-a-service (XaaS), robots and drones, and data analytics. It is forecast that, by 2022, more than 60% of the global gross domestic product will be digitized. By 2030 more than 70% of new value creation in the economy will be based on digitally-enabled platforms [2].

According to the WHO and the US Centers for Disease Control and Prevention (CDC), digital transformation technologies can play an essential role in fighting the COVID-19 pandemic.[2] For example, healthcare IoT, the Internet of Medical Things (IoMT) integrated into healthcare information technology (IT) systems, collects, analyzes, and transmits health data efficiently. Thus, it offers many services as

---

[1]https://www.bloomberg.com/news/articles/2020-04-08/world-economy-faces-5-trillion-hit-that-is-like-losing-japan.

[2]https://www.who.int/news-room/detail/03-04-2020-digital-technology-for-covid-19-response.

monitoring patients by the concerned healthcare professionals in a real-time manner from a remote location.

Utilization of digital transformation technologies to alleviate COVID-19 has the potential to provide benefits, improving efficiency and effectiveness of diagnosis, treatment, and prevention processes. After realizing how versatile they are, digital transformation technologies have recently attracted intense attention from both industry and academia, many prototypes and applications have been developed and started to be used. Although there are fragmented cases, the integration of these technologies, which provides a significant impact on tackling COVID-19, is needed to achieve a more dynamic and efficient healthcare environment. The aim of this study is to propose an integrated framework consisting of digital transformation technologies and to provide innovative solutions to tackle COVID-19. Correspondingly, the research question of the chapter is that how an integrated framework consisting of digital transformation technologies can provide innovative solutions to tackle COVID-19.

The remaining of the paper is structured as follows: first, the background of the study is given, followed by the results of the literature review related to digital transformation technologies used for alleviating COVID-19. Then, the proposed conceptual integrated digital transformation framework is given, and the services provided through the framework are described. After analyzing the benefits and challenges of this system, we conclude the paper with final remarks.

2   Background of the Study

Some of the emerging technologies creating digital transformation across all industries, as digital twin [3], Internet of Things (IoT) and connected devices [4], Artificial Intelligence [5, 6], Cyber-Physical Systems [5–7], Integration [5, 8, 9] Social media and platforms, Blockchain [10], Everything-as-a-Service (XaaS) [11], Mobile computing [12–14], Cloud computing [15–17], Robots and drones, Data Analytics [18, 19] and 3D printing [11] are summarized in Table 1.

3   Literature Review

Although present before COVID-19, several technologies gained attention to cope with the pandemic. In this section, we will investigate data-driven solutions, digital contact tracing, robotics, and virtual clinics. These solutions are maturing, yet a holistic framework is necessary to (1) understand the relationship between fragmented solutions, (2) figure out ways to combine fragmented solutions, (3) integrate data sources, and (4) introduce integrated digital solutions to tackle COVID-19.

Table 1 Emerging technologies used in digital transformation

Technology	Description
Digital Twin	Compiling and formatting data of physical reality into the virtual world. It is used for creating simulation models to monitor, diagnose, and forecast
Internet of Things (IoT) and connected devices	Using and seamlessly integrating low-cost sensors and processors provides the generation of big data and ubiquitous environments throughout operations
Artificial Intelligence (AI)	The development of software for accessing, combining, analyzing data, and converting it into valuable information for learning, explaining, and forecasting the operations, events, and trends
Cyber-Physical Systems	Combination of real-time data collected from the physical systems, software modeling, and statistics to predict the behaviors of the system under different scenarios for making decisions in real-time
Blockchain	Using distributed computers not owned by a single entity to manage immutable records of time-stamped data
Everything-as-a-Service (XaaS)	Tailoring the computing environments to reshape customer experiences
Social media and platforms	Bringing ecosystem parties together in a very accessible (low cost) way by providing interaction and giving feedback. I.e.: Uber
Robots and Drones	Using robotics technologies that have the capability of sensing and responding to the environment
Big Data Analytics	Collecting, storing, analyzing, and distributing big data to make better and faster decision making
3D printing	Printing the physical objects by using digital 3D models
Autonomous Vehicles	Developing vehicles that are driven utilizing built-in applications
Next-Generation Communication Network (5G)	Wireless communication supported by mobile networks has a better performance than 4G
Cloud Computing	Availability of data centers to many-users over Internet
Image processing	Using algorithms for processing digital images
Mobile Computing	Chatbots: applications for an online chat conversation Mobile applications: a software application developed to run on mobile devices
Virtual Reality	Simulation of the real world or a completely different world
GIS	Using spatial analytics, mapping, and location intelligence to determine the exact location

3.1 Data-Driven Solutions

As of August 5, 2020, confirmed COVID-19 cases amount up to 18,965,479 [1]. Many organizations, including the WHO and the European Centre for Disease

Prevention and Control (ECDC), publish near real-time data to COVID-19 [20]. Dashboards and frequently updated figures provide exploratory and explanatory **visual analysis** capabilities by aggregating and cleaning these data sources. Among the exploratory ones, there are John Hopkins' COVID-19 Dashboard [21] and Microsoft's Bing COVID-19 Tracker [22]. On the other hand, to assist decision-makers, the Financial Times provides explanatory figures updated frequently [23]. However, we do not know the real number of cases since widespread population testing is still not available in most countries. Thus, some studies investigate **prediction models**. These models are either diagnostic or prognostic [24]. Diagnostic models predict the existence of the disease while prognostic models predict outcomes such as mortality risk, length of hospital stay, and the necessity of intensive care. For example, Menni et al. [25] used self-reported data from a free smartphone application available in the UK. They have generated a linear diagnostic prediction model that considers age, sex, loss of smell and taste, severe or significant persistent cough, severe fatigue, and skipped meals.

Despite the limited amount of data available, there is considerable effort to **forecast** new COVID-19 cases in real-time. Chakraborty and Ghosh [26] proposed a novel hybrid approach to obtain real-time ten-day forecasts for Canada, France, India, South Korea, and the UK Centers for Disease Control and Prevention (CDC) [27] publishes weekly forecasts of total deaths and hospitalization for the USA. These forecasts are produced by numerous modeling groups that include Carnegie Mellon University, Columbia University, and Georgia Institute of Technology.

3.2 Digital Contact Tracing

The pandemic stimulated the development of mobile-based digital contact tracing systems. By keeping a history of contact proximity and duration, these systems immediately send notifications to close contacts of diagnosed cases prompting them to self-isolate [28]. Various protocols emerged to help implement such systems, among which are Apple and Google's combined Exposure Notification Framework [29, 30]. These protocols use Bluetooth or GPS technologies to track encounters between two devices. They differ in terms of their log processing and availability of specifications. Here, we only exemplify protocols that provide public specifications.

The majority of protocols respect the users' privacy by exchanging non-personally identifiable messages that change frequently. The goal is to prevent third parties from identifying individuals. The usage of centralized log processing still raises privacy concerns. BlueTrace [31] and PEPP-PT [32] are example protocols where infected individuals share all their encounters with a health authority. The health authority can decrypt the encounter history and obtain personally-identifiable information.

Decentralized protocols delegate log processing to local devices. All encounter data remains local, but a central server still exists to observe anonymous identifiers of infected users. DP-3T [33] and TCN [34] are example protocols that implement decentralized log processing. In the case of TCN, a trusted health authority can maintain the integrity of the positive test results. Two implementations of TCN are CoEpi [35] and CovidWatch [36]. CoEpi allows users to self-report infection, while CovidWatch allows reports through a trusted authority.

3.3 Robotics

Robots take on tedious and dangerous jobs, some of which are not suitable for human workers. COVID-19 stimulated further research and investment in robotics to fight infectious diseases. We now use robots for surface disinfection, diagnosis and screening, and delivering medication [37]. Moreover, Kimmig et al. [38] argue that robot-assisted surgery reduces the number of medically exposed staff and contamination with body fluids and surgical gasses. It also makes room for COVID-19 patients by reducing hospital stay for patients who urgently need complex surgery.

3.4 Virtual Clinics

In response to COVID-19, virtual consultations are on the rise to prevent physical presence at clinics. Using smartphones or webcam-enabled computers, healthcare providers can efficiently screen patients before they arrive in hospitals [39]. The decision to intake a patient would depend on symptoms combined with local epidemiologic information, and detailed travel and exposure history. Gilbert et al. [40] rapidly implemented virtual clinics in response to the COVID-19 crisis and surveyed patients who undertook virtual consultations via telephone or video call. Patient satisfaction scores were high. Among the reasons for high satisfaction were reduced travel times, reduced waiting times, and reduced travel impact on symptoms.

As stated in [41], although the proliferation of telemedicine systems can alleviate COVID-19, a more efficient and dynamic healthcare system can be realized through consolidation with other digital transformation technologies, like mobile applications, drones, smart wearables, and other IoT devices. In summary, several digital transformation technologies currently address the critical problems of COVID-19. However, there is a need to manage these digital transformation technologies from a holistic point of view and use them in conjunction with other digital transformation technologies. This study aims to fill this gap by proposing a conceptual framework architecture, described in the next section.

4 The Proposed Integrated Digital Transformation Framework

As stated in [41], they are a lack of integration of data sources of existing digital transformation technologies. The aim of this framework is to demonstrate how innovative solutions can be generated by integrating these verified data sources via a holistic digital transformation framework.

As depicted in Fig. 1, the proposed holistic digital transformation framework consists of four main components: Data Sources, Digital Transformation Technologies, Applications, and Users.

4.1 Data Sources

With the recent technological advancements in IoT and cloud computing domains, we are able to collect data from an extensive array of data sources. These data sources may have different forms, including unstructured, semi-structured, and structured. The unstructured form includes image, audio, video, and text data, the semi-structured form includes XML and JSON data, and the structured form includes data stored in relational database management platforms. The proposed framework aims to collect and blend these data forms in a single platform and transfer these data to the digital environment for further use. The data sources may include governmental data, including citizen and health records and surveillance cameras, Internet of Things (IoT) devices. Moreover, there are also personal private data sources, including mobile and wearable devices, social media, and Internet of Medical Things (IoMT) devices. The proposed framework aims to gather and integrate these verified data sources from the external environment and provide a resilient data pipeline for digital transformation technologies.

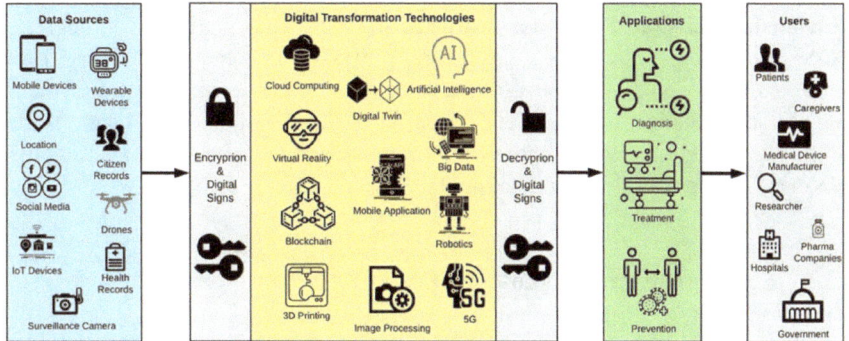

Fig. 1 The proposed framework architecture

4.2 Digital Transformation Technologies

Emerging digital technologies, including cloud computing, artificial intelligence, digital twin, virtual reality, image processing, big data, robotics, next-generation communication network, and 3D printing, are able to provide infrastructure for innovative applications to diagnose, treat and prevent pandemic by utilizing a wide range of data sources. However, these data sources may also include private information and health records, and there are strict government regulations and policies to transfer and share these data over the internet. The proposed framework needs to support data privacy by anonymization and encoding of private records such as name, surname, test results, and social security number. Hence, blockchain technology plays a crucial role in the proposed framework to transfer and share these private data anonymously and securely by using encryption and digital signs. Blockchain technology provides a solution for technical and social challenges, including privacy and quality of data. Accordingly, the utilization of blockchain technology in the healthcare domain presents promising solutions, including securing communications among stakeholders, efficient, accountable, transparent, and accurate delivery of private health records [42]. Moreover, blockchain technology enables us to integrate various kinds of private health records of individuals on a secure infrastructure. Even though some studies in the literature utilizes blockchain in the healthcare domain to solve a specific problem, including medical record management [43], medical [44], and financial transaction verification [45], these studies do not provide a comprehensive solution. To this end, the main focus of this study is proposing a holistic framework by integrating data sources and state-of-the-art digital transformation technologies on a blockchain-based secure infrastructure.

4.3 Applications

The proposed framework gathers and blends different various data sources in structured, semi-structured, and unstructured forms in a platform for interested parties, including researchers, healthcare professionals, and physicians. These interested parties may develop innovative applications by utilizing state-of-the-art digital transformation technologies. The result of these developed applications may be forwarded to interested parties in different forms to be used in diagnosis, treatment, and prevention applications for COVID-19.

4.3.1 Applications for Diagnosing COVID-19

With the state-of-the-art digital transformation technologies, we are enabled to develop applications for augmenting early diagnosis of COVID-19. The proposed framework assists data scientists and researchers to analyze provided a vast amount of datasets to analyze and train novel Artificial Intelligence (AI) and machine learning (ML) models to present more accurate and rapid diagnosis applications. These diagnosis applications are listed below:

- identifying people who had exposure to infected individuals
- automating the clinical testing process by using robots and Cyber-Physical Systems (CPS)
- observing patients' health indicators including stress level, cardiorespiratory variables, and daily activities in real-time with smart wearables, IoT devices, and mobile devices and notifying if there is any deviation from the baseline of the individual
- diagnosing COVID-19 via AI-powered medical imaging applications for analyzing computed tomography scans and X-rays
- supporting healthcare givers via telemedicine used in conjunction with AI, smart wearables, drones, and 5G cellular networks for evaluating, diagnosing and treating patients

4.3.2 Applications for Treating COVID-19

Developed treatment applications as a result of the utilization of emerging digital transformation technologies in the proposed framework are listed below:

- rapidly developing drugs and vaccines for COVID-19 by utilization of robots and CPS technologies
- helping hospitalized patients in transporting and serving food or medicines by robots
- developing unique treatment protocol for each patient and identifying patients with a higher possibility of developing severe symptoms based on their initial symptoms as a result of the utilization of AI and big data technology for analyzing patients' historical health records and response to current treatment
- offering extensive healthcare services as monitoring patients' health in a real-time manner and tracking their medicine use periodically by the integration of smart medical devices with IoMT technology.

4.3.3 Applications for Preventing COVID-19

These emerging digital transformation technologies and diverse set of data sources also enable us to develop prevention application for COVID-19 which are listed below;

- collecting proactive measures from the data sources and developing and using AI-powered applications for risk forecasting, simulation, and modeling of COVID-19 spread, to effectively manage it by the public health offices
- tracing and warning people about social distancing and wearing masks by utilizing IoT, mobile, and surveillance camera data sources. These data sources can also be utilized to determine potentially infected individuals who had close contact with a person who diagnosed COVID-19 in the last few days and notify these potential patients about regulations and guidelines they should adhere
- utilization of drones, robotics, and image processing technologies enables to determine crowded places, identifying high-risk zones according to social distancing and wearing mask rules
- thermal imaging, crowd surveillance, and broadcasting announcements by drones without any human intervention
- disinfecting surfaces by autonomous robots
- helping patients with mild COVID-19 symptoms who are under strict isolation at home
- preventing hospital-acquired infections by prompting alerts about any sanitation necessity risky for public safety by using IoT
- providing effective contact tracing as a result of the utilization of the mobile app, Bluetooth, and GPS data, which can be collected with the consent of the users. The information gathered as follows;

  – The real-time location and the previous locations of the person diagnosed with COVID-19
  – The list people exposed to the diagnosed person in the last few days to determine potentially infected people
  – The list of individuals exposed any potential COVID-19 patient

- notifying individuals immediately via the mobile application

  – if they exposed a person who has later diagnosed with COVID-19 and notifies the guidelines they should follow
  – when they enter the high-risk zone which assigned accurate risk scores to spaces

- preventing misinformation, myths, and conspiracy theories about COVID-19 by using AI on social media. For example, there was a myth on social media about spreading coronavirus through 5G infrastructure.

4.4 Users

The developed applications with digital transformation technologies can be used for patients, caregivers, medical device manufacturers, researchers, hospitals, pharma companies, and government. *Patients* can use these applications to track their health condition and their health indicators, including stress level, cardiorespiratory variables, and daily activities in real-time. *Caregivers* can use telemedicine applications for evaluating, diagnosing, and treating patients. Moreover, they can rapidly diagnose COVID-19 via AI-powered medical imaging applications for analyzing computed tomography scans and X-rays. *Medical device manufacturers and Pharma companies* can utilize these applications to improve their products and develop drugs, vaccines, and medical devices for COVID-19 by utilization of robots and CPS technologies. *Researchers* may get benefit from standardized and unified data sources to improve and accelerate their COVID-19 related researches. The developed applications and their collected data can contribute to the improvement of health data quality and quantity for COVID-19 related researches. *Hospitals* may disinfect surfaces, measure body temperature with thermal imaging, and broadcast announcements by autonomous robots without human intervention. *Governments* can trace the pandemic process in their countries with different indicators and develop emerging solutions to prevent COVID-19 and trace patients with their consent for ensuring that patients with mild COVID-19 symptoms are under strict isolation at home. Moreover, governments are able to determine crowded places to identify people who do not obey the rules, as social distancing and wearing a mask.

5 Benefits and Challenges of the Proposed System

Following the introduction of the proposed holistic framework and the analysis of principal components, we first discuss the benefits of producing integrated digital technologies to deal with contemporary problems caused by COVID-19. We then examine the challenges associated with building, maintaining, and exploiting the proposed framework.

5.1 Benefits of the Proposed Framework Architecture

5.1.1 Effective Strategic Management of COVID-19 Crisis

Public health agencies are responsible for making strategic decisions for tackling COVID-19 for their countries. Feeding integrated data from a multitude of sources to AI-powered applications yields better models, forecasts, and simulations. More reliable data products enhance the decision-making capability for public health agencies to manage the crisis effectively.

5.1.2 Reducing the Risk of Virus Transmission

After assigning accurate and real-time risk scores to zones, preventative applications can notify individuals immediately when entering a high-risk zone. Real-time detection of high-risk zones also enables local authorities to focus their efforts and take immediate action. Particularly in these zones and hospitals, autonomous robots can disinfect surfaces reducing the probability of staff members contracting the virus.

5.1.3 Detecting COVID-19 Carriers as Early as Possible

Since widespread population testing is not available in most countries, we do not know and thus isolate COVID-19 carriers if we do not implement strict lockdowns. A diagnosis application in the framework can combine user-reported symptoms, data from smart wearables, and travel and exposure history on smartphones. Telemedicine, together with remote testing or automated clinical testing procedures, can identify individuals who must self-isolate as early as possible.

5.1.4 Decreasing the Workload and the Stress Level of the Hospital Staff

The proposed framework supports healthcare givers to diagnose, treat, and prevent COVID-19 cases. It is possible to use telemedicine systems in conjunction with AI, smart wearables, remote testing, and autonomous robots. By reducing the workload and risk of virus contraction, integrated digital technologies also decrease the stress level of the hospital staff.

5.1.5 Reducing the Risk of an Overwhelmed Healthcare System

Early detection of carriers and decreased risk of virus transmission lead to a significant reduction in new case numbers, which require hospitalization. Moreover, better forecasts and simulations guide authorities' decisions on prioritizing healthcare investments and regions to send a limited amount of medical supplies (e.g., test kits and respirators). As a result, the risk of an overwhelmed healthcare system goes down.

5.1.6 Decreasing the Mortality Rates and Increasing the Treatment Success Rates

The use of AI, together with big data technologies, facilitate the development of unique treatment protocols for each patient. Treatment applications can accurately identify patients with a higher possibility of developing severe symptoms analyzing initial symptoms, response to current treatment, and historical health records. Predicting the necessity of intensive care early on would decrease mortality rates.

5.1.7 Reducing the Negative Impact of COVID-19 on the Economy

Digital transformation technologies can eliminate the need for lockdowns preventing economic shrinkage when they utilized in integrated forms. Furthermore, effectively preventing the spread would decrease the cost of treating patients.

5.1.8 Decreasing the Stress Level of People

There has been an abundance of misinformation, myths, and conspiracy theories about COVID-19 since the outbreak began. Combined with lockdowns, the inadequate response by public authorities, and overwhelmed states of hospitals, people's stress level spiked. As public health agencies can present better models and forecasts, we can expect an increase in trust and a decrease in misinformation.

5.2 Challenges of the Proposed Framework Architecture

5.2.1 Data Acquisition and Integration

In the proposed framework, heterogeneous data sources transmit data in different forms, such as text, image, audio, and video. This heterogeneity is a significant challenge since it can negatively affect the time necessary to develop AI-powered applications [46].

With the help of blockchain, disparate systems store patient health records. Every member of the chain is connected, and data is updated in all of the computers at once. Therefore, blockchain solves the inconsistency problem in which patients encounter inaccuracies in their health records, or their medical history is not available to all stakeholders. On the other hand, the proposed framework relies on trusted authorities to maintain data integrity. Selecting and keeping up with verified sources still create data accuracy and reliability challenges.

5.2.2 Privacy

Although blockchain technology highlights security through encryption, managing databases that contain having private health records, even in encrypted form, is a critical issue. Thus, the implementation of the proposed framework must adequately address access control.

5.2.3 The Lack of Historical Data

The lack of a substantial amount of unbiased training data affects the performance of the data analysis model. As a result of the unprecedented nature of the pandemic, the lack of historical data to train the models causes inefficient results.

5.2.4 Governance

In order for fragmented digital transformation technologies to work together, standards, protocols, and agreements need to emerge. Coordination can prove to be a challenge.

5.2.5 Expertise

There need to be people who have IT-skills to develop, implement, and maintain the components of this framework. Besides, for real-time scenarios to work out, domain experts and data scientists should be knowledgable of real-time big data processing platforms. The lack of people having expertise in these fields is a significant challenge for the effective implementation of the proposed framework.

5.2.6 Scalability

Since the framework uses a blockchain-based platform to manage digital trans-formation technologies, scalability stands as a challenge for the maintainability of the proposed framework.

5.2.7 Lack of Legislation

Appropriate legislation for governance rights, ownership of records, and distributed storage structure of the blockchain should be carefully defined. Since there are many stakeholders, data ownership is also an important issue needed to be solved.

5.2.8 The Lack of Infrastructure for 5G Network

A cellular network with a high bandwidth capability is necessary to maintain the implementation of the framework efficiently. The next-generation 5G networks can provide high-speed data transmission needed for the dynamic structure of the transactions performed through the system, such as collecting data via mobile applications or smart wearables, monitoring crowds via drones, or performing advanced data analytics. However, 5G networks have many challenges: still at the infancy stage, high cost of maintenance, and security issues for data confidentiality [47].

5.2.9   Cost of Setup and Operation

Since the proposed structure is a conceptual framework, setup and operation costs are not yet known. The decreased cost of IoT and storage devices and the availability of open-source technologies can reduce costs. Further cost reduction is possible if the implementation can use the present but fragmented technology solutions. We have an idea of the cost of lockdowns and the overwhelmed healthcare system, which can justify the implementation of the framework. Furthermore, the framework would continue to show benefits in a post-COVID-19 world.

5.2.10   Adoption and Trust

Blockchain-based infrastructures require a network of inter-connected computers for supplying the required computing power. Incentive mechanisms are necessary to encourage participants for adoption.

Most digital contact tracing apps fail because of the inability to build a critical mass of engaged users. If these applications fail to build a critical mass, they show highly inaccurate results. As suggested by Farronato et al. [48–50], the implementations can first provide immediate value to smaller communities to increase user engagement. Incentives might be necessary to increase adoption for other communities. Any implementation must consider the cultural context to minimize user resistance. Even so, the social implications of this framework should be managed effectively.

6   Conclusion

COVID-19 pandemic disease has cost lives and jobs. It has changed how we communicate, work, and socialize. Digital transformation technologies can play a crucial role in alleviating the COVID-19 outbreak. Although fragmented digital technologies try to address some of the contemporary problems created by the pandemic, a much more effective and dynamic healthcare system must use a combination of the digital transformation technologies in which data is integrated, private, and anonymized. In order to satisfy this necessity, a holistic digital transformation framework architecture is proposed in this study.

One of the main contributions of this study is to propose a privacy-first holistic digital transformation framework and its components. The second contribution is to identify innovative digital solutions developed for diagnosis, treatment, and prevention of the COVID-19 pandemic as a result of integrating data sources and state-of-art digital transformation technologies. The third contribution is to analyze the benefits and challenges of such a holistic digital transformation framework architecture developed for alleviating COVID-19. The proposed framework aims to provide a more efficient and dynamic healthcare system to reduce the negative impact on the economy, the risk of an overwhelmed healthcare system, the death toll, the risk of virus transmission, and healthcare givers' workload and stress levels.

References

1. Worldometer: COVID-19 coronavirus pandemic. https://www.worldometers.info/coronavirus/. Accessed 8 Aug 2020
2. World Economic Forum: Digital Transformation Initiative Maximizing the Return on Digital Investments (2018). http://www3.weforum.org/docs/DTI_Maximizing_Return_Digital_WP.pdf. Accessed 15 Aug 2020
3. Annunziata, M., Biller, S.: The industrial internet and the future of work. Mech. Eng. Mag. Sel. Artic. **137**(09), 30–35 (2015)
4. Gilchrist, A.: Industry 4.0: The Industrial Internet of Things. Apress, Berkeley (2016)
5. Kagermann, H., Helbig, J., Hellinger, A., Wahlster, W.: Recommendations for implementing the strategic initiative INDUSTRIE 4.0: securing the future of German manufacturing industry; final report of the Industrie 4.0 working group. Forschungsunion (2013)
6. Schuh, G., Anderl, R., Gausemeier, J., ten Hompel, M., Wahlster, W.: Industrie 4.0 Maturity Index. Manag. Digit. Transform. Companies, Munich Herbert Utz (2017)
7. Gökalp, E., Gökalp, M.O., Eren, P.E.: Industry 4.0 revolution in clothing and apparel factories: Apparel 4.0. In: Industry 4.0 from the MIS Perspective, pp. 169–183. Peter Lang, Bern, Switzerland (2018)
8. Schuh, G., Potente, T., Wesch-Potente, C., Weber, A.R., Prote, J.-P.: Collaboration mechanisms to increase productivity in the context of Industrie 4.0. Procedia CIRP **19**, 51–56 (2014)
9. Gökalp, E., Şener, U., Eren, P.E.: Development of an assessment model for industry 4.0: Industry 4.0-MM. In: Communications in Computer and Information Science, vol. 770, pp. 128–142 (2017)
10. Swan, M.: Blockchain: Blueprint for a New Economy. O'Reilly Media, Sebastopol (2015)
11. Schwab, K.: The Fourth Industrial Revolution. Currency, New York (2017)
12. Eren, P.E., Gökalp, E.: HealthGuide: A personalized mobile patient guidance system. In: Current and Emerging mHealth Technologies, pp. 167–187. Springer, Cham (2018)
13. Arpaci, I., Yardimci Cetin, Y., Turetken, O.: A cross-cultural analysis of smartphone adoption by Canadian and Turkish organizations. J. Glob. Inf. Technol. Manag. **18**(3), 214–238 (2015)
14. Arpaci, I., Yardimci Cetin, Y., Turetken, O.: Impact of perceived security on organizational adoption of smartphones. Cyberpsychology, Behav. Soc. Netw. **18**(10), 602–608 (2015)
15. Arpaci, I., Kilicer, K., Bardakci, S.: Effects of security and privacy concerns on educational use of cloud services. Comput. Human Behav. **45**, 93–98 (2015)
16. Arpaci, I.: Antecedents and consequences of cloud computing adoption in education to achieve knowledge management. Comput. Human Behav. **70**, 382–390 (2017)
17. Şener, U., Gökalp, E., Eren, P.E.: ClouDSS: A decision support system for cloud service selection. In: Lecture Notes in Computer Science (including subseries Lecture Notes in Artificial Intelligence and Lecture Notes in Bioinformatics), vol. 10537. LNCS (2017)

18. Coban, S., Gokalp, M.O., Gokalp, E., Eren, P.E., Kocyigit, A.: [WiP] Predictive maintenance in healthcare services with big data technologies. In: 2018 IEEE 11th Conference on Service-Oriented Computing and Applications (SOCA), pp. 93–98 (2018)
19. Gökalp, M.O., Kayabay, K., Akyol, M.A., Koçyiğit, A., Eren, P.E.: "Big Data in mHealth", in Current and Emerging mHealth Technologies, pp. 241–256. Springer, Cham (2018)
20. CSSEGISandData: COVID-19 Data Repository by the Center for Systems Science and Engineering (CSSE) at Johns Hopkins University (2020). https://github.com/CSSEGISandData/COVID-19/blob/master/README.md. Accessed 6 Aug 2020
21. J. H. University: COVID-19 Dashboard by the Center for Systems Science and Engineering (CSSE) at Johns Hopkins University (JHU) (2020). https://coronavirus.jhu.edu/map.html. Accessed 6 Aug 2020
22. Microsoft: COVID-19 Tracker (2020). https://www.bing.com/covid. Accessed 8 Aug 2020
23. F. Times: Coronavirus tracked (2020). https://www.ft.com/content/a2901ce8-5eb7-4633-b89c-cbdf5b386938. Accessed 6 Aug 2020
24. Wynants, L., et al.: Prediction models for diagnosis and prognosis of COVID-19: systematic review and critical appraisal. BMJ m1328, April (2020)
25. Menni, C., et al.: Real-time tracking of self-reported symptoms to predict potential COVID-19. Nat. Med. 26(7), 1037–1040 (2020). Jul. 2020
26. Chakraborty, T., Ghosh, I.: Real-time forecasts and risk assessment of novel coronavirus (COVID-19) cases: A data-driven analysis. Chaos, Solitons Fractals 135:109850 (2020)
27. C. for D. C. and Prevention: COVID-19 Forecasting: Background Information (2020). https://www.cdc.gov/coronavirus/2019-ncov/cases-updates/forecasting.html. Accessed 6 Aug 2020
28. Ferretti, L. et al.: Quantifying SARS-CoV-2 transmission suggests epidemic control with digital contact tracing. Science 368(6491), eabb6936 (2020). 8 May 2020
29. Apple: ExposureNotification Framework (2020). https://developer.apple.com/documentation/exposurenotification. Accessed 24 Jul 2020
30. Google: Exposure Notifications API (2020). https://developers.google.com/android/exposure-notifications/exposure-notifications-api. Accessed 24 Jul 2020
31. Bay, J., et al.: BlueTrace : A privacy-preserving protocol for community-driven contact tracing across borders (2020)
32. Pepp-pt: Documentation for Pan-European Privacy-Preserving Proximity Tracing (PEPP-PT) (2020). https://github.com/pepp-pt/pepp-pt-documentation. Accessed 8 Aug 2020
33. Troncoso, C. et al.: Decentralized Privacy-Preserving Proximity Tracing (2020)
34. TCNCoalition: Specification and reference implementation of the TCN Protocol (2020). https://github.com/TCNCoalition/TCN. Accessed 8 Aug 2020
35. CoEpi: CoEpi: Community Epidemiology in Action (2020). https://www.coepi.org/. Accessed 8 Aug 2020
36. Sydney Von Arx, H.X., Becker-Mayer, I., Blank, D., Colligan, J., Fenwick, R., Hittle, M., Ingle, M., Oliver Nash, M., Nguyen, V., Petrie, J., Schwaber, J., Szabo, Z., Veeraghanta, A., Voloshin, H.X., White, T.: Slowing the Spread of Infectious Diseases Using Crowdsourced Data (2020)
37. Yang, G.-Z., et al.: Combating COVID-19—the role of robotics in managing public health and infectious diseases. Sci. Robot. 5(40), eabb5589 (2020). Mar. 2020
38. Kimmig, R., Verheijen, R.H.M., Rudnicki, M., and for SERGS Council: Robot assisted surgery during the COVID-19 pandemic, especially for gynecological cancer: a statement of the Society of European Robotic Gynaecological Surgery (SERGS). J. Gynecol. Oncol. 31(3), e59 (2020)
39. Hollander, J.E., Carr, B.G.: Virtually Perfect? Telemedicine for Covid-19. N. Engl. J. Med. 382(18), 1679–1681 (2020). Apr. 2020
40. Gilbert, A.W., et al.: Rapid implementation of virtual clinics due to COVID-19: report and early evaluation of a quality improvement initiative. BMJ Open Qual. 9(2), e000985 (2020). May 2020

41. Chamola, V., Hassija, V., Gupta, V., Guizani, M.: A comprehensive review of the COVID-19 pandemic and the role of IoT, drones, AI, bblockchain, and 5G in managing its impact. IEEE Access **8**(April), 90225–90265 (2020)
42. Gökalp, E., Gökalp, M.O., Çoban, S., Eren, P.E.: Analysing opportunities and challenges of integrated blockchain technologies in healthcare. In: EuroSymposium on Systems Analysis and Design, pp. 174–183 (2018)
43. Xia, Q., Sifah, E.B., Asamoah, K.O., Gao, J., Du, X., Guizani, M.: MeDShare: trust-less medical data sharing among cloud service providers via blockchain. IEEE Access **5**, 14757–14767 (2017)
44. Yue, X., Wang, H., Jin, D., Li, M., Jiang, W.: Healthcare data gateways: found healthcare intelligence on blockchain with novel privacy risk control. J. Med. Syst. **40**(10), 218 (2016)
45. Basu, A., Subedi, P., Kamal-Bahl, S.: Financing a cure for diabetes in a multipayer environment. Value Heal. **19**(6), 861–868 (2016)
46. Gokalp, M.O., Kayabay, K., Akyol, M.A., Eren, P.E., Kocyigit, A.: Big data for industry 4.0: a conceptual framework. In: 2016 International Conference on Computational Science and Computational Intelligence, pp. 431–434, Dec (2016)
47. Li, D.: 5G and intelligence medicine—how the next generation of wireless technology will reconstruct healthcare? Precis. Clin. Med. **2**(4), 205–208 (2019)
48. Farronato, C., Iansiti, M., Bartosiak, M., Denicolai, S., Ferretti, L., Fontana, R.: How to get people to actually use contact-tracing apps. Harvard Bus. Rev. (2020)
49. Gökalp, E., Gökalp, M.O., Çoban, S.: Blockchain-based chain management: understanding the determinants of adoption in the context of organizations. Inf. Syst. Manag. 1–22 (2020) https://doi.org/10.1080/10580530.2020.1812014
50. Çaldağ, M.T., Gökalp, E.: Exploring Critical success factors for blockchain-based intelligent transportation systems. Emerg. Sci. J. **4**, 27–44 (2020) https://doi.org/10.28991/esj-2020-SP1-03

Application of Modern Technologies on Fighting COVID-19: A Systematic and Bibliometric Analysis

Irsa Azam and Muhammad Usman

Abstract COVID-19 pandemic drastically increased the demand for essential medical healthcare equipment's, medicines along with the strict lockdown conditions to prevent disease transmission. As a result it becomes challenging for healthcare professionals to provide In-person treatment. This poses pressure to perform medical practices besides taking care of the spread of the novel coronavirus. Studies have declared that information technologies have the potential to fulfill customized requirements of COVID-19 pandemic. Thinking about advance technologies and its benefits, this study is going to provide a through literature about the application of advance technologies in real-time. The study also highlights that Telemedicine and Telehealth are the most widely used technological term in the COVID-19 research literature. The first aspect of the study is to search the literature that allowed us to understand the series of advance technologies emerging recently and how they are helpful during this pandemic? The second stream of literature focused on classifying the top 12 technologies under different scenarios as an outcome of the current situation. Study mapped the literature by selecting the final 83 articles to understand the benefits and application of these technologies in different areas of healthcare system.

Keywords COVID-19 · Telehealth · Telemedicine · Artificial intelligence · Internet of things · Pandemic · Robotics

I. Azam (✉) · M. Usman
Department of Management Sciences, University of Gujrat, Punjab, Pakistan
e-mail: Irsa.azam63@gmail.com

M. Usman
e-mail: drusman@uog.edu.pk

© The Author(s), under exclusive license to Springer Nature Switzerland AG 2021   167
I. Arpaci et al. (eds.), *Emerging Technologies During the Era of COVID-19 Pandemic*, Studies in Systems, Decision and Control 348,
https://doi.org/10.1007/978-3-030-67716-9_11

1 Introduction

The world is witnessing the deadly coronavirus (COVID-19) outbreak. To date, the center for disease control and prevention (CDC) reported 1.7 million cases of COVID-19 with 104,396 deaths [1]. The WHO (World Health Organization) declared this outbreak a pandemic on the date of March 11, 2020. Moreover, WHO also called for mutual support for the health sector to prepared them to fight against this pandemic [2, 3]. Similarly likely, on March 13, 2020 the executive officer of the president of US indicate this pandemic a global emergency. As the growth of the incidents of COVID-19 continues to rise, the healthcare system is struggling to prepare and fulfill the increasing clinical demands [3, 4].

COVID-19 has dramatically affected healthcare facilities and treatment systems worldwide [2]. The Healthcare system is under great pressure to provide primary healthcare services to patients. In order to fulfill the growing demands of healthcare equipment's and to mitigate the spread of coronavirus, hospitals need to improve the efficiency of their medical system [5]. In this global health emergency, medical personnel and scientific researchers are looking for ways to address the challenges of this deadly virus in more efficient way [6]. In this emergency situation, science and technology are playing a key role [7]. Studies also examined that this pandemic has not just impacted on physical health but it also severely impacting the mental health of people. In the presence of the fear of this deadly virus people are choosing to stay at home and practice social distancing [8].

Researchers are continuously suggesting that advanced technology has the potential to mitigate the disease by naturally controlling the disaster [9]. Despite the fact that some countries are not accepting latest technologies, it the only option to fight with this pandemic [10–12]. Countries are trying to adopt innovative technologies almost in all the department of the healthcare system to get relief from this emergency situation [13, 14]. This alarming situation raised one question in everyone's mind, about how public healthcare services are immediately responding to infectious disease with the help of latest technologies. Therefore, this present study focus on the ways of utilizing modern technologies by the healthcare system and the practical benefits of these technologies in fight against the pandemic. This study will provide a thorough review of the application of advanced technologies used by different healthcare systems in different departments. The study will not only provide insight into the application of technologies but also it will provide a deep understanding of their reaction to coronavirus pandemics.

2 Methodology

For this systematic search, the study developed a search strategy to identify relevant literature. This search strategy was tailored to Scopus database. A Boolean operator keyword technique is utilized by using "COVID-19" AND Technologies, "COVID-19" AND industry 4.0 to collect data. Study included journal articles

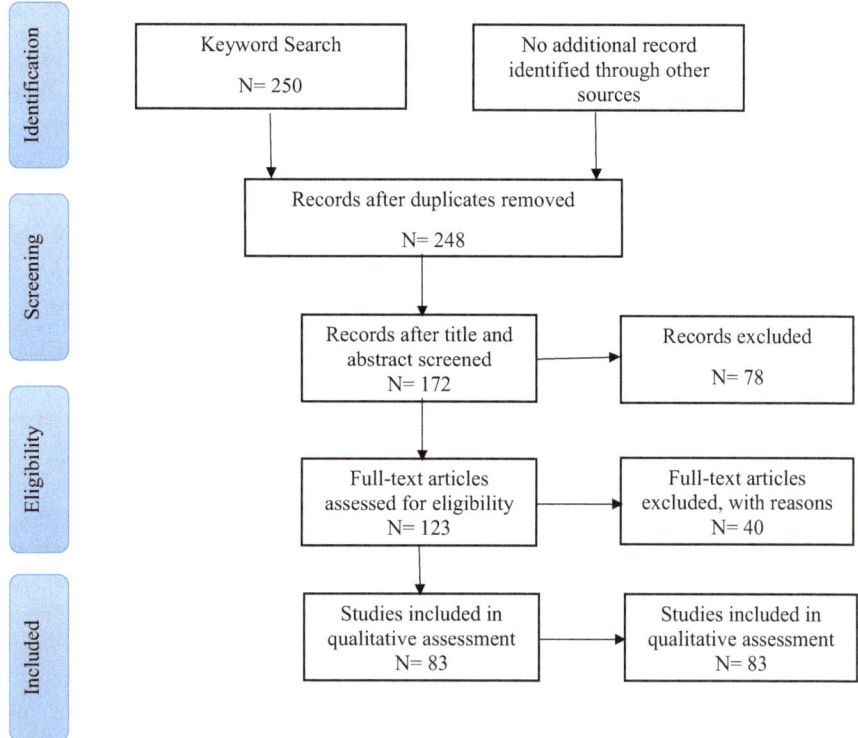

Fig. 1 PRISMA framework

published in English only. The selection criteria were based on the Fig. 1 PRISMA Statement [15]. The research mainly focused on the mapping existing literature on COVID-19 in the field of technological innovation in healthcare system. Therefore all the other articles targeting other areas are excluded. A total of 78 articles were excluded at this stage. The studies of 172 records were extracted at this stage. For maintaining the quality of the review all duplications of the articles were checked deeply. A careful evaluation of each research paper was carried out and 89 articles were removed at a later stages. Finally, a total of 83 articles were selected for final conducting the study. These 83 articles were fulfilled all the basic criteria of the study.

3   Results

COVID-19 teaches us many lessons on becoming resilient in future with the supply of resources [16]. The study found many advanced technologies which is successfully utilized in healthcare system of different countries. These technologies can help in the proper control and management of COVID-19 pandemic [17, 18]. All these technologies discussed in this paper can also help in the detection and diagnosis of COVID patients and also rectify other health-related problems.

3.1 Telemedicine and Telehealth Service During COVID-19

The concept of telemedicine can be defined as the process of providing care to a patient by a physician or other healthcare professional at a different venue [19]. It is the Use of Information Technology platforms, including voice, audio, text and digital data exchange as help to diagnosis, prescription and follow up evaluation [20]. Researchers are continuously trying to suggest the measures to overcome the possible pitfalls of telemedicine [21, 22].

Globally, the existing healthcare system has severely challenged by COVID-19 pandemic. The transmission speed of the virus makes the hospital to avoid face to face consultation with patients [23]. Therefore, many routine checkups including dental procedures, neurologic examination, and cardiology monitoring have been suspended and only emergency surgeries are allowed to perform. All over the world, government-imposed serious restriction on social interactions and travel and proposed remote working concept [24]. The concept of telemedicine is not new, some countries were already adopting the practices of telehealth especially palliative care for seriously ill patients. Research evidence declared that those patients who were receiving palliative care at their homes were very much satisfied with time saving and convenient virtual care [25]. The emergence of COVID-19 makes the telemedicine critically essential tool to mitigate the spread of coronavirus and save personal protection equipment [25].

After reviewing the literature and performing bibliometric analysis, the study found that telemedicine and telehealth are most widely used terms in COVID-19 research shown in Fig. 2. A large number of studies have been conducted in this area as compared to any other technological terms. This clearly indicate the important role being played by telemedicine and telehealth in healthcare service. Telemedicine research is conducted in almost all the departments of healthcare systems including Dentistry, Neurology, Orthopedic, Cardiologist, Dermatologist and eye care [26] (Table 1).

3.2 3D Printing Technology

The world is facing a huge shortage of personal protection equipment (PPE) during COVID-19. Countries are looking for creative solutions to fulfill this shortage [27]. 3D printing is appearing to be the best solution for this problem. It is playing huge role by making low-cost face shields and changing full-face snorkel masks [28]. The 3D printable adopter, having a transparent face shield is used as an effective barrier against infection. The Polylactic Acid material used for making adopters is

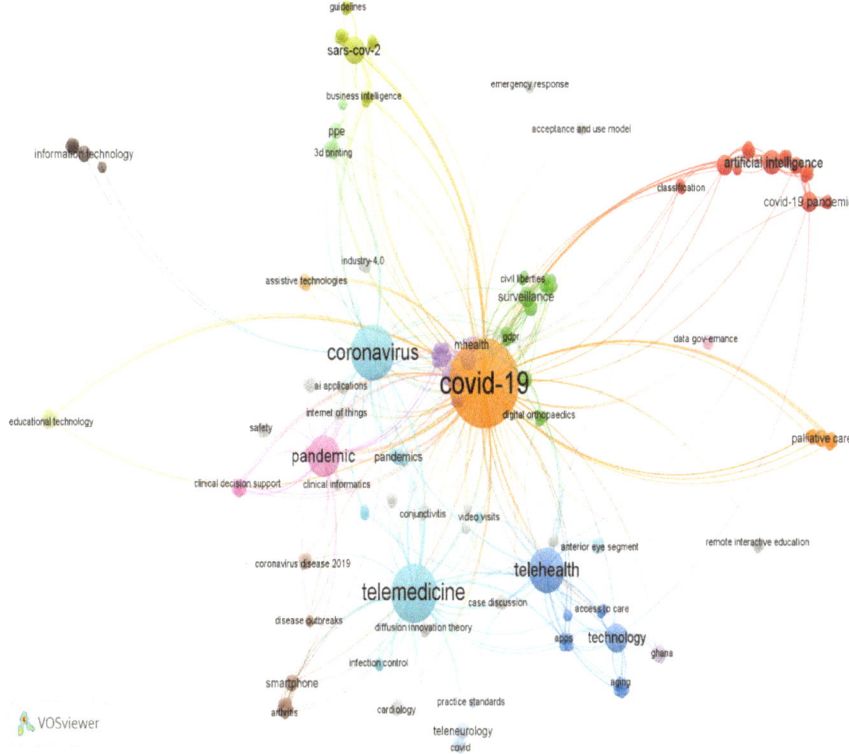

Fig. 2 The most used keywords in the context of emerging technologies during COVID-19

cost-effective, available at a very cheap price. The front transparent sheet is made with the help of polyester and used for laser printing [27].

The facemask produced with the help of 3D printing technology can be used to test a large number of people in just 30 min. Previously used face masks including surgical and N95 were considered unfriendly for the environment. However, these masks are made with NanoHack 3D which is environment friendly and can be reused [2].

3.3 Artificial Intelligence (AI)

Researchers identified that Artificial Intelligence (AI) is also one of the powerful tool to fight against COVID-19. AI has the capacity to accurately predict the outbreak and also helps in minimizing the spread of Coronavirus. It can also guide about the COVID patient future health condition, which one is going to develop the severe respiratory disease [29]. If the wrong information is circulating in social

Table 1 Top 12 most widely used technologies in COVID-19 Pandemic

Implementation of technology	Area	Functions	References
Telemedicine [44–46]	Teledentistry Teleneurology Teledermatology Digital Orthopedics Telecardiologist Telehealth eyecare	Remote dental screening, making diagnosis, providing consultation and proposing treatment plan Virtual neurologic examinations by using well established robust information technology WebEx virtual conference call system to perform daily telemedicine rounds Close communication with patients Delivering care in the most appropriate venue, whether that be at home, the clinic, or the hospital Use video communication software and implanted devices for remote cardiology monitoring Triage for anterior eye, enhancing compliance, subjective reflection and visual acuity, imaging, contact lens fitting Blood pressure and oxygen saturation	[23] [19, 47] [48] [49–52] [53, 54] [24, 55, 56]
3D-Printing Technology	COVID-19 intensive care unit (ICU)	Doctor use the headlight to protect themselves from respiratory droplets, blood, sputum and other fluids. Useful for ENT clinic, emergency room, operating room, ENT ward	[27, 57–59]
5G, 6G Mobile Data	Wireless awareness, Forecasting, Connectivity Health Status Code	Facilitate the scientists in forecasting the next global epidemic from the recent epidemic Role of B5G or 6G in delivering faster download speeds Future possibilities of speeds up to one terabit per second Patient data including state of health, travel history, and number of contacted people is used to tell a person about requirement of quarantine or not Code is assigned by giving red, yellow or green color	[35, 60, 61]

(continued)

Table 1 (continued)

Implementation of technology	Area	Functions	References
Telehealth [62–65]	Zoom, Microsoft Teams Palliative care Cisco WebEx virtual conference call Remote non-clinical services	Provide additional visual information, diagnostic clues and therapeutic presence Remote assessments of patients presenting COVID-19, Provision of proper care to patient Use to Perform daily telemedicine rounds via a store and forward consultative service Provision of training, administrative meetings and continuing medical education	[66–68] [25, 69–71] [72–75]
Artificial Intelligence	2D Deep Convolutional Neural Network 3-Dimensional Deep Learning Model Smartphone Applications Business Intelligence	Disease tracking Prediction outcome of patient's health condition Computational Biology and Medicines perspective, Protein structure predictions Prevention and follow-up for COVID-19 patients BI can provide real-time data on how the epidemic is spreading and where are the clusters Use of first aid calls to identify the place of contagion occurring	[29, 66, 76, 77]
Internet of Thing (IoT)	Internet of healthcare thing (IoHT)/Internet of medical thing (IoMT) Mobile Applications	Alerts and tracks any types of diseases to improve the safety of the patient Digitally captures the data and information of the patient without any human interaction Biometric measurements like blood pressure, heartbeat and glucose level Facilitates the use of patient medical information without the physical manipulation of patient records	[78, 79] [80, 81]
Robotics	Robots	Hospital used robots to deliver medicines, food and measure the temperature of coronavirus patients	[34, 82]
Drone technology	Remote Areas	Transportation of COVID-19 test samples to laboratory centers The autonomous drones are used to haul samples taken from suspected persons in the remote	[40]

(continued)

Table 1 (continued)

Implementation of technology	Area	Functions	References
		parts of the country to testing facilities in the inspection area	
Automatic Machine	Solar-powered automated hand washing machine	The machine, which uses solar and minimalizes the risk of self-contamination when washing hands	[40]
GPS, Wifi, Bluetooth	Digital Contact Tracing App Infected Patient Contact Tracing Digitized security controls	Use of electronic information to identify exposures of an individual to infection Identification of close contact for giving instructions about patient (self-quarantine, respiratory hygiene/cough etiquette) Advice for receiving early care patient develop symptoms Hospital used digitalized administrative processes such as using infrared thermometer to detect temperature without body contact	[41, 83]
5G cloud partnership	CT and X-ray synchronization system ECG data monitoring and ultrasonic images 5G cloud-based smart robot 5G IT-based infrared temperature monitoring Surveillance System Honghu Hybrid System (HHS)	accurate detection of CT and other images in the screening of alleged COVID-19 incidents Enhancing the medical treatment potential for urgent and serious cases Conduct remote health care, body temperature monitoring, sanitization, washing, and medication distribution Used in traffic centers to track passenger conditions in a variety of cities. Monitor social distancing, mask wearing, and body temperature Information on symptoms, psychological status, contact history, social behavior, and the physical environment	[36]
Wearable Biosensor	Patient Arm	physiological parameters for monitoring patients, detecting disease progression, skin temperature, heart rate, respiratory rate, oxygen saturation, perspiration and activity of ambulatory subjects in a 24/7 basis	[23, 32]

media AI can automatically detect them and delete them. AI is also used in developing Robots which further helps in performing an online examination of people and sanitization job. CT scan is produced with the help of this technology which healthcare system used for the detection of pneumonia caused by a virus [30, 31].

Many countries have started different quarantine measures for restricting the movement of asymptomatic individuals having COVID-19 exposure to the community. These individual needs to stay at home with quarantine facilities. Moreover, they were also asked to perform symptom surveillance and fever during those 14 days. However, studies reported that there are 50% infected people who had no fever until the development of full-blown disease [4]. So relying on body temperature only is not sufficient. Therefore, the development of AI-based wearable technology (Biosensors) is used for the continuous monitoring of COVID-19 patient. This topic is not much discussed in the literature but this measure can enhance patient compliance with the monitoring system [32, 33].

3.4 Robotics

The outbreak of COVID-19 made caused a severe workload on medical staff of hospitals. Robotics are good alternatives for providing similar care as a doctor does. According to [34] China had placed 14 Robots into their hospital, which were used to clean and disinfect, deliver medicine and food, measure patient temperature, and also provides entertainment to patients. It is a great alternative to use and decrease the direct interaction between patients and doctors. They can perform the assigned tasks very carefully. They can perform daily work, assist doctors with remote diagnosis, and reduce the fear of spreading disease [34].

During the lockdown, a police Robot placed in different patrolling areas can carefully examine that people are following the orders of lockdown or not. Similarly likely embedded Robots in hospitals looks after the work of doctors. They used to make sure that doctors are performing their duties without any disruption. This technology is very helpful in decreasing the spread of the deadly virus [2].

3.5 Mobile Data (5G, 6G) and Cloud Partnership

For dealing with fast-growing coronavirus, Wuhan builds cabin hospital in its most difficult time. That hospital was built to deal with emergency patients, confirmed with mild symptoms. The main purpose of cabin hospitals is to check patients with oral medicine, mobile computed tomography, and CT scanning and some other functions. The basic purpose of the hospital is to build a wireless medical healthcare management system where services will be provided remotely. Therefore, hospital worked on 5G wireless network system. The wireless router consists of local network unit was configured to get access the internet, address the need of internet access, reports different types of data and also the share of a file between cabin hospital and outside of hospital. They found that 5G network signals are more stable without packet drop, having bandwidth speed 10 time faster than 4G, and intranet bandwidth exceed 50 M and efficiently able to meet the demands placed on the network. The role of B5G or 6G is significant, in delivering faster download speeds, with future possibilities of speeds up to one terabit per second [35].

3.6   Cloud Partnership

Studies indicate that hospitals can use 5G based portal for free diagnosis of COVID-19 and efficient medical care. If there is any limitation in this portal (lack of radiologist), it can be addressed by using CT scan and X-ray with the focus on 5G cloud partnership. 5G cloud partnership is helpful in using the accurate detection of CT and other image during COVID-19 alleged screening. A 5G cloud based smart robots can also be developed for performing some medical tasks. Some cities are already practicing 5G IT based infrared temperature monitoring in different traffic centers for tracking passenger condition [36].

3.7   Internet of Things (IoT)

This advance technology helps in making sure that all the coronavirus infected patients are quarantine. It is also used for proper monitoring of quarantine patients. If a patient found to be at high risk, it can quickly be tackled by using an internet based network. This technology can also help in the reduction of the workload of the medical staff and ultimately their efficiency also improves. Patients can enjoy the benefits of superior treatment services with a small number of mistakes. IoT is considered to be most widely used technology in almost all sectors including health, education and business [37, 38]. This platform is useful to identify a person who is going to contact COVID patient [39]

3.8   Drone Technology

Drone technology can be used to check the proper implementation of quarantine and mask-wearing. Ghana became the very first country in the world who make the use of drone technology for the transfers of COVID-19 test samples to laboratory centers. The process includes the transportation of haul sample taken from the suspected person of remote area into the testing centers of two major cities [40]. Then the results of the test are delivered through short Message Service. This technology makes the haulage of medical supplies to remote clinics possible by decreasing the transportation cost.

3.9   Solar-Powered Automated Handwashing Machine

The continuous pressure of washing hands with soap under running water makes the Ghanaian inventor able to invent hand washing machine. This automated handwashing machine uses solar energy and reduces the risk of self-contamination. This innovation got a certificate from the standard authority for starting mass production. The machine used a remote sensor that allows for a 15 s gap before between the discharge of handwashing soap and water to hands into the barrel. A single machine can be used by a minimum 150 people before refill [40].

3.10   GPS, WiFi and Bluetooth

Many countries including Australia, Canada, Germany, France and United Kingdom have changed the traditional contact tracing system and adopted digital tracing. GPS and WiFi sensors are used to identify the location of the app users and that geolocation information is utilized to dictate the proximity to an infected person. Barcoding is also used for contact tracing these days. China is using this technology for tracking information. Another strategy is the use of Wi-Fi fingerprints which make use of the received signal strength of each Wi-Fi network to create a fingerprint for each location [41, 42].

4   Conclusion

The study provides a thorough insight on the list of technologies which the healthcare system is utilizing in performing multiple tasks. The results are relevant with [43] study. Health sector is using this technology in making customized face masks, gloves and also used them in collecting information about the proper treatment of COVID patients. The study discussed 12 advance technologies which is applied in different areas of healthcare system and which provide relief during the war against the deadly virus. The application of these technologies can give updates on daily basis, about specific region in a specific given time of epidemic. The proper application of advance technologies can provide a lot of innovative ideas and solutions in order to better deal with this virus. Each technology contains its specific characteristics. The generalizability of telemedicine and telehealth made them the most popular term found in the bibliometric analysis of the literature given in Fig. 2. The proper application examples of these technologies can encourage those who are still depending on face to face treatment in this difficult time. The study can also provide a guideline to practitioners on the benefits and application areas of these technologies.

References

1. Chavis, A., Bakken, H., Ellenby, M., Hasan, R.: Coronavirus disease-2019 and telehealth: prevention of exposure in a medically complex patient with a mild presentation. J. Adolesc. Heal. 2019–2021 (2020)
2. Javaid, M., Haleem, A., Vaishya, R., Bahl, S., Suman, R., Vaish, A.: Industry 4.0 technologies and their applications in fighting COVID-19 pandemic. Diabetes Metab. Syndr. Clin. Res. Rev. **14**(4), 419–422 (2020)
3. Reeves, J.J., et al.: Rapid response to COVID-19: Health informatics support for outbreak management in an academic health system. J. Am. Med. Informatics Assoc. **27**(6), 853–859 (2020)
4. Guan, W., et al.: Clinical characteristics of coronavirus disease 2019 in China. N. Engl. J. Med. **382**(18), 1708–1720 (2020)
5. Anthony, B. Jr.: J. Med. Syst. **44**(7), 132. https://www-ncbi-nlm-nih-gov.offcampus.lib. washington.edu/pmc/articles/PMC4762820/ (2020)
6. Ntshalintshali, S.D., Mnqwazi, C.: Affordable digital innovation to reduce SARS-CoV-2 transmission among healthcare workers. South African Med. J. **110**(7), 605–606 (2020)
7. Kumar, A., Gupta, P.K., Srivastava, A.: A review of modern technologies for tackling COVID-19 pandemic. Diabetes Metab. Syndr. Clin. Res. Rev. **14**(4), 569–573 (2020)
8. Arpaci, I., et al.: Analysis of Twitter Data Using Evolutionary Clustering during the COVID-19 Pandemic. Comput. Mater. Contin. **65**(1), 193–204 (2020)
9. Gates, B.: Responding to Covid-19—a once-in-a-century pandemic? N. Engl. J. Med. **382**(18), 1677–1679 (2020)
10. Farrugia, G., Plutowski, R.W.: Innovation lessons from the COVID-19 pandemic. Mayo Clin. Proc. **95**(8), 1574–1577 (2020)
11. Chayomchai, A., Phonsiri, W., Junjit, A., Boongapim, R., Suwannapusit, U.: Factors affecting acceptance and use of online technology in Thai people during COVID-19 quarantine time. Manag. Sci. Lett. **10**(13), 3009–3016 (2020)
12. Pahuja, M.,Wojcikewych, D.: Systems barriers to assessment and treatment of COVID-19 positive patients at the end of life. J. Palliat. Med. **XX**(Xx), 10–12 (2020)
13. Katapally, T.R.: A global digital citizen science policy to tackle pandemics like COVID-19. J. Med. Internet Res. **22**(5), 1–7 (2020)
14. Shen, Y., et al.: Emergency responses to COVID-19 outbreak: experiences and lessons from a General Hospital in Nanjing, China. Cardiovasc. Intervent. Radiol. **43**(6), 810–819 (2020)
15. Moher, D., et al.: Preferred reporting items for systematic reviews and meta-analyses: the PRISMA statement. PLoS Med. **6**(7) (2009)
16. Iyengar, K., Mabrouk, A., Jain, V.K., Venkatesan, A., Vaishya, R.: Learning opportunities from COVID-19 and future effects on health care system. Diabetes Metab. Syndr. Clin. Res. Rev. **14**(5), 943–946 (2020)
17. Kitchin, R.: Civil liberties or public health, or civil liberties and public health? Using surveillance technologies to tackle the spread of COVID-19. Sp. Polity 1–20 (2020)
18. Leite, H., Hodgkinson, I.R., Gruber, T.: New development: 'Healing at a distance'— telemedicine and COVID-19. Public Money Manag. **40**(6), 483–485 (2020)
19. Grossman, S.N., et al.: Rapid implementation of virtual neurology in response to the COVID-19 pandemic. Neurology **94**(24), 1077–1087 (2020)

20. M. V., S. T., A. C.: Using telemedicine during the COVID-19 pandemic. Indian Pediatr (2020). http://www.embase.com/search/results?subaction=viewrecord&from=export&id= L631770422
21. Iyengar, K., Jain, V.K., Vaishya, R.: Pitfalls in telemedicine consultations in the era of COVID 19 and how to avoid them. Diabetes Metab. Syndr. Clin. Res. Rev. **14**(5), 797–799 (2020)
22. Smith, W.R., Atala, A.J., Terlecki, R.P., Kelly, E.E., Matthews, C.A.: Implementation guide for rapid integration of an outpatient telemedicine program during the COVID-19 pandemic. J. Am. Coll. Surg. **231**(2), 216–222.e2 (2020)
23. Ghai, S.: Teledentistry during COVID-19 pandemic. Diabetes Metab. Syndr. Clin. Res. Rev. **14**(5), 933–935 (2020)
24. Nagra, M., Vianya-estopa, M., Wol, J.S.: Since January 2020 Elsevier has created a COVID-19 resource centre with free information in English and Mandarin on the novel coronavirus COVID- 19. The COVID-19 resource centre is hosted on Elsevier Connect, the company' s public news and information, January (2020)
25. Calton, B., Abedini, N., Fratkin, M.: Telemedicine in the time of coronavirus. J. Pain Symptom Manage. **60**(1), e12–e14 (2020)
26. Cohen, B.H., Busis, N.A., Ciccarelli, L.: Coding in the world of COVID-19: non-face-to-face evaluation and management care. Continuum (Minneap. Minn) **26**(3), 785–798 (2020)
27. Viera-Artiles, J., Valdiande, J.J.: 3D-printable headlight face shield adapter: personal protective equipment in the COVID-19 era. Am. J. Otolaryngol. Head Neck Med. Surg. (May), 102576 (2020)
28. Ishack, S., Lipner, S.R.: Applications of 3D printing technology to address COVID-19—related supply shortages. Am. J. Med. **133**(7), 771–773 (2020)
29. Mortazavi, S.A.R., Mortazavi, S.M.J., Parsaei, H.: COVID-19 pandemic: how to use artificial intelligence to choose non-vulnerable workers for positions with the highest possible levels of exposure to the novel coronavirus. J. Biomed. Phys. Eng. **10**(3), 383–386 (2020)
30. Coombs, C.: Will COVID-19 be the tipping point for the intelligent automation of work? A review of the debate and implications for research. Int. J. Inf. Manage. (June), 102182 (2020)
31. Cuffaro, L., Di Lorenzo, F., Bonavita, S., Tedeschi, G., Leocani, L., Lavorgna, L.: Dementia care and COVID-19 pandemic: a necessary digital revolution. Neurol. Sci. **41**(8), 1977–1979 (2020)
32. Wong, C.K., et al.: Artificial intelligence mobile health platform for early detection of COVID-19 in quarantine subjects using a wearable biosensor: protocol for a randomised controlled trial. BMJ Open **10**(7), e038555 (2020)
33. Ke, Y.Y., et al.: Artificial intelligence approach fighting COVID-19 with repurposing drugs. Biomed. J. xxxx, 1–8 (2020)
34. Zeng, Z., Chen, P.J., Lew, A.A.: From high-touch to high-tech: COVID-19 drives robotics adoption. Tour. Geogr. **22**(3), 724–734 (2020)
35. Zhou, B., Wu, Q., Zhao, X., Zhang, W., Wu, W., Guo, Z.: Construction of 5G all-wireless network and information system for cabin hospitals. J. Am. Med. Informatics Assoc. **27**(6), 934–938 (2020)
36. Gong, M., Liu, L., Sun, X., Yang, Y., Wang, S., Zhu, H.: Cloud-based system for effective surveillance and control of COVID-19: useful experiences from Hubei, China. J. Med. Internet Res. **22**(4), 1–9 (2020)
37. Kertesz, A., Varadi, Sz., Gultekin Varkonyi, G.: Legal Issues of IoT Services using Fogs and Clouds. Springer (2020)
38. Al-Emran, M., Al-Maroof, R., Al-Sharafi, M.A., Arpaci, I.: What impacts learning with wearables? An integrated theoretical model. Interact. Learn. Environ. 1–21 (2020)
39. Nicol, G.E., Piccirillo, J.F., Mulsant, B.H., Lenze, E.J.: Action at a distance: geriatric research during a pandemic. J. Am. Geriatr. Soc. **68**(5), 922–925 (2020)
40. Sibiri, H., Zankawah, S.M., Prah, D.: Coronavirus diseases 2019 (COVID-19) response: highlights of Ghana's scientific and technological innovativeness and breakthroughs. Ethics, Med. Public Heal. **14**(May), 0–5 (2020)

41. Kleinman, R.A., Merkel, C.: Digital contact tracing for covid-19. CMAJ **192**(24), E653–E656 (2020). https://doi.org/10.1503/cmaj.200922
42. Cheng, W., Hao, C.: Case-initiated COVID-19 contact tracing using anonymous notifications. JMIR mHealth uHealth **8**(6), e20369 (2020)
43. Singh, R.P., Javaid, M., Kataria, R., Tyagi, M., Haleem, A., Suman, R.: Significant applications of virtual reality for COVID-19 pandemic. Diabetes Metab. Syndr. Clin. Res. Rev. **14**(4), 661–664 (2020)
44. Vilendrer, S., et al.: rapid deployment of inpatient telemedicine in response to COVID-19 across three health systems. J. Am. Med. Informatics Assoc. **27**(7), 1102–1109 (2020)
45. C.S. Report: Phthiriasis Palpebrarum Presenting as Anterior Blepharitis. Indian J. Public Health **62**(3), 2018–2020
46. Martínez-García, M., et al. (2020) Tracing of COVID-19 patients by telemedicine with telemonitoring. Rev. Clínica Española (English Ed. (2020)
47. Huri, E., Hamid, R.: Technology-based management of neurourology patients in the COVID-19 pandemic: is this the future? A report from the International Continence Society (ICS) institute. Neurourol. Urodyn. **39**(6), 1885–1888 (2020)
48. Rismiller, K., Cartron, A.M., Trinidad, J.C.L.: Inpatient teledermatology during the COVID-19 pandemic. J. Dermatolog. Treat. **31**(5), 441–443 (2020)
49. Bini, S.A., et al.: Digital orthopaedics: a glimpse into the future in the midst of a pandemic. J. Arthroplasty **35**(7), S68–S73 (2020)
50. Evans, K.D., Yang, Q., Liu, Y., Ye, R., Peng, C.: Sonography of the lungs: diagnosis and surveillance of patients with COVID-19. J. Diagnostic Med. Sonogr. **36**(4), 370–376 (2020)
51. Negrini, S., et al.: Telemedicine from research to practice during the pandemic. 'Instant paper from the field' on rehabilitation answers to the COVID-19 emergency. Eur. J. Phys. Rehabil. Med. **56**(3), 327–330 (2020)
52. G. Rh, G. J, H. R, S. Mf: Orthopaedic forum. Bone **58**, 429–437 (2008)
53. Vandekerckhove, P., Vandekerckhove, Y., Tavernier, R., De Jaegher, K., de Mul, M.: Leveraging user experience to improve video consultations in a cardiology practice during the COVID-19 pandemic: initial insights. J. Med. Internet Res. **22**(6), e19771 (2020)
54. Grange, E.S., et al.: Responding to COVID-19: the UW medicine information technology services experience. Appl. Clin. Inform. **11**(2), 265–275 (2020)
55. Rastogi, S., Singh, N., Pandey, P.: Telemedicine for Ayurveda consultation: Devising collateral methods during the COVID-19 lockdown impasse. J. Ayurveda Integr. Med. xxxx, 5–7 (2020)
56. Tashkandi, E., et al.: Virtual management of patients with cancer during the COVID-19 pandemic: web-based questionnaire study. J. Med. Internet Res. **22**(6), e19691 (2020)
57. Callahan, C.J., et al.: Open development and clinical validation of multiple 3D-printed nasopharyngeal collection swabs: rapid resolution of a critical COVID-19 testing bottleneck. J. Clin. Microbiol. **58**(8), 1–10 (2020)
58. Chaturvedi, S., Gupta, A., Krishnan, V.S., Bhat, A.K.: Design, usage and review of a cost effective and innovative face shield in a tertiary care teaching hospital during COVID-19 pandemic. J. Orthop. **21**(May), 331–336 (2020)
59. Singh, S., Prakash, C., Ramakrishna, S.: Three-dimensional printing in the fight against novel virus COVID-19: technology helping society during an infectious disease pandemic. Technol. Soc. **62**(June), 101305 (2020)
60. Ben Pan, X.: Application of personal-oriented digital technology in preventing transmission of COVID-19, China. Ir. J. Med. Sci. 2318 (2020)
61. Smith, C.D., Mennis, J.: Incorporating geographic information science and technology in response to the COVID-19 pandemic. Prev. Chronic Dis. **17**, E58 (2020)
62. Vokinger, K.N., Nittas, V., Witt, C.M., Fabrikant, S.I., von Wyl, V.: Digital health and the COVID-19 epidemic: an assessment framework for apps from an epidemiological and legal perspective. Swiss Med. Wkly. **150**(May), w20282 (2020)

63. Wherton, J., Shaw, S., Papoutsi, C., Seuren, L., Greenhalgh, T.: Guidance on the introduction and use of video consultations during COVID-19: important lessons from qualitative research. BMJ Lead. 1–5 (2020)

64. Xie, B., Charness, N., Fingerman, K., Kaye, J., Kim, M.T., Khurshid, A.: When going digital becomes a necessity: ensuring older adults' needs for information, services, and social inclusion during COVID-19. J. Aging Soc. Policy **32**(4–5), 460–470 (2020)

65. Li, Y., et al.: Current treatment approaches for COVID-19 and the clinical value of transfusion-related technologies. Transfus. Apher. Sci. (May), 102839 (20200

66. Hart, J.L., Turnbull, A.E., Oppenheim, I.M., Courtright, K.R.: Family-centered care during the COVID-19 era. J. Pain Symptom Manage. **60**(2), e93–e97 (2020)

67. Maese, J.R., Seminara, D., Shah, Z., Szerszen, A.: What a difference a disaster makes: the telehealth revolution in the age of COVID-19 pandemic. Am. J. Med. Qual. 1–3 (2020)

68. Ros, M., Neuwirth, L.S.: Increasing global awareness of timely COVID-19 healthcare guidelines through FPV training tutorials: Portable public health crises teaching method. Nurse Educ. Today **91**(May), 104479 (2020)

69. Madden, N., et al.: Telehealth uptake into prenatal care and provider attitudes during the COVID-19 pandemic in New York City: a quantitative and qualitative analysis. Am. J. Perinatol. **37**(10), 1005–1014 (2020)

70. Poncette, A.S., et al.: Improvements in patient monitoring in the intensive care unit: survey study. J. Med. Internet Res. **22**(6), e19091 (2020)

71. Rubulotta, F., et al.: Technologies to optimize the care of severe COVID-19 patients for health care providers challenged by limited resources. Anesth. Analg. **131**(2), 351–364 (2020)

72. Mercadante, S., Adile, C., Ferrera, P., Giuliana, F., Terruso, L., Piccione, T.: Palliative care in the time of COVID-19. J. Pain Symptom Manage. **60**(2), e79–e80 (2020)

73. Chae, S.H., Kim, Y., Lee, K.S., Park, H.S.: Development and clinical evaluation of a web-based upper limb home rehabilitation system using a smartwatch and machine learning model for chronic stroke survivors: prospective comparative study. JMIR mHealth uHealth **8** (7), e17216 (2020)

74. Puro, N.A., Feyereisen, S.: Telehealth availability in US Hospitals in the face of the COVID-19 pandemic. J. Rural Heal. **00**, 1–7 (2020)

75. Roy, B., et al.: Teleneurology during the COVID-19 pandemic: a step forward in modernizing medical care. J. Neurol. Sci. **414**(May), 116930 (2020)

76. Vaishya, R., Javaid, M., Khan, I.H., Haleem, A.: Artificial Intelligence (AI) applications for COVID-19 pandemic. Diabetes Metab. Syndr. Clin. Res. Rev. **14**(4), 337–339 (2020)

77. Sechi, G.M., et al.: Business intelligence applied to emergency medical services in the lombardy region during sars-COV-2 epidemic. Acta Biomed. **91**(2), 39–44 (2020)

78. Naik, B.N., Gupta, R., Singh, A., Soni, S.L., Puri, G.D.: Real-time smart patient monitoring and assessment amid COVID-19 Pandemic—an alternative approach to remote monitoring. J. Med. Syst. **44**(7), 2–3 (2020)

79. Swayamsiddha, S., Mohanty, C.: Application of cognitive Internet of Medical Things for COVID-19 pandemic. Diabetes Metab. Syndr. Clin. Res. Rev. **14**(5), 911–915 (2020)

80. Singh, R.P., Javaid, M., Haleem, A., Suman, R.: Internet of things (IoT) applications to fight against COVID-19 pandemic. Diabetes Metab. Syndr. Clin. Res. Rev. **14**(4), 521–524 (2020)

81. Iyengar, K., Upadhyaya, G.K., Vaishya, R., Jain, V.: COVID-19 and applications of smartphone technology in the current pandemic. Diabetes Metab. Syndr. Clin. Res. Rev. **14** (5), 733–737 (2020)

82. Yan, A., Zou, Y., Mirchandani, D.A.: How hospitals in mainland China responded to the outbreak of COVID-19 using IT-enabled services: an analysis of hospital news webpages. J. Am. Med. Inform. Assoc. **27**(May), 991–999 (2020)

83. Almeida, B. de A., et al.: Personal data usage and privacy considerations in the covid-19 global pandemic. Cienc. e Saude Coletiva **25**, 2487–2492 (2020)

Mid-Term Forecasting of Fatalities Due to COVID-19 Pandemic: A Case Study in Nine Most Affected Countries

Sneha Rai and Mala De

Abstract The outbreak of COVID-19 pandemic has presented the entire world with an unrivalled challenge of public health leaving a remarkable impact on the social, economic, and financial lives of humanity. Though a major portion of the globe is under lockdown due to this deadly virus, the number of causalities is still growing rapidly. Therefore, it is very important to predict the number of infected and fatality cases for the future to overcome the consequences, save the lives of people, and plan accordingly. This paper proposes a data-driven analysis based on univariate Holt's double exponential smoothening method with parameter optimization and polynomial curve fitting technique for one month ahead forecasting of the death cases in nine profoundly affected countries across the world namely India, USA, Italy, UK, China, France, Iran, Spain, and Germany. In contrast to the complex deep learning-based predictors, the proposed Holt's model is simple yet efficient enough to give outstanding prediction performance for all the countries under this study and can be further used for the prediction of infections and causalities for the rest of the countries in future. The future estimation of the number of death cases will act as a beneficial tool for the successful allocation of the medical resources and as an early warning to the policymakers and health officials as well as to the residents of the country to boost their self-awareness.

Keywords Coronavirus · Fatality rate · Mid-term forecasting · Optimized Holt's model · Polynomial curve fitting method

1 Introduction

Coronavirus disease (COVID-19) which was first identified in December 2019 in Wuhan, China is declared as a global pandemic by WHO on March 11. COVID-19 pandemic is the highest global crisis faced by the entire world since World War-II.

S. Rai (✉) · M. De
National Institute of Technology Patna, Patna, Bihar 800005, India
e-mail: sneharai0212@gmail.com

© The Author(s), under exclusive license to Springer Nature Switzerland AG 2021
I. Arpaci et al. (eds.), *Emerging Technologies During the Era of COVID-19 Pandemic*, Studies in Systems, Decision and Control 348,
https://doi.org/10.1007/978-3-030-67716-9_12

This pandemic is moving like a wave all around the globe and as of May 21, 2020, the outbreak has resulted in 5,105,902 infected cases with 330,004 reported deaths and 2,035,445 recovered cases worldwide [1–3]. 215 countries or territories around the world have been affected so far by this deadly virus. Apart from the health crisis, COVID-19 has taken away the peace of the people by devastating the social, economic, and political crisis in every country under its effect [4]. The whole world is competing to reduce the havoc created with the spread of the virus by slowing down the contamination rate through increased testing and treatment of the patients, promoting self prevention and social distancing, quarantining people, contact tracing, stopping gatherings, etc. [5]. Social media platforms such as Facebook, Instagram, YouTube, and Twitter allow the spread of an enormous amount of fake information and rumors which manipulate the human mind [6]. A self-report instrument is developed in [7] to test the different psychological problems such as fear, panic, or phobia experienced by the people during this pandemic situation. The epicenter of the new crisis and death toll for COVID-19 has shifted in several months from China initially to Europe and then to USA [8–10]. The rapid increase in the contagion and deaths caused by COVID-19 is putting a significant load on the health care system of every highly affected country in the world [11]. Thus, it has become cutting edge research to develop the future trends of the novel COVID-19 causalities at this moment.

The present study is therefore motivated to develop a prediction model which can accurately predict the daily death toll in nine most affected countries like India, USA, Italy, UK, China, France, Iran, Spain, and Germany and could be used to predict the number of infections and causalities that will occur in future around the world. This prediction is extremely required by the government to develop a proper health care structure that is highly effected regions to minimize the damage of mankind due to this deadly virus by utilising the available resources in the best possible way. For building the prediction model, many conventional and modern forecasting techniques are available whose prediction accuracy mainly depends on the data availability and the reliability of the selection of different attributes used for forecasting [1, 12–15]. Depending upon the length of the forecast interval, forecasting may be classified as (i) short-term (ii) mid-term, and (iii) long-term. In this study, a univariate time series method has been proposed to generate forecasts of death cases in multiple countries for a few weeks to a month ahead which comes under midterm forecasting. Since the variation of the COVID-19 cases is highly non-linear in nature, therefore the decisions based on the linear time series forecasting methods like ARIMA [16, 17] and regression techniques [18–20] would be highly crucial. In many studies, the wavelet-based prediction model [21, 22] has shown good prediction accuracy for non-stationary data but fine analysis, it becomes computationally intensive and it becomes a tedious job to select the proper wavelet for the specific purpose and implement it correctly. A population model developed in [23] for healthy and infected people using fractional-order derivatives identify free immigration as an important factor to increase the infection of the current outbreak. In [24], it has been demonstrated that the predictions using more complex models (like SEIR) may not be more dependable as compared to the simpler models.

This work proposes a Holt's double exponential smoothening model which is a univariate time series method and is independent of the other attributes affecting the forecasts and performs direct prediction of the data by considering the historical data as model input [25–27]. The added advantage of Holt's model is that the smoothening coefficients representing the level, trend, and seasonal component of the data can be optimized to get the high prediction accuracy of the model. Unlike the advanced deep learning-based methods such as long short-term memory (LSTM) and convolution neural network (CNN) [15], the Holt's method with optimized parameters gives accurate and consistent prediction results for all the countries under this study with very less computational time and memory space requirement. For establishing the superiority of the proposed model, the results obtained by Holt's model are compared with the prediction results of the polynomial curve fitting model which is a basic mathematical technique to find a mathematical function that can be used to model a data [28]. In [29], the polynomial curve fitting model has been used to forecast energy production by the renewable sources and it has been found better than the linear regression model in terms of accuracy and model fit. It is a simple and basic method of forecasting but has one limitation that the prediction does not take into account the significant fluctuations in the historical data (like trend and seasonality factors) [30], and hence it is more accurate for short-term forecasts and may not yield good prediction performance for long-term forecasts. The suitability of the proposed Holt's model for nine profoundly affected countries signifies that the model will be applicable to any country for short-term, mid-term, and long term predictions of the cases related to infections and deaths due to COVID-19 pandemic.

The rest of the paper is organised as follows: Sect. 2 introduces the proposed prediction methods for one month ahead (mid-term) prediction of COVID-19 death cases for India, USA, Italy, UK, France, China, Iran, Spain, and Germany. The country-wise dataset description and their temporal variation are given in Sect. 3. Section 4 presents the results and discussion along with the comparison of the two prediction models discussed in Sect. 2. Finally, Sect. 5 concludes the paper.

2   Methodology

This work proposes the Holt's Double Exponential Smoothening method with optimized coefficients and Polynomial fitting method to predict the total number of deaths due to COVID-19 across the nine most affected countries across the world. The complete forecasting procedure is shown in Fig. 1. The data for daily death cases is collected for nine most affected countries namely India, USA, Italy, UK, France, China, Spain, Iran, and Germany. Now, as the rise in the daily death cases in all the affected countries is extremely high, therefore a suitable prediction model is required to predict the causalities (deaths, infections, etc.) which can occur in the future at a faster rate and higher accuracy. Due to this reason, the univariate time series prediction method is used here which requires only the past input data of the variable to be predicted and takes very less processing time to provide precise results. The entire dataset is divided into two parts: training and testing dataset.

The models are trained by the past data and the prediction results obtained are compared to select the best prediction model for all the countries. Finally, the selected model is used for one month ahead prediction of death cases in all the nine countries as mentioned above.

2.1 *Holt's Exponential Smoothening Method*

Holt's double exponential method is a univariate time series forecasting method used to predict the data having a trend. This method depends upon two smoothening coefficients, one for the level component and the other for the trend component. The mathematical equations describing the level and trend component are given below:

$$\text{Level} : A_t = \alpha \times y_t + (1 - \alpha) \times (A_{t-1} + T_{t-1}) \tag{1}$$

$$\text{Trend} : T_t = \beta \times (A_t - A_{t-1}) + (1 - \beta) \times T_{t-1} \tag{2}$$

Fig. 1 The stages of forecasting

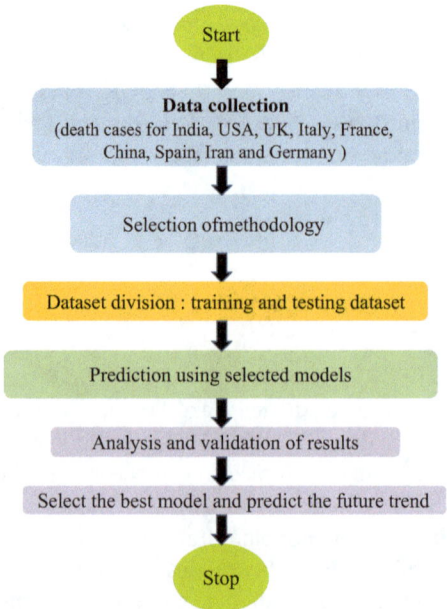

The final forecast is the sum of the level and trend component of the predicted data. The final forecast equation can be written as:

$$\text{Forecast}: F_{t+m} = A_t + T_t \times m \tag{3}$$

The initial value of the level and trend component can be calculated using the following equations:

$$A_1 = y_1, \quad T_1 = \frac{(y_2 - y_1) + (y_4 - y_3)}{2} \tag{4}$$

where; α and β are the level and trend smoothening coefficients, A_t is the level component, T_t is the trend component, y_t is actual load, t is time period, F_{t+m} is the forecasted load for m periods ahead.

2.2 Polynomial Curve Fitting

The polynomial curve fitting method is a simple mathematical technique to model a data having a non-linear trend by assigning the best fit curve along with the entire range of the data spread. The polynomial function which fits the data more accurately is decided using the least square method that minimizes the sum of residuals of the actual and the plotted points. Initially, polynomial interpolation is performed to find the lowest degree polynomial which passes through the maximum given data points, and the obtained polynomial is extrapolated to estimate the data beyond the actual range of the observed data. The polynomial with the nth degree can be described by the given mathematical equation:

$$p = p_1 x^n + p_2 x^{n-1} + p_3 x^{n-2} + \cdots + p_n + p_{n+1} \tag{5}$$

where; $p(x)$ is the polynomial of degree n, $p_1, p_2, \ldots, p_{n+1}$ are the polynomial coefficients and the length of polynomial p is $n + 1$. The prediction performance of the model is determined by the different performance parameters discussed below.

2.3 Performance Parameters

The prediction accuracy of the developed model is determined using the mean absolute percentage error (MAPE), root mean square error (RMSE), and the Pearson correlation coefficient (r). The mathematical formulas for the calculation of all the three metrics are given below:

$$\text{MAPE} = \frac{1}{N} \left| \sum_{i=1}^{N} \frac{(y_i - \tilde{y}_i)}{y_i} \right| \times 100 \tag{6}$$

where; N is the number of testing data samples, y_i is actual data and $\tilde{y}_i$ is predicted data.

The RMSE between the actual and the predicted data is calculated as:

$$\text{RMSE} = \sqrt{\frac{\sum_{i=1}^{N} (y_i - \tilde{y}_i)^2}{N}} \tag{7}$$

Pearson correlation coefficient (r) is used to determine the relationship between the actual and predicted variable. It is calculated as:

$$r_{xy} = \frac{n \sum x_i y_i - \sum x_i \sum y_i}{\sqrt{\left(n \sum x_i^2 - (\sum x_i)^2\right)} \sqrt{\left(n \sum y_i^2 - (\sum y_i)^2\right)}} \tag{8}$$

where r_{xy} is the Pearson correlation coefficient between x and y, n is the number of observations used for testing, x_i and y_i are the values of x and y for the ith observation.

3 Case Study

This section describes the datasets used for testing and validating the proposed prediction models. This paper deals with the daily death cases due to COVID-19 in the nine most affected countries across the world namely India, USA, Italy, China, UK, Spain, France, Iran, and Germany. The proposed Holt's model is trained by the daily death cases used as input and the prediction results are compared with the polynomial model to obtain the best-suited prediction model for the given test system. The dataset consists of 218 daily observations ranging from 31/12/19 to 04/08/20. Here, 85% (188 data points) of the total data is used for training, and the remaining 15% (30 data points) is used for testing and validation of the model. The temporal variation of death cases of ten countries is shown in Fig. 2. From Fig. 2 it can be seen that the number of deaths has increased exponentially from around March 2020 for all the shown countries except China, as it has entered the recovery phase.

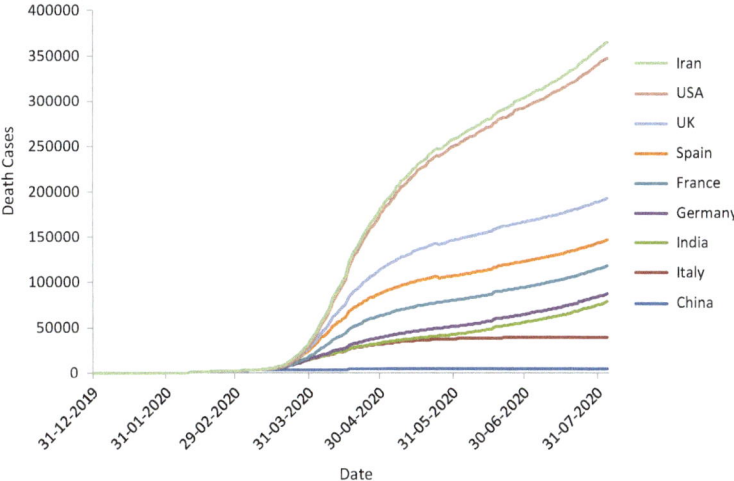

Fig. 2 Temporal variation of death cases

4 Results and Discussion

The Holt's double exponential smoothening method with optimized parameters is used to predict the future death cases due to COVID-19 in nine most affected countries of the world. The performance of the optimized Holt's model is compared with the fundamental statistical polynomial extrapolation method to determine the most accurate prediction with the least errors.

Optimized Holt's method

Holt's method is a univariate method for the prediction of the temporal data. It comprises three components: level, trend, and seasonality. In this paper, Holt's double exponential smoothening method is used to predict the data having a level and trend components. Two smoothening coefficients α and β (0–1) and the three Eqs. (1), (2) and (3) mentioned in Sect. 2.1 are used to develop the prediction model. Initially, the value of level and trend coefficients α and β are taken as 0.5 each and the corresponding MAPE and RMSE values are computed for all the testing scenarios. To improve the prediction accuracy of the model, the coefficients α and β are optimized to give the minimum MAPE value. The objective function is mentioned below:

$$\text{Minimize MAPE}(\alpha, \beta) = \left| \sum_{(i=1)}^{N} \frac{y_i - \tilde{y}_i}{y_i} \right| \times 100 \qquad (9)$$

where y_i is the actual data, $\tilde{y}_i$ is the predicted data and N is the number of testing samples used. In this work, Generalized Reduced Gradient (GRG) nonlinear solver is used for parameter optimization. Initially, the solver calculates the objective function value which is MAPE by taking the parameters α and β as 0.5. Now, as the input value changes, the GRG nonlinear solver calculates the slope of the objective function and finds the optimum solution when the partial derivatives become 0. Finally, it calculates the optimum value of the smoothening coefficients which yields the minimum MAPE and RMSE for all the given inputs. The optimized values of α and β and the corresponding MAPE and RMSE values before and after are given in Table 1.

It can be observed that the performance of Holt's model in terms of MAPE and RMSE is improved many times after the optimization of the smoothening parameters (α and β) for each country under study. The comparison of the prediction performance by Holt's model and polynomial curve fitting method is given in Table 2. The degree of the polynomial (n) is the minimum degree which gives the least computed error or there is no significant decrease in the error as the degree is increased. The results show that the prediction performance of Holt's method with optimized parameters is far better than the statistical polynomial curve fitting method in all the cases. This proves the accuracy of the optimized Holt's model over the polynomial model for the non-linear temporal testing data. The MAPE for all the nine mentioned countries is reduced by 80% by using Holt's model as observed from Table 2. The minimum and maximum MAPE obtained by Holt's model is between 0-3% whereas the polynomial fitting model is around 5–25% in all the cases which again signifies that the predicted values are very close to the actual data resulting in greater prediction accuracy and high goodness of fit.

The real-time prediction plots showing the comparison of the model forecasts for all the countries under study are shown in Fig. 3. The plots indicate that the predicted data is much closer to the observed data points in the case of Holt's model in all the nine countries used for this study. The red line for Holt's model in the plots indicates the least deviation of the predicted data from the actual data for all the nine countries as visualized from (i) to (viii) of Fig. 3. It can be seen from the comparison plot of China in Fig. 3(iv), that the prediction curve is plotted for the

Table 1 Forecasting using Holt's model

S. No.	Country	Before optimization				After optimization			
		α	β	MAPE	RMSE	α	β	MAPE	RMSE
1	India	0.5	0.5	7.87	3326.80	0.15	1.0	2.21	767.43
2	Italy	0.5	0.5	0.46	194.64	0.14	0.23	0.34	148.22
3	USA	0.5	0.5	0.99	2189.82	0.36	1.0	0.71	1261.93
4	UK	0.5	0.5	1.73	924.58	0.08	0.22	0.91	100.21
5	China	0.5	0.5	0.18	12.88	1.0	0.06	0.082	5.32
6	France	0.5	0.5	0.30	112.13	0.67	1.0	0.75	286.83
7	Iran	0.5	0.5	3.74	734.06	0.18	1.0	1.89	408.55
8	Spain	0.5	0.5	0.54	178.08	0.99	0.85	0.11	51.73
9	Germany	0.5	0.5	0.67	73.76	0.14	1.0	0.382	42.88

Table 2 Prediction performance by different models

S. No	Country	Optimized Holt's model			Polynomial fitting model			
		MAPE	RMSE	r	MAPE	RMSE	r	n
1.	India	2.21	767.43	0.990	2.45	1095.72	0.986	3
2.	Italy	0.34	148.22	0.989	24.61	8.84E + 03	0.982	1
3.	USA	0.71	1261.93	0.997	7.37	1.04E + 04	0.988	3
4.	UK	0.91	100.21	0.992	6.82	3.72E + 03	0.977	3
5.	China	0.082	5.32	0.989	14.81	805.32	0.974	4
6.	France	0.75	286.83	0.980	22.26	7.92E + 03	0.968	2
7.	Iran	1.89	408.55	0.995	4.90	9.95E + 03	0.991	2
8.	Spain	0.11	51.73	0.985	10.15	2.62E + 03	0.977	2
9.	Germany	0.382	42.88	0.988	13.78	1.11E + 03	0.970	3

n = degree of the polynomial

entire range of data including the training and testing dataset i.e. total 136 data points, unlike other countries where the prediction plot is shown for the testing dataset only. This is because of the entire different trend in the variation of death cases in China as it has entered into the recovery phase when all other countries are facing the exponential rise in the death cases, this can also be proved from Fig. 2 which shows the time-series variation of death cases for different countries. The comparative MAPE values obtained by both the prediction models for the nine most affected countries can also be visualized by the box plot shown in Fig. 4. The two boxes in the plot show the range of MAPE values obtained for all the nine countries using Holt's and Polynomial prediction models. The red line in each box represents the median of the MAPE values. From the two boxes of the box plot, it can be observed that the MAPE values are lower and in a comparable range in the case of Holt's model as compared to the polynomial model of prediction. For Holt's model, the MAPE is in the range of 0–3% whereas, for the polynomial model, the range of MAPE is between 5 and 25% which concludes that the Holt's model with optimized parameters gives the high prediction accuracy for all the nine countries tested under this study and hence can be used further for prediction of various causalities across the whole world due to this pandemic.

The one month ahead prediction of death cases in all the nine most affected countries mentioned above using Optimized Holt's double exponential smoothening method is shown in Fig. 5.

As a case study, the actual and predicted death counts in all the nine countries for two days (15/07/20 and 08/08/20) are given in Table 3. The data of 15/07/20 is used for testing the proposed model and 08/08/20 is the independent out of sample data for which the prediction has been done. It has been found that the death count predicted by the proposed method is very close to that of the actual death count irrespective of the country under observation. On an average, the accuracy for the

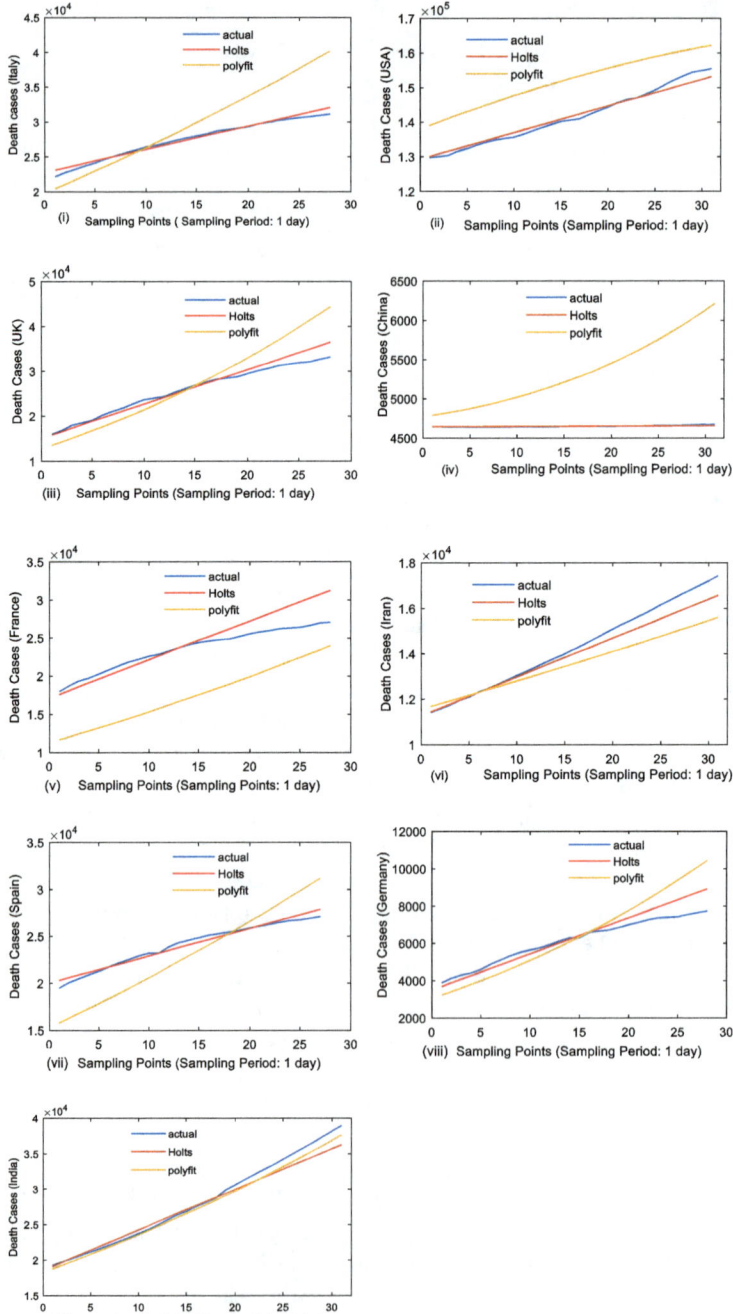

Fig. 3 Real-time forecasts (for 30 days) of Holt's and Polynomial curve fitting model for 9 countries namely (i) Italy, (ii) USA, (iii) UK, (iv) China, (v) France, (vi) Iran, (vii) Spain, (viii) Germany, and (ix) India

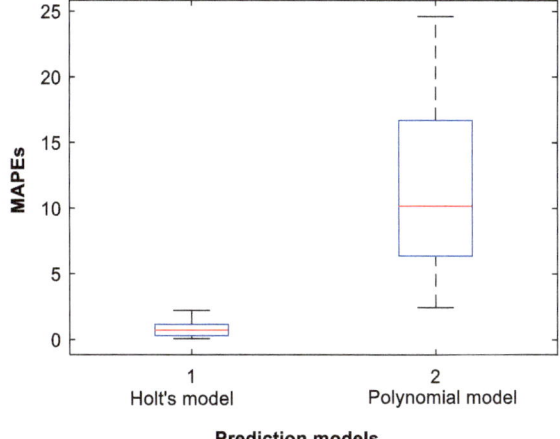

Fig. 4 Box plot showing MAPE values using two prediction models

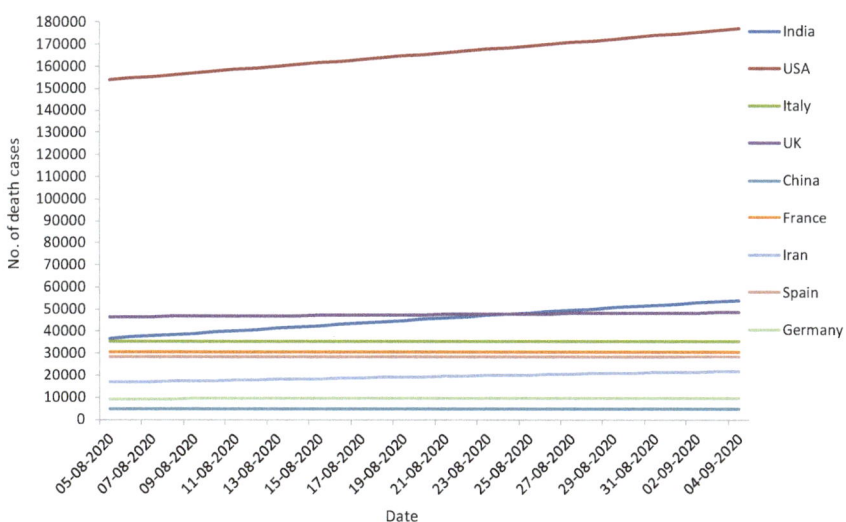

Fig. 5 One month ahead forecasts of death cases using Holt's model

two mentioned dates as per Table 3 is 99.57% and 97.85% respectively. This confirms that the proposed model gives remarkably good prediction results for the in sample (testing) as well as for out of sample future data for various countries under this study.

Table 3 Comparison between actual and predicted death counts using Proposed Holt's model

Country	Death count on 15-July-2020		Death count on 08-August-2020	
	Predicted	Actual	Predicted	Actual
India	24,792	24,309	38,515	42,518
Italy	34,977	34,984	35,226	35,190
USA	137,686	136,466	156,183	161,356
UK	44,931	44,968	46,591	46,511
China	4646	4642	4657	4681
France	30,032	30,029	30,377	30,324
Iran	13,138	13,211	17,250	18,132
Spain	28,410	28,413	28,464	28,503
Germany	9089	9071	9259	9195
Average prediction accuracy (%)	99.57		97.85	

5 Conclusion

This paper proposes a new prediction technique for forecasting of fatalities caused due to the outbreak of COVID-19 pandemic using optimized Holt's double exponential smoothening method. The model is developed using the datasets collected from the nine most affected countries around the world namely India, USA, UK, Italy, China, France, Spain, Iran, and Germany. The accuracy of the proposed method is evaluated by training the model using the past 188 daily death cases data and tested for 30 days ahead prediction within the sample which was further used to forecast one month ahead out of sample death cases in nine countries mentioned above. The suitability of the proposed method is validated through the comparison of forecasts obtained by the polynomial curve fitting method. The average MAPE for all the given countries calculated by Holt's method is around 1% and by Polynomial fitting model is around 12%, which concludes that the Holt's model gives the best forecasting performance for all different scenarios of input datasets. The proposed method has been used to estimate the total death counts for a future date (08/08/20) with an average accuracy close to 98%. As the model performs well in the mid-term, the same work can be extended for forecasting the number of infections and death cases in all the countries for a mid-term, short-term, and long-term period too. The prediction of fatalities can help in the social, economical, and financial planning of the governments in advance so that the least possible devastation of lives can occur.

Acknowledgements This work is supported by the Science & Engineering Research Board, Department of Science and Technology, Government of India, under grants number ECR/2017/001027.

References

1. Chakraborty, T., Ghosh, I.: Real-time forecasts and risk assessment of Novel Coronavirus (COVID-19) cases: a data-driven analysis. Chaos, Solitons Fractals **135**, 109850 (2020)
2. Saez, M., Tobias, A., Varga, D., Barcelo, M.A.: Effectiveness of the measures to flatten the epidemic curve of COVID-19: the case of Spain. Sci. Total Env. **727**, 138761 (2020)
3. Wang, C., Horby, P.W., Hayden, F.G., Gao, G.F.: A novel Coronavirus outbreak of global health concern. Lancet **395**(10223), 470–473 (2020)
4. Liu, L.: Emerging study on the transmission of the Novel Coronavirus (COVID-19) from urban perspective: evidence from China. Cities **103**, 102759 (2020)
5. Zhao, Z., Li, X., Liu, F., Zhu, G., Ma, C., Wang, L.: Prediction of the COVID-19 spread in African countries and implications for prevention and control: a case study in South Africa, Egypt, Algeria, Nigeria, Senegal and Kenya. Sci. Total Env. **729**, 138959 (2020)
6. Arpaci, I., Ishehabi, S., Al-Emran, M., Khasawneh, M., Mahariq, I., Abdeljawad, T., Hassanien, A.E.: Analysis of twitter data using evolutionary clustering during the COVID-19 pandemic. In: Cmc-Computers Materials & Continua (2020). Accessed 13 Sep 2020. https://doi.org/10.32604/cmc.2020.011489
7. Arpaci, I., Karataş, K., Baloğlu, M.: The development and initial tests for the psychometric properties of the COVID-19 Phobia Scale (C19P-S). Personality Individ. Differ. **164**, 110108 (2020). https://doi.org/10.1016/j.paid.2020.110108
8. Reis, R.F., Quintela, B.M., Campos, J.O., Gomes, J.M., Rocha, B.M., Lobosco, M., Santos, R.W.: Characterization of the COVID-19 pandemic and the impact of uncertainties, migration strategies, and underreporting of cases in South Korea, Italy, and Brazil. Chaos, Solitons Fractals (2020). https://doi.org/10.1016/j.chaos.2020.109888
9. Elavarsan, R.M., Pugazhendhi, R.: Restructured society and environment: A review on potential technological strategies to control the COVID-19 pandemic. Sci. Total Env. (2020). https://doi.org/10.1016/j.scitotenv.2020.138858
10. Li, Q., Guan, X., Wu, P., Wang, X., Zhou, I., Tong, Y.: Early transmission dynamics in Wuhan, China, of novel coronavirus-infected pneumonia. N top N Engl. J. Med. (2020)
11. Qarnain, S.S., Muthuvel, S., Bathrinath, S.: Review on government action plans to reduce energy consumption in buildings amid COVID-19 pandemic outbreak. Materials Today: Proceedings (2019). https://doi.org/10.1016/j.matpr.2020.04.723
12. Moghram, I., Rahman, S.: Analysis and evaluation of five short-term load forecasting techniques. IEEE Trans. Power Syst. **4**(4), 1484–1491 (1989)
13. Aly Hamed, H.H.: A proposed intelligent short-term load forecasting hybrid models of ANN, WNN, and KF based on clustering techniques for smart grid. Electr. Power Syst. Res. **182** (2020)
14. Ceperic, E., Ceperic, V., Member, S., Baric, A.: A Strategy for short-term load forecasting by support vector regression machines. IEEE Trans. Power Syst. **28**(4), 4356–4364 (2013)
15. Tian, C., Ma, J., Zhang, C., Zhan, P.: A deep neural network for short-term load forecast based on LSTM and convolution neural network. Energies **11**, 3493 (2018)
16. Singh, S., Parmar, K.S., Kumar, J., Makkhan, S.J.S.: Development of new hybrid model of discrete wavelet decomposition and ARIMA models in application to one month forecast the causalities cases of COVID-19. Chaos, Solitons Fractals (2020). https://doi.org/10.1016/j.chaos.2020.109866
17. Box, G.E., Jenkins, G.M., Reinsel, G.C., Ljung, G.M.: Time series analysis: forecasting and control. Wiley, Hoboken (2015)
18. Alex, D., Timothy, C.: A regression-based approach to short term system load forecasting. IEEE Trans. Power Syst. **5**(4), 1535–1550 (1990)
19. Hagan, M.T., Behr, S.M.: The time series approach to short term load forecasting. IEEE Trans. Power Syst. **2**(3) (1987)

20. Ribeiro, M.H.D.M., Silva, R.G., Mariani, V.C., Coelho, L.S.: Short-term forecasting COVID-19 cumulative confirmed cases: perspectives for Brazil. Chaos Solitons Fractals. https://doi.org/10.1016/j.chaos.2020.109853 (2020)
21. Alfieri, L., Falco, P.D.: Wavelet-based decompositions for probabilistic load forecasting. IEEE Trans. Smart Grid 12(2), 1367–1376 (2020)
22. Percival, D.B., Walden, A.T.: Wavelet Methods for Time Series Analysis, 4. Cambridge University Press, Cambridge (2000)
23. Shah, K., Abdeljawad, T., Mahariq, I., Jarad, F.: Qualitative analysis of a mathematical model in the time of COVID-19. BioMed Res. Int. May 27 (2020). https://www.hindawi.com/journals/bmri/2020/5098598
24. Roda, W.C., Varughese, M.B., Han, D., Li, M.Y.: Why is it difficult to accurately predict the COVID-19 epidemic. Infect. Dis. Modell. 5, 71–281 (2020). https://doi.org/10.1016/j.idm.2020.03.001
25. Gob, R., Lurz, K., Pievatolo, A.: Electrical load forecasting by exponential smoothing with covariates. Appl. Stoch. Models Bus. Ind. 29(6), 629–645 (2013)
26. Taylor, J.W.: Triple seasonal method for short-term electricity demand forecasting. Eur. J. Oper. Res. 204(1), 139–152 (2010)
27. Gupta, S., Raghuvanshi, G.S., Chanda, A.: Effect of weather on COVID-19 spread in US: a prediction model for India in 2020. Sci. Total Env. 728, 138860 (2020)
28. Stein, E., Schmidt, M.: Advances in adaptive computational methods in mechanics. Studies Appl Mech. (1998)
29. Kafazi, I.E., Bannari, R., Aboutafail, M.O., Guerrero, J.M.: Energy production: a comparison of forecasting methods using the polynomial curve fitting and linear regression. IEEE 978 (2017)
30. Johnson, A.T.: Curve Fitting. Chapter 13. Elsevier (1991)

Problematic Use of Digital Technologies and Its Impact on Mental Health During COVID-19 Pandemic: Assessment Using Machine Learning

Anshika Arora, Pinaki Chakraborty, and M. P. S. Bhatia

Abstract Many nations have imposed lockdowns due to the COVID-19 pandemic as a measure to prevent the spread of disease among its population. These lockdowns have confined people at their homes which is leading them to use digital technologies such as Internet, social media, smartphones, more than ever before. The problematic use of these digital technologies may impact their mental and emotional health. This chapter discusses the role of machine learning to assess addiction to various digital technologies and its impact on mental and emotion health and on sleep quality during the COVID-19 pandemic. Three case studies are provided to demonstrate how machine learning can be used to assess these addictions and related disorders during the pandemic. Gaussian mixture clustering is implemented to group people with similar Twitter usage patterns to identify addictive Twitter usage during the pandemic. The results convey that 11.71% of users show addictive Twitter usage patterns and 4.05% of users show highly addictive Twitter usage patterns while 2.70% of users show dangerously addictive usage patterns. "Sadness" and "anger" are the dominating emotions among these users in contrast to "happiness" which is the dominating emotion among non-addictive users. A similar approach is used to cluster students with similar smartphone usage patterns and nomophobia scores to identify nomophobic behavior during the pandemic. The results show that 4.5% of students are at extremely high risk whereas 73% of students are at high risk. A review of studies identifies the emergence of machine learning for assessment of mental and emotional health during the COVID-19 pandemic. A case study on sleep quality assessment using data from wearable sensors convey that sleep quality of students

A. Arora (✉) · P. Chakraborty · M. P. S. Bhatia
Department of Computer Science and Engineering, Netaji Subhas University of Technology, New Delhi 110078, India
e-mail: anshika.arora.trf19@nsut.ac.in

P. Chakraborty
e-mail: pinaki.chakraborty@nsut.ac.in

M. P. S. Bhatia
e-mail: mpsbhatia@nsut.ac.in

I. Arpaci et al. (eds.), *Emerging Technologies During the Era of COVID-19 Pandemic*, Studies in Systems, Decision and Control 348,
https://doi.org/10.1007/978-3-030-67716-9_13

has been reduced significantly during the pandemic with a maximum decrease of 90.90%.

Keywords Digital technologies · Behavioral addiction · Machine learning · Mental health

1 Introduction

Addiction to digital technologies is a type of non-substance addictions which is a subset of behavioral addictions. The term technological addiction encompasses wide range of addictions such as Internet addiction [1], social media addiction [2, 3], smartphone addiction [4], online gaming addiction, and online gambling addiction. When people spend significant amounts of time using these digital technologies, and Internet-related activities developing dependency on these technologies for social interaction, entertainment and information retrieval, it leads to negative outcomes including problematic and addictive behaviors. Furthermore, technological addiction to an extent causes mental impairments [5–8], emotional instability [5, 9] and sleep disorders [10–12].

The coronavirus disease 2019 (COVID-19) epidemic has spread across the world. Countries are imposing nationwide lockdowns to control the spread of the disease. The lockdowns are affecting the daily lives of people around the world leading them to use digital technologies, including the Internet, social media, and smartphones more than ever before. According to *Statista* [13], worldwide 40% of Internet users were using their laptops more than ever before during the pandemic whilst 32% and 22% of worldwide Internet users increased their usage of desktops and tablets, respectively. Smart TV and media streaming service usage was also surged with around 30% of worldwide Internet users using these services more than ever. An average of 14% of Internet users worldwide were utilizing game consoles more than ever before as a direct result of the COVID-19 outbreak and its associated lockdowns. Global survey reports of increased use of in-home media consumption by the Internet users revealed, on an average 51% of worldwide respondents were watching films and shows on streaming services such as Netflix more. More than 45% of worldwide Internet users spent longer on social media and messaging services. A large proportion (35%) of worldwide Internet users agreed on spending more time on computer and video games and listening to more streaming services such as Apple music and Spotify. However, a small proportion of worldwide Internet users professed to have read more magazines and newspapers (16% and 14% respectively). Interestingly, at least 50% of respondents in most countries said that they were watching more news coverage. Also, 14% of respondents agreed to have created and uploaded more videos on YouTube and Tik Tok.

The emergence Artificial Intelligence (AI) and Machine Learning (ML) based methods in various applications in the context of COVID-19 has been reported [14–

16]. A review of publications emphasizing on the use of AI in the COVID-19 research domains was conducted by Hossain et al. [15]. It identified various research domains including epidemiological characteristics, diagnostics, prevention and control, psychological conditions leveraging the techniques of AI during this pandemic.

The pervasiveness of interactive functions such as chat rooms and forums on the Internet make it more addictive in nature compared to the less interactive functions and cause significant impairment to their family, social, financial, and professional lives and health. [1]. Signs of tolerance, withdrawal, and craving existed among the addictive Internet users [17] and the negative consequences of excessive use of digital technologies included preoccupation with the Internet, feeling of excitement or euphoric when online, going online to escape other problems, obsessive-like characteristics, socializing online more than in person, staying online longer than planned and losing track of time [18, 10]. Moreover, the most vulnerable group to addictive usage of digital technologies is students due to the accessibility of the Internet and the flexibility of their schedules [19] which also affects their academic performance [20]. Moreover, Social networking is widespread and convenient method to stay in touch with family and friends amidst social distancing directives during the COVID-19 pandemic, nevertheless timeless use of these technologies is further impacting the mental health of people causing anxiety, depression, sleep problems and others. Several instruments have been devised to assess the addiction to digital technologies. For instance, Media and Technology Usage and Attitudes Scale (MTUAS) [21] is a long, 66-item scale for assessment of extent of media and technology usage. However, it cannot assess different types of technological addictions such as Internet Addiction, social media addiction and smartphone addiction. Therefore, the research problem is to assess addiction to different types of digital technologies during the COVID-19 pandemic and its impact on mental health employing various machine learning techniques for reliable and accurate results.

This chapter discusses about the applications of ML for prediction of addictive use of digital technologies and its impact on mental and emotional health during the COVID-19 pandemic. Section 2 discusses about Internet addiction with excessive use of Internet during the pandemic. It discusses about the existing approaches to assess Internet addiction. Section 3 discusses about excessive use of social media during the pandemic and existing approaches to evaluate social media addiction followed by a case study to assess addictive Twitter usage during the COVID-19 pandemic using machine learning. Section 4 discusses about the assessment of smartphone addiction and its associated psychological disorders. It presents a case study to evaluate excessive usage of smartphone during the pandemic using machine learning. Section 5 discusses about how addiction to digital technologies affect mental health of people and reviews studies employing machine learning for assessment of mental and emotional health during the COVID-19 pandemic followed by a case study to assess sleep quality of undergraduate students. Finally, Sect. 6 concludes the chapter.

2 Internet Addiction

There exist multiple terms used to refer Internet addiction namely, Problematic Internet Use (PIU), Excessive Internet Use, Compulsive Internet Use, Pathological Internet Use, and Internet Addiction Disorder (IAD) describing the negative effects of excessive use of Internet on personal lives of Internet users. However, everyone using the Internet is not necessarily a victim of problematic use. Griffiths [22] has argued that many of the users who are excessively using the Internet may be using it as a medium to fuel other addictions and cannot be referred to as Internet addicts. He put forward a need to distinguish between addictions to the Internet and addictions on the Internet. Wąsiński and Tomczyk [23] provided a clear definition of Internet addiction stating, "it may become problematic only for those who are unable to control their online activities". Addicted individuals abandon their everyday activities and devote their time to the activities that they discover on the Internet. In simple terms, Problematic Internet Use refers to the use of the Internet that creates interreference in one's personal daily life in the form of psychological disruptions, social withdrawal, sleep deprivation, reduced school or work performance in a person's life [10, 20, 24, 25].

According to American Psychiatric Association's Diagnostic and Statistical Manual of Mental Disorders-Fourth Edition (DSM-IV) [17], Internet addiction is a blanket term covering an extensive variety of behaviors and impulse control problems. These behaviors are categorised into five specific subtypes as follows:

- *Cyber relationship addiction*: It includes excessive involvement in online relationships and addiction to social networking platforms to the point where one prefers virtual relationships over real-life friends.
- *Information overload*: It is associated with compulsive web surfing eventually leading to lower work productivity
- *Net compulsions*: It is associated with obsessive online gambling, shopping or online trading often resulting in financial crisis.
- *Computer addiction*: It is associated with obsessive computer game-playing.
- *Addiction to inappropriate material*: It is associated with the compulsive use of websites with adult content.

2.1 Excessive Internet Usage During COVID-19 Pandemic

As per the reports of *OpenVault's Broadband Insights* [26], the average broadband consumption has increased to 402.5 GB for the first quarter of 2020 as compared to 273.5 GB during the same time last year which sums up to 47% increase. It also results in 17% rise over the fourth quarter of 2019. Overall Internet usage surged by 47% in first quarter of 2020 largely due to the COVID-19 pandemic. There are various applications and services responsible for this growth. It has been reported by *Infinera* [27] that 50,000 years' worth of media streaming was observed in just one day, on April 4. This is due to the increased reliance of people on streaming services. Netflix, a media streaming service has seen a 22% growth in subscribers. Peloton, a collaborative workout company's user membership growth rose by 66%. The use of social media has increased considerably leading to a 27% increase in daily Facebook traffic flow and a 26% growth in quarterly sessions on LinkedIn. Also, 25% rise in monthly downloads has been seen by TikTok. Moreover, messaging applications such as WhatsApp have been retrieving twice as many video and voice calls. Social video applications have also seen a flow in acceptance, with Bunch receiving 1 million downloads in just seven days and House party, a social video and gaming application seeing a 70% increase in monthly signups. Nintendo, a video gaming company has seen 41% surge in monthly profit.

2.2 Assessment of Internet Addiction: Conventional Approach

The earliest pragmatic research conducted on addictive Internet use was by Young in 1996 [1]. Young [1] developed an 8-item questionnaire to assess the traces of addiction among Internet users. The questionnaire was based on DSM-IV [17] criteria for pathological gambling. Later, in 2008, Young [5] developed a 20-item scale assessing frequency of Internet use, frequency of interaction on Internet, its various negative impacts on one's personal life and its impact on mental and emotional health (anxiety, depression, mood, nervousness). The assessment is predominantly known as Internet Addiction Test (IAT). In 1997, Brenner [28] devised a 32-item instrument called the Internet-Related Addictive Behavior Inventory (IRABI) with all the items as dichotomous. In 1999, Pratarelli et al. [18] formulated a set of analytical criteria to examine possible constructs underlying Internet addiction. A of 93-item questionnaire was formulated, with items related to categorical demographic and Internet use and dichotomous items. Another attempt in formulating a reliable Internet addiction scale was made by Beard and Wolf [24] in 2001. They tried to modify the Young's questionnaire, based on concerns with the objectivity and reliance on self-report by removing vague terms and clarifying some terminologies. Furthermore, Rotunda et al. [10] devised a tool names as the Internet Use Survey (IUS) which analyses three components comprising demographic data and Internet usage; negative consequences and experience associated with Internet use; personal history and psychological characteristics of participants. The authors focussed on the need to consider contextual and dispositional factors associated with frequent Internet use rather than inaccurately assuming it as excessive, pathological, or addictive. Later, in 2003, Shapira et al. [29] anticipated that problematic Internet use be conceptualized as a form of impulse control disorder. They modified their diagnostic criteria for problematic Internet use containing broader aspects. A bigger study was conducted by Greenfield [30] called the Virtual Addiction Survey (VAS). The VAS focussed on descriptive information items including frequency and duration of Internet use, specific purpose of Internet usage, and clinical items such as disinhibition, loss of time, and behavior online. Lately, Chen et al. [31] developed a Chinese Internet addiction scale in 2003 which is commonly known as Chen Internet Addiction Scale (CIAS). The scale is based on a questionnaire exploring weekly on-line hours, habitual domains, and experience of Internet usage. Later, the scale was revised, and CIAS-R was developed with modification of item wording and addition and elimination of some items. More recently, Demetrovics et al. [32] devised an 18-item scale and named it as Problematic Internet Use Questionnaire (PIUQ) which contains three subscales: obsession, neglect, and control disorder.

2.3   Assessment of Internet Addiction: Machine Learning Based Approach

In the last decade, automation has been introduced for detection of Internet addiction and problematic Internet use by several researchers utilizing various data mining and machine learning techniques. Di et al. [33] utilized Support Vector Machines for automated detection of Internet addiction disorder among Chinese college students using data from CIAS and multiple personality questionnaires. Their model achieved a high performance with accuracy of 96.32% in detecting Internet addiction Nandhini et al. [34] implemented multiple machine learning classification algorithms including Naive Bayes, JRip, ZeroR, J48, RepTree using data from a survey to evaluate Internet addiction disorder among Indian students. Ji et al. [35] proposed a model for detection of Internet addiction utilizing a rule-based model, extended classifier which combines Reinforcement Learning and Evolutionary Computation. They used responses to CIAS as their dataset. Ioannidis et al. [36] employed machine learning techniques such as Logistic Regression, Random Forests and Naïve Bayes for detection of problematic Internet use utilizing the IAT data.

3   Social Media Addiction

Social Networking Sites are virtual platforms providing wide-ranging services to its users such as interaction with real-life friends, and other people with similar interests to maintain online relationships. Instant messaging applications refer to communication platforms which deliver more engaging, one-to-one interactions in contrast to social networking platforms which promote one-to-many conversations.

But since both provide virtual interaction among users, both can be considered as social media platforms. Over the last decade, social media use has become increasingly prevalent in daily activities and hence, various researchers are studying the prevalence and effects of social media addiction. Social media addiction belongs to the category of cyber relationship addiction among various Internet addiction categories as described in Sect. 2.

Social Network Mental Disorder (SNMD) is a disorder among social media users associated with undue use of social media applications accompanied with a loss of the sense of time and withdrawal including feelings of anxiety, depression, anger, or tension on inaccessibility of the social networking applications [37]. SNMDs are social-oriented and tend to happen to users who are dependent on social networking platforms for interacting with others. These people generally lack offline interactions and hence, seek cyber-relationships for compensation.

3.1 Excessive Use of Social Media During Covid-19 Pandemic

The impact of COVID-19 outbreak and its associated lockdowns has been seen as increased global in-home media consumption including the social media by the Internet users worldwide. According to *Statista* [13], around 45% of worldwide Internet users spent longer on messaging services such as WhatsApp and Facebook messenger whilst 44% of worldwide users agreed on spending longer on social media including Facebook, Instagram and Twitter during the pandemic.

Participants from Philippines, Italy, China and Brazil were most likely to be spending more time on social networking platforms. Participants from Spain, Italy, China and Philippines were most likely to be spending more time on social networking platforms. Several cities in the United States have instructed their residents to stay at home during the pandemic. A survey was conducted in U.S. in March 2020 to ask users whether they believe that they will use selected social media more if restricted at home due to the pandemic. YouTube, Facebook and Instagram were popular social media platforms that users were estimating to increase their usage during being at home.

According to *Statista* [13], various social media platforms have been used during the pandemic, with Facebook being the most used with 78.1% of U.S. adults using the platform. The second-most used platform was Instagram, with 49.5% of U.S. adults using it.

India went into a country-wide lockdown on March 25, 2020, which was extended until May 17, 2020. A survey was conducted to explore the impact of lockdown on media usage across India and it was observed that there was a spike in usage of social networking applications in the first phase of the lockdown. The usage stabilized in the further phases of lockdown. Respondents reported to have been using social networking applications for as high as five hours. In contrast, users spent just over three hours using social media platforms in the weeks before the coronavirus lockdown.

3.2 Assessment of Social Media Addiction: Conventional Approach

A set of Berner's addiction scales were developed in the past decade to evaluate addiction to various social networking sites. To evaluate addiction to Facebook, Bergen Facebook Addiction Scale (BFAS) [2] was constructed consisting of 18 items. The scale comprised six items, each one based on individual core aspect of addiction namely, salience, mood modification, tolerance, withdrawal, conflict, and relapse. Later, Bergen Social Media Addiction Scale (BSMAS) [3] was constructed to cover the addictive usage of all social network sites which is alternatively known as Bergen Social Networking Addiction Scale (BSNAS). BSMAS and BFAS compose same addiction evaluation criteria and structure of items though BSMAS using the word "social media" instead of "Facebook" where social media refers to commonly used platforms such as Facebook, Twitter, Instagram, etc. Bergen's addiction scales have been related to addiction's negative outcomes (e.g., poor sleep quality, anxiety, depression). In 2015, Idubor [38] investigated Social Media Usage and Addiction Levels among university students of Nigeria. He developed an instrument named as Social Media Utilisation and Addiction Questionnaire (SMUAQ) which comprised respondents' personal information and two major sections evaluating social media utilization and level of addiction. More recently, Liu et al. [39] developed a Chinese social media addiction scale comprising seven conventional dimensions of behavioral addiction, i.e. compulsive use, withdrawal, negative consequence, mood alteration, salience, tolerance, and relapse, and two additional dimensions which are preference for online social interaction and continued use (continuous use of social media despite being aware of its negative consequences). Addiction evaluation instruments to evaluate addiction to specific social media platforms also exist in literature. Twitter Addiction Scale (TAS) was developed as a customized version of Internet Addiction Test by replacing the word "Internet" with "Twitter" [40]. The Instagram Addiction Scale (TIAS) [41] was devised to measure addiction behavior on Instagram. TIAS consists of two subscales which are Instagram Feed Addiction and Instagram Stories Addiction.

3.3 Assessment of Social Media Addiction: Machine Learning Based Approach

Recent studies show the evidences of employing machine learning techniques for prediction and assessment of social media addiction. Leong et al. [42] proposed a neural network-based model for prediction of social media addiction and validated their model on data of 615 Facebook users. Shuai et al. [43] attempted to predict Social Network Mental Disorder Detection (SNMDD) by exploiting features extracted from social network data using machine learning techniques. In their work they detected three types of SNMD namely, Cyber Relationship Addiction, Net compulsion, and Information Overload using semi supervised learning techniques. They further extend their study [37] by introducing SNMD-based Tensor Model (STM) to improve the accuracy.

3.4 Case Study I: Machine Learning for Analysis of Addictive Use of Twitter During COVID-19 Lockdown in India

This case study aims to determine the effects of the pandemic and the lockdown on the behavioral health of people based on their tweets during the first two phases of lockdown in India. This case study analyses usage patterns of twitter users focusing on addictive usage.

Tweets have been collected during the first two phases of lockdowns (25 March 2020–14 April 2020 and 15 April 2020–3 May 2020) in India. The tweets have been collected based on two categories as follows:

- *Emotion-based tweets* to get the information of users who frequently express their feelings and emotions on social media platforms such as Twitter. Four basic emotions: happiness, sadness, anger/disgust, and surprise/fear have been considered in this work. Scraping of emotion-based tweets has been done using the hashtags of emotion words for each emotion. The emotion words are collected from existing literature [44, 45]. Lexical variants of these emotion words have also been considered.

- *Situation-based tweets* for analysing the active twitter users expressing their views on current situation. Scraping of these tweets has been done based on hashtags related to the current situation in India. Further, subjectivities of situation-based tweets are assessed and tweets with subjectivity equal to 0 have been discarded for the further process. This is done to exclude informative user profiles such as news portal profiles and e-commerce profiles for optimal analysis of addictive usage on Twitter.

After removing the duplicate tweets, total data comprises 7664 tweets.

The user profiles with more than 3 tweets among the collected tweets by same username during 40 days of lockdown period are selected for analysis of their usage patterns. 222 such users have been identified. To get the information on daily usage patterns of the selected user profiles all the tweets during the 40 days of lockdown

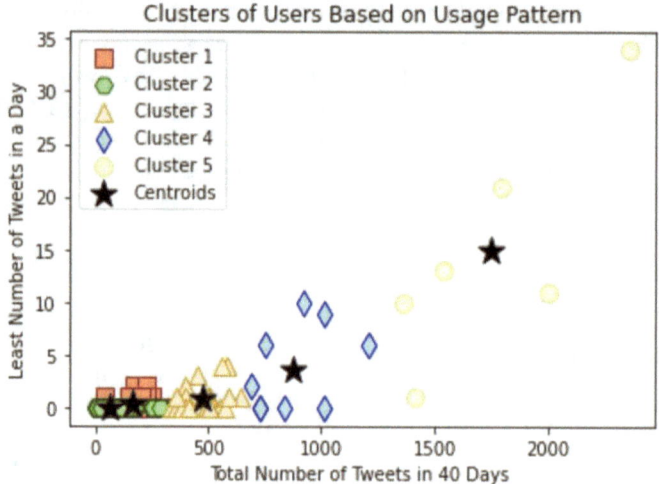

Fig. 1 Distribution of clusters based on twitter usage patterns

period from each user profile have been scrapped. Also, emotion of a user has been labelled based on the most used emotion word (including hashtags) among all the tweets posted by a user during this period.

A non-supervised machine learning based technique named as Gaussian Mixture Model Clustering has been used to cluster users based on similar usage pattern. 15 features of users' tweeting patterns have been sent as input to the clustering algorithm. These features include number of tweets posted in each of the least active 7 days, number of tweets posted in each of the most active 7 days and total number of tweets posted during the 40 days study period.

The result of clustering is depicted in Fig. 1, which gives the distribution of clusters. Optimal number of clusters is chosen based on Bayesian Information Criterion (BIC) and gradient of BIC scores. The optimal number of clusters is chosen to be five. BIC score corresponding to five clusters is 1825.

For each cluster, average number of tweets posted by users, and overall emotion of users during the study period have been analysed. The results of this analysis have been presented in Table 1. Based on this analysis, the usage of users in these clusters are identified as normal, frequent, addictive, highly addictive and dangerously addictive.

It can be observed that 11.71% of users show addictive usage patterns and 4.05% of users show highly addictive usage patterns with sadness as a dominating emotion among these users while 2.70% of users show dangerously addictive usage patterns with anger as a dominating emotion among these users.

Table 1 Analysis of clusters based on twitter usage patterns and overall emotion

	Clusters				
	Cluster 1	Cluster 2	Cluster 3	Cluster 4	Cluster 5
Number of students	142 (63.96%)	39 (17.57%)	26 (11.71%)	9 (4.05%)	6 (2.70%)
Mean of average number of tweets posted daily	1.52	4.09	11.53	21.39	42.56
Overall emotion of most of users	Happiness [Happiness (60%), Sadness (23%), Anger/ Disgust (3%), Surprise/ Fear (14%)]	Happiness [Happiness (51%), Sadness (36%), Anger/ Disgust (3%), Surprise/ Fear (10%)]	Sadness [Happiness (27%), Sadness (42%), Anger/ Disgust (23%), Surprise/ Fear (8%)]	Sadness [Happiness (11%), Sadness (56%), Anger/ Disgust (33%), Surprise/ Fear (0%)]	Anger [Happiness (0%), Sadness (33%), Anger/ Disgust (67%), Surprise/Fear (0%)]
Type of usage	Normal	Frequent	Addictive	Highly addictive	Dangerously addictive

4   Smartphone Addiction

Smartphones usage has grown to the extent where the phones have become an integral part of everyone's lives. Today, smartphones are used throughout the day for multiple reasons, including communication, productivity, utilities, and even entertainment, social networking, and gaming. Smartphone addiction is a shifting technological addiction as the mobile phones have evolved to smartphones encompassing varied Internet features and applications [46]. Smartphone addiction refers to overuse of smartphones with corresponding functional impairments. Several terminologies have been used in the literature to refer smartphone addiction such as Problematic Mobile Phone use (PMPU), Problematic Smartphone Usage (PSU) mobile phone dependency, mobile phone addiction, and smartphone use disorder. Studies have been reported assessing the effect of smartphone addiction on daily lives of people. Hawi et al. [46] present strong evidence on the adverse effect of smartphone addiction on academic performance of university students by studying link between smartphone multitasking and the decline in academic performance. Recent investigations have explored novel psychological variables in association with smartphone addiction. These variables are nomophobia and Fear of Missing Out (FOMO).

Nomophobia is a phobia of millennials which is abbreviation of No Mobile Phone Phobia. It is defined as a disorder of the modern world which describes the discomfort in the form of nervousness, distress and/or anxiety caused by the non-availability of a smartphone which also includes fear of not being able to access information and communicate or losing connection with people [47–49]. Nomophobia occurs as a result of overuse and dependency on smartphones for social networking and information retrieval and arises from feeling of not being able to do calling, messaging, losing internet connectivity and access to social networking, and losing access to online information [50] leading to nomophobic people being anxious in such situations. Nomophobia has been studied to impact psychological, social, academic, and professional lives of smartphone users [24, 51]. Such people tend to keep their phone switched on 24 h a day and even take their phone to bed causing sleep disorders. Arora and Chakraborty [52] reviewed the studies on nomophobia to identify the techniques used to detect nomophobia. It was found in their review that most of the studies used self-reporting-based techniques to diagnose nomophobia with Nomophobia Questionnaire (NMP-Q) framed by Yildirim and Correia [50] being the most used instrument. The NMP-Q is a 20-item questionnaire devised for evaluation of nomophobia where higher scores correspond to higher nomophobia severity.

FOMO involves the fear of missing out on pleasing and enjoyable experiences of smartphone usage. It involves the apparent need to persistently stay connected with the social network [53]. FOMO has been associated with anxiety and depression severity [54–57] and with symptoms of smartphone addiction severity [55, 56, 58].

4.1 Excessive Smartphone Usage During COVID-19 Pandemic

According to Statista [13], media device usage has been increased worldwide among Internet users due to coronavirus outbreak. As a result of this, around 70% of Internet users worldwide were using their smartphones more, though this varied significantly by country. People were using their smartphones excessively during the COVID-19 pandemic. In China and the Philippines around 86% people were using their phones more.

4.2 Assessment of Smartphone Addiction: Conventional Approach

The first attempt to assess smartphone addiction was done by Bianchi and Phillips in 2005 [59]. They devised a 27-item scale named Mobile Phone Problem Use Scale (MPPUS) to evaluate problematic usage of mobile phones. Another attempt in devising a questionnaire for problematic usage of mobile phones was done in 2008 by Billieux et al. [60]. They named their instrument as Problematic Mobile

Phone Use Questionnaire (PMPUQ). The PMPUQ intends to assess actual as well as potential problematic usage of mobile phones by assessing 30 items developed to target dangerous use of mobile phone, financial problem because of mobile phone use, and dependence on mobile phone. Lately, Kwon et al. [4] in 2013 constructed the Smartphone Addiction Scale (SAS) which is a 33-item scale devised for smartphone addiction assessment which consisted of six factors including overuse, withdrawal, cyberspace orientated relationship, daily life disturbance, positive anticipation, and tolerance. Later, they developed a short version of SAS namely Smartphone Addiction Scale-Short Version (SAS-SV) which comprises 10 items which were selected using content validity [61]. In 2016, Csibi et al. [62] developed Smartphone Application-Based Addiction Scale (SABAS) is a 6-items. It is assessment of smartphones' application-based addictions. In another recent study, Marty-Dugas et al. [63] Smartphone Use Questionnaires (SUQ-G&A) differentiates general smartphone usage from absent-minded smartphone usage. It uses scores of two 10-item scales named as, general (SUQ-G) and absent-minded (SUQ-A). SUQ-G emphases on specific uses of smartphone such as frequency of checking social media applications while SUQ-A deals with mindless usage such as frequency of checking mobile phone without realizing the purpose.

4.3 Assessment of Smartphone Addiction: Machine Learning Based Approach

Shin et al. [64] use a range of mobile phone usage data and identify several features in order to develop a machine learning based model for automated prediction of problematic use of smartphones. The algorithms used by them include Naïve Bayes, Support Vector Machines and AdaBoost and their model achieved and accuracy of 89.6% for detection of problematic smartphone use. Lawanont et al. [65] built a smartphone addiction recognition system based on smartphone usage data. Based on this data, they implemented a classification model utilizing Naïve Bayes, Decision Tree, K-Nearest Neighbor, and Support Vector Machines for recognition of likelihood of having smartphone addiction. In a recent study, Ellis et al. [66] utilized k-means clustering algorithm to cluster the users with similar smartphone usage behavior. They used the objective behavior data retrieved from Apple's Screen Time application which automatically logs a series of behavioral metrics related to screen time over a period of seven days. In another more recent study, Elhai et al. [67] utilized supervised machine learning algorithms to detect PSU severity among Chinese undergraduate students. They also correlated FOMO, depression and anxiety symptoms with PSU severity using the data from responses to SAS-SV, Depression Anxiety Stress Scale-21, FOMO Scale, and Ruminative Responses Scale and concluded that FOMO had the largest relative contribution in modelling PSU severity. A more technically advanced research was conducted by Kim et al. [68]. They proposed a model based on Deep Belief Networks, K-Nearest

Neighbor, and Support Vector Machines for prediction of smartphone addiction levels in individuals. They use EEG signals of participants for emotion analysis of individuals and concluded that the risk group was more emotionally stable than the non-risk group especially in expression of emotion "Fear".

4.4 Case Study II: Assessment of Nomophobia Among University Students During COVID-19 Pandemic Using Machine Learning

Due the COVID-19 pandemic, to maintain distancing Universities have been shifting from classroom-based teaching to online education. The online teaching approaches such as video lectures, sharing of study materials through distant learning services are leading students to use digital technologies including smartphones frequently and for long hours. As nomophobia occurs due to overuse and dependency on smartphones, this case study intends to identify the nomophobic behavior among undergraduate university students during the pandemic.

The materials used to collect data are NMP-Q [50] and an android mobile application which automatically collects usage data from the smartphones of the students. The attributes collected by the app are total phone usage in last 7 days (hours), total night-time usage in last 7 days (hours) and total number of times the phone has been unlocked in last 7 days. Using the app and the questionnaire data of 111 students was collected.

Gaussian Mixture clustering has been implemented to identify students at different levels of risk based on the similarity in their NMP-Q scores and smartphone usage pattern. The result of clustering is depicted in Fig. 2 which gives the

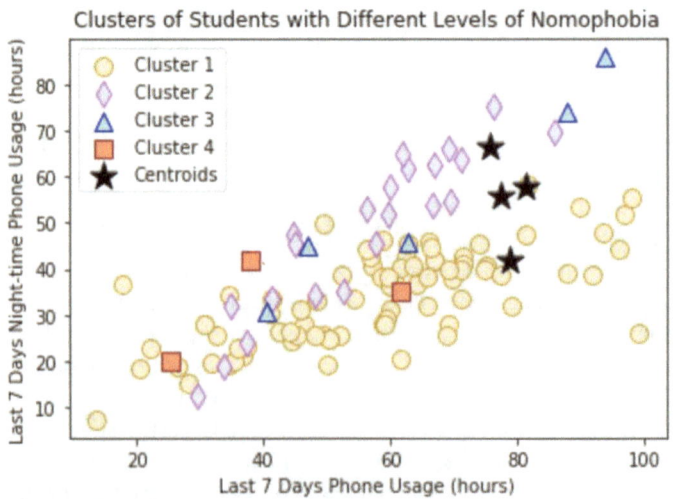

Fig. 2 Distribution of clusters based smartphone usage patterns

Table 2 Analysis of clusters based on mobile usage patterns

	Clusters			
	Cluster 1	Cluster 2	Cluster 3	Cluster 4
Number of students	81 (73%)	22 (19.8%)	5 (4.5%)	3 (2.7%)
Mean of average daily smartphone usage (hours)	8.22	7.98	9.47	5.95
Mean of average daily night-time usage (hours)	4.62	5.01	8.03	6.89
Level of risk	High	Medium	Extremely high	Low

distribution of clusters. Optimal number of clusters is chosen based on Bayesian Information Criterion (BIC) and gradient of BIC scores. The optimal number of clusters is chosen to be four. BIC score corresponding to four clusters is 4362.

Students in each cluster have been analysed for average hours of daily smartphone usage and average hours of daily night-time smartphone usage. Based on the results, the levels of risk of students in the clusters are identified as Extremely High, High, Medium and Low. Table 2 gives the result of analysis of clusters.

It can be observed that 4.5% of students are at extremely high risk whereas 73% of students are at high risk.

5   Impact on Mental and Emotional Health and Sleep

There exist empirical studies relating Internet addiction, social media addiction and smartphone addiction to mental health impairments such as anxiety [5, 8, 11, 12], depression [6, 8, 11, 69], stress [6, 7, 12, 70], aggressive behaviors [71], emotional instability [5, 9, 72] and sleep disorders [10–12, 70]. In fact, mental instability, emotional instability and sleep disruption are among the four key factors associated with excessive Internet/technology use as studied by several researchers [2, 3, 5, 10, 18]. These key factors are as follows:

- *Absorption*: it deals with over engrossment with the Internet to an extent often leading to staying on the Internet longer than planned due to the loss of track of time while using the Internet. It also deals with the tendency to socialise more on the internet than in person and emotional instability while using the Internet.
- *Negative Consequences*: It deals with the effects of excessive use of the Internet on one's personal life. This happens when one prefers to be on the Internet than spending time with loved ones often leading to loss of loved ones, missing appointments, loneliness, eventually leading to anxiety and mental instability.
- *Sleep*: It deals with sleep pattern disruption due to excessive use of the Internet to an extent that one schedules his sleep around Internet use. It involves tendencies such as sleeping during the day and being online at night and having less than five hours of sleep because of Internet use.

- *Deception*: It deals with the tendency of lying to others online about one's identity or amount of time spent online.

Several studies relate mental health problems such as anxiety and depression to problematic internet use [73–75], social media exposure [76]. problematic online gaming [77, 78], excessive smartphone usage [79] during the COVID-19 outbreak. The problematic and timeless use of digital technologies during the pandemic are impacting the mental health of people causing anxiety, depression, sleep problems and multiple psychological commotions. The next section reviews studies employing machine learning techniques for assessment and analysis of mental and emotional disruptions during the pandemic.

5.1 Mental and Emotional Health

According to the study of Hossain et al. [15], a cluster of articles highlight a growing use of AI in COVID-19-related psychological research. In another research [80], the authors study the prevalence, need and applications of Artificial Intelligence technologies for mental health research during the COVID-19 pandemic. Venigalla et al. [81] developed a web portal for analysing the emotions of Twitter users of India during the COVID-19 pandemic. They considered six basic emotions (Anger, Disgust, Happiness, Surprise, Fear and Sadness). They believe that Machine Learning techniques could be used to improve the classification and plan to utilize these techniques in their future studies.

Table 3 lists the details of studies assessing and analysing mental and emotional health variables via machine learning techniques during the COVID-19 pandemic around the world. The articles have been retrieved from MEDLINE database and Google Scholar. The articles have been retrieved using the keywords (mental health OR emotional health) AND (machine learning) AND (COVID-19 OR coronavirus). The articles assessing mental or emotional health using other techniques, but machine learning have not been selected for the review. The studies have been reviewed to determine the following.

- Various machine learning techniques used for evaluation of mental and emotional health.
- Types of datasets used by researchers to assess mental and emotional health.
- Variables predicted for assessment of mental and emotional health.

5.2 Sleep

The adverse effects of overuse of digital technologies due to excessive screen time on sleep has also been reported in literature [89]. Vernon et al. [90] study the disruptions in daily activities and sleep activities caused by use of mobile phones in

Table 3 Studies utilizing machine learning for mental and emotional health assessment during the pandemic

S. No.	Reference	Data	Machine larning techniques	Psychological prediction
1.	Lauren et al. [82]	Social media (twitter tweets)	Not mentioned	Psychological stress
2.	Zhou et al. [83]	Social media (twitter tweets)	Logistic, regression, linear discriminant analysis, Gaussian Naïve Bayes	Depression
3.	Guntuku et al. [84]	Social media (twitter tweets)	Not mentioned	Mental health based on sentiment, anxiety, stress and loneliness
4.	Ćosić et al. [85]	Multimodal sensor-based data	Spiking neural networks, support vector machine, random forest,	Mental health disorders
5.	Khattar et al. [86]	Questionnaire	Association rule mining	Overall mental and emotional health
6.	Tummers et al. [87]	The CORD-19 dataset	K-means clustering	Intellectual disability
7.	Li et al. [88]	Social media (Weibo posts)	Online ecological recognition	Scores of emotional indicators (anxiety, depression, indignation, and Oxford happiness)

adolescent and consider this as a big challenge. In this study conducted on hight school students, it was found that poor sleep quality associated with late-night texting or calling was linked to a decline in mental health, including depression and low self-esteem. It was concluded that students who used their cell phones frequently in the evenings were at greater risk for depression the following year. It is also reported that excessive use of smartphones during the day, increase the likelihood of a sleeping disorder, stress, anxiety [12] leading to feeling of less physically active after experiencing these symptoms. It has been reported that Adults with Internet Addiction disorder have frequent difficulty initiating and maintaining sleep, non-restorative sleep, daytime functional impairment [91]. The addiction to digital technologies disrupts sleep quality by delayed bedtimes and reduced total sleep duration. The underlying mechanisms of these adverse associations between excessive screen time and sleep quality include time displacement (i.e., time spent on screens replaces time spent sleeping and other activities), psychological stimulation based on media content, and the effects of light emitted from devices on circadian timing, sleep physiology, and alertness.

Case Study III: Sleep Quality Assessment During COVID-19 Pandemic

Rajkumar [92] reviewed the existing literature on the COVID-19 outbreak pertinent to mental health and concluded that symptoms of anxiety and depression and self-reported stress were common psychological reactions to the COVID-19 pandemic which might be associated with disturbed sleep. People are using digital technologies more than ever before which is leading to sleep deprivation in the form of late in-bed time, less duration of sleep, and increased sleep onset latency.

Many wearables have been proposed and developed in the recent past to assess sleep with inbuilt sensors and the intervention of machine learning and deep learning algorithms with wearable technology for sleep quality assessment is evident to give impressive results [93]. This case study aims to assess daily sleep quality of Indian undergraduate students using data from smartwatches during the COVID-19 pandemic. Data of twelve undergraduate students has been collected for seven consecutive days before and after the imposition of lockdown in India and their daily sleep quality is assessed using the methodology proposed by Arora et al. [93]. Samsung Galaxy smartwatch and Xiomi MI smartband have been used for the collection of data.

The results obtained convey that daily sleep quality has been significantly decreased for maximum number of students with different values. The maximum decrease is observed to be 90.90% and the lowest decrease is observed to be 11.11%.

6   Research Model

This chapter studies the applicability of machine learning for assessment addiction to digital technologies and its impact on mental and emotional health and sleep. The research design includes study of addiction to various types of digital technologies, its assessment using conventional methods and machine learning techniques, its associated psychological disorders, and its impact on mental and emotional health further leading to various mental and emotional disorders and impacting sleep. Applicability of machine learning in assessment of these disorders has also been studied. Figure 3 presents the research model.

The model is justified with case studies analysing twitter addiction and nomophobia during the COVID-19 pandemic using unsupervised machine learning contributing to the existing literature in the domain of application of machine learning in mental and behavioral health.

The materials used for performing case studies are presented in Table 4.

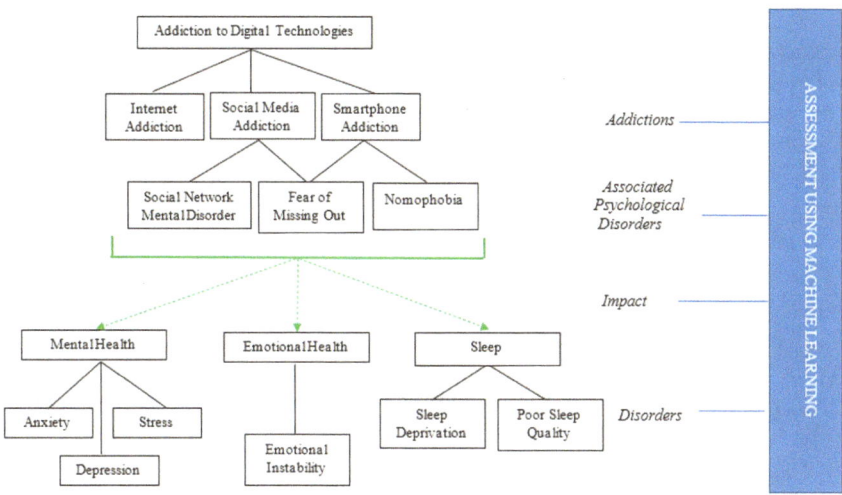

Fig. 3 Research model

Table 4 Materials

Materials	Purpose
GetOldTweets3 (https://pypi.org/project/getoldtweets3)	Scraping tweets
Tweet-pre-processor (https://pypi.org/project/tweet-preprocessor)	Cleaning tweets
TextBlob (https://textblob.readthedocs.io/en/dev)	Subjectivity analysis
NLTK (https://www.nltk.org)	Removing stop words before identifying frequently used words
Activity tracker (self-developed mobile application)	Accessing smartphone usage
Samsung Galaxy smartwatch and Xiomi MI smartband	Sleep data collection
Scikit-learn (https://scikit-learn.org/stable)	Gaussian mixture clustering
Matplotlib (https://pypi.org/project/matplotlib)	Plotting clusters

7 Discussion and Conclusion

Addiction to digital technologies encompasses wide range of behavioral addictions such as Internet addiction, social media addiction, smartphone addiction, online gaming addiction, etc. The use of digital technologies has increased with high rate during the COVID-19 pandemic as people are confined at homes due to lockdowns imposed to prevent the spread of disease. This increases their dependency on these technologies for media streaming, information retrieval, social communications and so on. The high dependency on these digital technologies may lead to timeless, problematic and addictive usage and development of associated psychological disorders such as social network mental disorder, nomophobia, and fear of missing out. This chapter discusses the role of machine learning to assess addiction to various digital technologies during the COVID-19 pandemic. The chapter discusses about the existing approaches (conventional and machine learning based) of assessment of Internet addiction, social media addiction and smartphone addiction and their associated psychological disorders. Three case studies are provided to demonstrate how machine learning can be used to assess these addictions and related disorders during the COVID-19 pandemic. Gaussian mixture clustering is implemented to group people with similar Twitter usage patterns to identify addictive Twitter usage during the pandemic. The results show that 11.71% of users show addictive Twitter usage patterns and 4.05% of users show highly addictive Twitter usage patterns while 2.70% of users show dangerously addictive usage patterns. "Sadness" and "anger" are the dominating emotions among these users in contrast to "happiness" which is the dominating emotion among non-addictive users. A similar approach is used to cluster students with similar smartphone usage patterns and nomophobia scores to identify nomophobic behavior during the pandemic. The results show that 4.5% of students are at extremely high risk whereas 73% of students are at high risk. The results convey that a moderate proportion of users are addicted to digital technologies during the pandemic and emotions play a role in this addictive usage. The chapter also discusses how these behavioral addictions impact mental health of users leading to stress, anxiety, depression, emotional instability and sleep deprivation. A review of studies identifies the emergence of machine learning for assessment of mental and emotional health during the COVID-19 pandemic. A case study on sleep quality assessment using data from wearable sensors convey that sleep quality of students has been reduced significantly during the lockdown with a maximum decrease of 90.90%.

Prior studies indicate that addictive behaviors towards digital technologies are affected by gender and cultural differences [94–96] and Internet addiction relate to more serious issues such as cyberbullying [97], hence motivating researchers to study the problem and provide solutions to it. Researchers suggest users' social media posts may affect human psychology and behavior during the COVID-19 pandemic [98]. A study analysing social media posts during the pandemic found that there exist high frequency of words like "death", "test", "spread", and "lockdown" which suggests that people fear of being infected and death [99]. This research contributes to the existing literature of study of human behavior during the pandemic analysing their social media usage patterns, smartphone usage patterns and sleep patterns.

The chapter discusses the applicability of machine learning methods for accurate assessment of addiction to digital technologies and its impact on mental and emotional health and sleep. This chapter opens avenues for prediction of mental health disorders using supervised machine learning techniques. Also, overall behavioral health of people using data from questionnaires, social media, smartphones, and smartwatches can also be studied in future. Deep learning techniques for these predictions should also be explored.

References

1. Young, K.: Internet addiction: the emergence of a new clinical disorder. CyberPsychology & Behav. **3**, 237–244 (1996)
2. Andreassen, C.S., Torsheim, T., Brunborg, G.S., Pallesen, S.: Development of a Facebook addiction scale. Psychol. Rep. **110**(2), 501–517 (2012)
3. Andreassen, C.S., Pallesen, S., Griffiths, M.D.: The relationship between addictive use of social media, narcissism, and self-esteem: Findings from a large national survey. Addict. Behav. **64**, 287–293 (2017)
4. Kwon, M., Lee, J.Y., Won, W.Y., Park, J.W., Min, J.A., Hahn, C., Gu, X., Choi, J.H., Kim, D.J.: Development and validation of a smartphone addiction scale (SAS). PLoS ONE **8**(2), e56936 (2013)
5. Young, K.S., Griffin-Shelley, E., Cooper, A., O'mara, J., Buchanans, J.: Online infidelity: A new dimension in couple relationships with implications for evaluation and treatment. Sex. Addict. & Compulsivity: J. Treat. Prev. **7**(1–2), 59–74 (2000)
6. Seki, T., Hamazaki, K., Natori, T., Inadera, H.: Relationship between internet addiction and depression among Japanese university students. J. Affect. Disord. **256**, 668–672 (2019)
7. Jacobsen, W.C., Forste, R.: The wired generation: Academic and social outcomes of electronic media use among university students. Cyberpsychology, Behav., Soc. Netw. **14**(5), 275–280 (2011)
8. Scimeca, G., Bruno, A., Cava, L., Pandolfo, G., Muscatello, M. R. A., Zoccali, R.: The relationship between alexithymia, anxiety, depression, and internet addiction severity in a sample of Italian high school students. Sci. World J. (2014)
9. Xiuqin, H., Huimin, Z., Mengchen, L., Jinan, W., Ying, Z., Ran, T.: Mental health, personality, and parental rearing styles of adolescents with Internet addiction disorder. Cyberpsychology, Behav., Soc. Netw. **13**(4), 401–406 (2010)
10. Rotunda, R.J., Kass, S.J., Sutton, M.A., Leon, D.T.: Internet use and misuse: Preliminary findings from a new assessment instrument. Behav. Modif. **27**(4), 484–504 (2003)
11. Salicetia, F.: Internet addiction disorder (IAD). Procedia - Soc. Behav. Sci. **191**, 1372–1376 (2015)
12. Thomée, S., Härenstam, A., Hagberg, M.: Mobile phone use and stress, sleep disturbances, and symptoms of depression among young adults-a prospective cohort study. BMC Public Health **11**(1), 66 (2011)
13. https://www.statista.com/
14. Nguyen, T. T.: Artificial intelligence in the battle against coronavirus (COVID-19): a survey and future research directions (Preprint) (2020)
15. Hossain, M. M., Sarwar, S. A., McKyer, E. L. J., Ma, P.: Applications of Artificial Intelligence Technologies in COVID-19 Research: A Bibliometric Study (2020)
16. Wynants, L., Van Calster, B., Bonten, M.M., Collins, G.S., Debray, T.P., De Vos, M., Haller, M.C., Heinze, G., Moons, K.G., Riley, R.D., Schuit, E.: Prediction models for diagnosis and prognosis of covid-19 infection: systematic review and critical appraisal. BMJ 369 (2020)
17. American Psychiatric Association: Diagnostic and statistical manual of mental disorders (DSM-IV), 4th edn. American Psychiatric Association, Washington, DC (1994)
18. Pratarelli, M.E., Browne, B.L., Johnson, K.: The bits and bytes of computer/Internet addiction: A factor analytic approach. Behav. Res. Methods, Instrum., & Comput. **31**(2), 305–314 (1999)
19. Moore, D. W.: The emperor's virtual clothes: The naked truth about Internet culture. Algonquin Books (1995)
20. Chang, M.K., Law, S.P.M.: Factor structure for Young's Internet Addiction Test: A confirmatory study. Comput. Hum. Behav. **24**(6), 2597–2619 (2008)
21. Rosen, L.D., Whaling, K., Carrier, L.M., Cheever, N.A., Rokkum, J.: The media and technology usage and attitudes scale: An empirical investigation. Comput. Hum. Behav. **29**(6), 2501–2511 (2013)

22. Griffiths, M.: Internet addiction-time to be taken seriously? Addict. Res. **8**(5), 413–418 (2000)
23. Wąsiński, A., Tomczyk, Ł.: Factors reducing the risk of internet addiction in young people in their home environment. Child Youth Serv. Rev. **57**, 68–74 (2015)
24. Beard, K.W., Wolf, E.M.: Modification in the proposed diagnostic criteria for Internet addiction. Cyberpsychology & Behav. **4**(3), 377–383 (2001)
25. Siciliano, V., Bastiani, L., Mezzasalma, L., Thanki, D., Curzio, O., Molinaro, S.: Validation of a new Short Problematic Internet Use Test in a nationally representative sample of adolescents. Comput. Hum. Behav. **45**, 177–184 (2015)
26. https://openvault.com/complimentary-report-Q120/
27. https://www.infinera.com/
28. Brenner, V.: Psychology of computer use: XLVII. Parameters of Internet use, abuse and addiction: the first 90 days of the Internet Usage Survey. Psychol. Rep., **80**(3), 879–882
29. Shapira, N.A., Lessig, M.C., Goldsmith, T.D., Szabo, S.T., Lazoritz, M., Gold, M.S., Stein, D.J.: Problematic internet use: proposed classification and diagnostic criteria. Depress. Anxiety **17**(4), 207–216 (2003)
30. Greenfield, D.N.: Psychological characteristics of compulsive Internet use: A preliminary analysis. Cyberpsychology & Behav. **2**(5), 403–412 (1999)
31. Chen, S. H., Weng, L. J., Su, Y. J., Wu, H. M., Yang, P. F.: Development of a Chinese Internet addiction scale and its psychometric study. *Chin. J. Psychol.* (2003)
32. Demetrovics, Z., Szeredi, B., Rózsa, S.: The three-factor model of Internet addiction: The development of the Problematic Internet Use Questionnaire. Behav. Res. Methods **40**(2), 563–574 (2008)
33. Di, Z., Gong, X., Shi, J., Ahmed, H.O., Nandi, A.K.: Internet addiction disorder detection of Chinese college students using several personality questionnaire data and support vector machine. Addict. Behav. Rep. **10**, 100200 (2019)
34. Nandhini, C., Krishnaveni, K.: Evaluation of internet addiction disorder among students. Indian J. Sci. Technol. **9**(19), 1–5 (2016)
35. Ji, H. M., Chen, L. Y., Hsiao, T. C.: Real-time detection of internet addiction using reinforcement learning system. In: Proceedings of the Genetic and Evolutionary Computation Conference Companion, pp. 1280–1288, July (2019)
36. Ioannidis, K., Chamberlain, S.R., Treder, M.S., Kiraly, F., Leppink, E.W., Redden, S.A., Stein, D.J., Lochner, C., Grant, J.E.: Problematic internet use (PIU): associations with the impulsive-compulsive spectrum. An application of machine learning in psychiatry. J. Psychiatr. Res. **83**, 94–102 (2016)
37. Shuai, H.H., Shen, C.Y., Yang, D.N., Lan, Y.F.C., Lee, W.C., Philip, S.Y., Chen, M.S.: A comprehensive study on social network mental disorders detection via online social media mining. IEEE Trans. Knowl. Data Eng. **30**(7), 1212–1225 (2017)
38. Idubor, I.: Investigating social media usage and addiction levels among undergraduates in University of Ibadan, Nigeria. J. Educ., Soc. Behav. Sci., 291–301 (2015)
39. Liu, C., Ma, J.: Development and validation of the Chinese social media addiction scale. Personality Individ. Differ. **134**, 55–59 (2018)
40. Kircaburun, K.: Effects of gender and personality differences on twitter addiction among Turkish undergraduates. J. Educ. Pract. **7**(24), 33–42 (2016)
41. Sholeh, A., Rusdi, A.: A new measurement of instagram addiction: psychometric properties of the instagram addiction scale (TIAS). Feedback, **737**:499
42. Leong, L.Y., Hew, T.S., Ooi, K.B., Lee, V.H., Hew, J.J.: A hybrid SEM-neural network analysis of social media addiction. Expert Syst. Appl. **133**, 296–316 (2019)
43. Shuai, H. H., Shen, C. Y., Yang, D. N., Lan, Y. F., Lee, W. C., Yu, P. S., Chen, M. S.: Mining online social data for detecting social network mental disorders. In: Proceedings of the 25th International Conference on World Wide Web, pp. 275–285, April (2016)
44. Shaver, P., Schwartz, J., Kirson, D., O'connor, C.: Emotion knowledge: further exploration of a prototype approach. J. Pers. Soc. Psychol. **52**(6), 1061 (1987)

45. Wang, W., Chen, L., Thirunarayan, K., Sheth, A. P.: Harnessing twitter "big data" for automatic emotion identification. In: 2012 International Conference on Privacy, Security, Risk and Trust and 2012 International Confernece on Social Computing, pp. 587–592, IEEE, September (2012)
46. Hawi, N.S., Samaha, M.: To excel or not to excel: Strong evidence on the adverse effect of smartphone addiction on academic performance. Comput. Educ. **98**, 81–89 (2016)
47. Yildirim, C., Correia, A. P.: Understanding nomophobia: A modern age phobia among college students. In: International Conference on Learning and Collaboration Technologies, pp. 724–735. Springer, Cham, August (2015)
48. King, A.L.S., Valença, A.M., Silva, A.C.O., Baczynski, T., Carvalho, M.R., Nardi, A.E.: Nomophobia: Dependency on virtual environments or social phobia? Comput. Hum. Behav. **29**(1), 140–144 (2013)
49. King, A. L. S., Valença, A. M., Silva, A. C., Sancassiani, F., Machado, S., Nardi, A. E.: "Nomophobia": impact of cell phone use interfering with symptoms and emotions of individuals with panic disorder compared with a control group. Clin. Pract. Epidemiol. Ment. Health: CP & EMH. **10**:28 (2014)
50. Yildirim, C., Correia, A.P.: Exploring the dimensions of nomophobia: Development and validation of a self-reported questionnaire. Comput. Hum. Behav. **49**, 130–137 (2015)
51. Caplan, S.E.: Problematic Internet use and psychosocial well-being: development of a theory-based cognitive–behavioral measurement instrument. Comput. Hum. Behav. **18**(5), 553–575 (2002)
52. Arora, A., Chakraborty, P.: Diagnosis, prevalence and effects of nomophobia-A review. Psychiatry Res. **288**, 112975 (2020)
53. Przybylski, A.K., Murayama, K., DeHaan, C.R., Gladwell, V.: Motivational, emotional, and behavioral correlates of fear of missing out. Comput. Hum. Behav. **29**(4), 1841–1848 (2013)
54. Elhai, J.D., Levine, J.C., Dvorak, R.D., Hall, B.J.: Fear of missing out, need for touch, anxiety and depression are related to problematic smartphone use. Comput. Hum. Behav. **63**, 509–516 (2016)
55. Wolniewicz, C.A., Tiamiyu, M.F., Weeks, J.W., Elhai, J.D.: Problematic smartphone use and relations with negative affect, fear of missing out, and fear of negative and positive evaluation. Psychiatry Res. **262**, 618–623 (2018)
56. Elhai, J.D., Yang, H., Fang, J., Bai, X., Hall, B.J.: Depression and anxiety symptoms are related to problematic smartphone use severity in Chinese young adults: Fear of missing out as a mediator. Addict. Behav. **101**, 105962 (2020)
57. Rozgonjuk, D., Levine, J.C., Hall, B.J., Elhai, J.D.: The association between problematic smartphone use, depression and anxiety symptom severity, and objectively measured smartphone use over one week. Comput. Hum. Behav. **87**, 10–17 (2018)
58. Rozgonjuk, D., Elhai, J.D., Ryan, T., Scott, G.G.: Fear of missing out is associated with disrupted activities from receiving smartphone notifications and surface learning in college students. Comput. Educ. **140**, 103590 (2019)
59. Bianchi, A., Phillips, J.G.: Psychological predictors of problem mobile phone use. CyberPsychology & Behav. **8**(1), 39–51 (2005)
60. Billieux, J., Linden, M., Rochat, L.: The role of impulsivity in actual and problematic use of the mobile phone. Appl. Cogn. Psychol.: Off. J. Soc. Appl. Res. Mem. Cogn. **22**(9):1195–1210 (2008)
61. Kwon, M., Kim, D.J., Cho, H., Yang, S.: The smartphone addiction scale: development and validation of a short version for adolescents. PLoS ONE **8**(12), e83558 (2013)
62. Csibi, S., Demetrovics, Z., Szabo, A.: Hungarian adaptation and psychometric characteristics of Brief Addiction to Smartphone Scale (BASS) [In Hungarian]. Psychiatria Hungarica **31**(1), 71–77 (2016)
63. Marty-Dugas, J., Ralph, B.C., Oakman, J.M., Smilek, D.: The relation between smartphone use and everyday inattention. Psychol. Conscious.: Theory, Res., Pract. **5**(1), 46 (2018)

64. Shin, C., Dey, A. K.: Automatically detecting problematic use of smartphones. In: Proceedings of the 2013 ACM International Joint Conference on Pervasive and Ubiquitous Computing, pp. 335–344, September (2013)
65. Lawanont, W., Inoue, M.: A development of classification model for smartphone addiction recognition system based on smartphone usage data. In: International Conference on Intelligent Decision Technologies, pp. 3–12. Springer, Cham, June (2017)
66. Ellis, D.A., Davidson, B.I., Shaw, H., Geyer, K.: Do smartphone usage scales predict behavior? Int. J. Hum Comput Stud. **130**, 86–92 (2019)
67. Elhai, J.D., Yang, H., Rozgonjuk, D., Montag, C.: Using machine learning to model problematic smartphone use severity: The significant role of fear of missing out. Addict. Behav. **103**, 106261 (2020)
68. Kim, S.K., Kang, H.B.: An analysis of smartphone overuse recognition in terms of emotions using brainwaves and deep learning. Neurocomputing **275**, 1393–1406 (2018)
69. Young, K.S., Rogers, R.C.: The relationship between depression and Internet addiction. Cyberpsychology & Behav. **1**(1), 25–28 (1998)
70. Thomée, S., Härenstam, A., Hagberg, M.: Computer use and stress, sleep disturbances, and symptoms of depression among young adults–a prospective cohort study. BMC psychiatry **12**(1), 176 (2012)
71. Ko, C.H., Yen, J.Y., Liu, S.C., Huang, C.F., Yen, C.F.: The associations between aggressive behaviors and Internet addiction and online activities in adolescents. J. Adolesc. Health **44**(6), 598–605 (2009)
72. Zamani, B.E., Abedini, Y., Kheradmand, A.: Internet addiction based on personality characteristics of high school students in Kerman. Iran. Addiction & health **3**(3–4), 85 (2011)
73. Király, O., Potenza, M. N., Stein, D. J., King, D. L., Hodgins, D. C., Saunders, J. B., Griffiths, M. D., Gjoneska, B., Billieux, J., Brand, M., Abbott, M. W.: Preventing problematic internet use during the COVID-19 pandemic: Consensus guidance. Compr. Psychiatry, 152180 (2020)
74. Li, Y., Wang, Y., Jiang, J., Valdimarsdóttir, U. A., Fall, K., Fang, F., Song, H., Lu, D., Zhang, W.: Psychological distress among health professional students during the COVID-19 outbreak. Psychol. Med., 1–3 (2020)
75. Kakunje, A., Mithur, R., Kishor, M.: Emotional well-being, mental health awareness, and prevention of suicide: Covid-19 pandemic and digital psychiatry. Arch. Med. Health Sci. **8**(1), 147 (2020)
76. Gao, J., Zheng, P., Jia, Y., Chen, H., Mao, Y., Chen, S., Wang, Y., Fu, H., Dai, J.: Mental health problems and social media exposure during COVID-19 outbreak. PLoS ONE **15**(4), e0231924 (2020)
77. King, D. L., Delfabbro, P. H., Billieux, J., Potenza, M. N.: Problematic online gaming and the COVID-19 pandemic. J. Behav. Addict. (2020)
78. Amin, K. P., Griffiths, M. D., Dsouza, D. D.: Online Gaming During the COVID-19 Pandemic in India: Strategies for Work-Life Balance. Int. J. Ment. Health Addict., 1–7 (2020)
79. Sun, S., Lin, D., Operario, D.: Need for a population health approach to understand and address psychosocial consequences of COVID-19. Psychol. Trauma: Theory, Res., Pract., Policy **12**(S1), S25 (2020)
80. Hossain, M. M., McKyer, E. L. J., Ma, P.: Applications of artificial intelligence technologies on mental health research during COVID-19 (2020)
81. Venigalla, A. S. M., Vagavolu, D., Chimalakonda, S.: Mood of India During Covid-19–An Interactive Web Portal Based on Emotion Analysis of Twitter Data. arXiv preprint arXiv:2005.02955 (2020)
82. Hung, M., Lauren, E., Hon, E. S., et al.: Social Network Analysis of COVID-19 Sentiments: Application of Artificial Intelligence. J Med Internet Res (inpress, 3 August 2020)
83. Zhou, J., Zogan, H., Yang, S., Jameel, S., Xu, G., Chen, F.: Detecting community depression dynamics due to COVID-19 pandemic in Australia. arXiv preprint arXiv:2007.02325 (2020)

84. Guntuku, S. C., Sherman, G., Stokes, D. C., Agarwal, A. K., Seltzer, E., Merchant, R. M., Ungar, L. H.: Tracking mental health and symptom mentions on twitter dCOVID-19. J. Gen. Intern. Med., 1–3 (2020)

85. Ćosić, K., Popović, S., Šarlija, M., Kesedžić, I., Jovanovic, T.: Artificial intelligence in prediction of mental health disorders induced by the COVID-19 pandemic among health care workers. Croat. Med. J. **61**(3), 279 (2020)

86. Khattar, A., Jain, P. R., Quadri, S. M. K.: Effects of the disastrous pandemic COVID 19 on learning styles, activities and mental health of young Indian students-A machine learning approach. In: 2020 4th International Conference on Intelligent Computing and Control Systems (ICICCS), 1190–1195. IEEE (2020)

87. Tummers, J., Catal, C., Tobi, H., Tekinerdogan, B., Leusink, G.: Coronaviruses and people with intellectual disability: an exploratory data analysis. J. Intellect. Disabil. Res. **64**(7), 475–481 (2020)

88. Li, S., Wang, Y., Xue, J., Zhao, N., Zhu, T.: The impact of COVID-19 epidemic declaration on psychological consequences: a study on active Weibo users. Int. J. Environ. Res. Public Health **17**(6), 2032 (2020)

89. LeBourgeois, M.K., Hale, L., Chang, A.M., Akacem, L.D., Montgomery-Downs, H.E., Buxton, O.M.: Digital media and sleep in childhood and adolescence. Pediatrics **140** (Supplement 2), S92–S96 (2017)

90. Vernon, L., Modecki, K.L., Barber, B.L.: Mobile phones in the bedroom: Trajectories of sleep habits and subsequent adolescent psychosocial development. Child Dev. **89**(1), 66–77 (2018)

91. Kim, K., Lee, H., Hong, J.P., Cho, M.J., Fava, M., Mischoulon, D., Kim, D.J., Jeon, H.J.: Poor sleep quality and suicide attempt among adults with internet addiction: A nationwide community sample of Korea. PLoS ONE **12**(4), e0174619 (2017)

92. Rajkumar, R. P.: COVID-19 and mental health: A review of the existing literature. Asian J. Psychiatry, 102066 (2020)

93. Arora, A., Chakraborty, P., Bhatia, M. P. S.: Analysis of Data from Wearable Sensors for Sleep Quality Estimation and Prediction Using Deep Learning. Arab. J. Sci. Eng., 1–20 (2020)

94. Baloğlu, M., Şahin, R., Arpaci, I.: A review of research in problematic internet use: gender and cultural differences. Curr. Opin. Psychol. (2020)

95. Arpaci, I., Unver, T. K.: Moderating role of gender in the relationship between big five personality traits and smartphone addiction. Psychiatr. Q., 1–9 (2020)

96. Arpaci, I.: Relationships between early maladaptive schemas and smartphone addiction: The moderating role of mindfulness. Int. J. Ment. Health Addict., 1–15 (2020)

97. Arpaci, I., Abdeljawad, T., Baloğlu, M., Kesici, Ş., Mahariq, I.: Mediating effect of internet addiction on the relationship between individualism and cyberbullying: cross-sectional questionnaire study. J. Med. Internet Res. **22**(5), e16210 (2020)

98. Arpaci, I., Karataş, K., Baloğlu, M.: The development and initial tests for the psychometric properties of the COVID-19 Phobia Scale (C19P-S). Pers. Individ. Differ., 110108 (2020)

99. Arpaci, I., Alshehabi, S., Al-Emran, M., Khasawneh, M., Mahariq, I., Abdeljawad, T., Hassanien, A.E.: Analysis of twitter data using evolutionary clustering during the COVID-19 pandemic. CMC-Comput., Mater. & Contin. **65**(1), 193–204 (2020)

The Role of Technology Acceptance in Healthcare to Mitigate COVID-19 Outbreak

Adi A. AlQudah⬧, Said A. Salloum⬧, and Khaled Shaalan⬧

Abstract The recent decade has included huge achievements in the development for information technologies in healthcare. Now, these technologies can be employed as part of the response to the COVID-19 pandemic. Information technologies in healthcare are crucial to store, manage and exchange the clinical data. On the other hand, the success or failure of a specific technology relies on the acceptance to use that technology. There is a need to assess the user's technology acceptance prior to the development or improvements for that technology. The study objective is to systematically review the studies that empirically had evaluated the acceptance of technology in healthcare through the technology acceptance model (TAM), its extensions and integrated models based on it. Also, the study will highlight the various studied technologies in healthcare arena, and how these technologies can be utilized to provide the health services, as a respond to the on-going pandemic. PRISMA guidelines were used to perform the review; and the search process has been completed using six digital libraries: Google Scholar, PubMed, IEEE Xplore, Springer Link, ACM, and Science Direct. Out of 1768 studies, a total of 99 empirical studies were found to be eligible and included in this study. A thorough statistical analysis was achieved, to understand the situation of technology acceptance as in the recent decade. The analysis included the key factors, as they were extensively utilized to clarify the technology acceptance, along with the key confirmed hypotheses to build robust and valid technology acceptance models in healthcare. It was found that electronic records, tele-medicine and mobile health solutions have attracted the most of researchers in the last ten years. Where the acceptance of those solutions was explored, through various user types and settings, within different countries particularly Taiwan and the United States; who are leading this research domain.

A. A. AlQudah · S. A. Salloum (✉) · K. Shaalan
Faculty of Engineering & IT, The British University in Dubai, Dubai, United Arab Emirates
e-mail: ssalloum@sharjah.ac.ae

S. A. Salloum
Research Institute of Sciences & Engineering, University of Sharjah, Sharjah, United Arab Emirates

© The Author(s), under exclusive license to Springer Nature Switzerland AG 2021   223
I. Arpaci et al. (eds.), *Emerging Technologies During the Era of COVID-19 Pandemic*, Studies in Systems, Decision and Control 348,
https://doi.org/10.1007/978-3-030-67716-9_14

Keywords COVID-19 · Technology acceptance model · Healthcare · Systematic review · PRISMA

1 Introduction

Apart from the health perspective, the impact of coronavirus has expanded to every single side of our lives. Basically, the key objective for the development of technology solutions healthcare was to ease the delivery of health services, instead of necessity [1]. As per the healthcare professionals, the key precaution to reduce the spread rate is to maintain the social distancing rules [2]. So, the initial reaction from the governments was to minimize or even stop all clinical services [3]; which opened the door to make the use of information technologies essential. Globally, healthcare organizations are trying to use different technologies to provide their medical services while patients are stayed at homes, i.e. tele-medicine technologies, especially that many cases can be managed and treated effectively from a distance [4–6]. Also, other solutions can help to reduce the number of visits during this pandemic or even in the future including mobile health services [7], [8], and health portals [9–11].

To ensure the success of any information technology, it is important to have the user's acceptance for that technology [12]. Low level of acceptance for a specific information technology may result to failure, or at least slowness in the execution of that technology [13–15]. In healthcare, the absence of technology acceptance has negative influence on the key objectives, i.e. patients' data management and storage [16]. Technology acceptance indicates the positive psychological status towards the usage intention of innovative technology solutions [17–19]. Technology acceptance is constant process due to the continuous changes in the requirements of users, and significant at any time of the technology life cycle; including the design and after implementation phases [20]. It is obvious that information technologies are continuously getting extended in healthcare domain [21]. Information technologies facilitate the quality of clinical services, and maximize patient's safety. As well, information technologies is playing vital role to enhance healthcare staff's work efficiency and effectiveness [22].

Through the years, the Technology Acceptance Model (TAM) [23–29], its extensions [30, 31] and modifications have been employed, to explain acceptance of various information technologies in healthcare domain [32–35]. Where these technologies Include healthcare websites [10], mobile applications [16], tele-medicine solutions [36], electronic health records [37]. The objective of this study is to provide an overview for the studied technologies in healthcare, with respect to the technology acceptance. Through conducting a systematic review, the study aims to explain the situation of technology acceptance literature in healthcare. The study is looking to clarify the current direction of technology acceptance literature, and how it can be improved to facilitate in the battle with COVID-19 virus or other similar crises in future.

Moreover, it is not possible to deny the other conducted reviews, to survey the technology acceptance in healthcare arena [22], [38–45]. On the other hand, this review is novel for several reasons. First, the study will include only studies with empirical assessment for TAM, its extensions, and modifications. Second, the review will discuss those technologies with respect to their role to respond to the current pandemic (COVID-19). Third, different information technologies will be reviewed in the study, and not only one technology, e.g. Electronic Health Records. Fourth, studies with various settings and user types were considered. Fifth, the reviewed studies were published in the last ten years (2010–2019), to provide new summary about the literature. Finally, this study will discuss different implications, and future directions that are vary from other reviews.

2 Novel Coronavirus (COVID-19)

With its high levels of spread and unique composition, COVID-19 is considered the most known pandemic disease [2, 46]. Nevertheless, other respiratory viruses from the same family of Coronaviruses have arisen with severe harms. Porcine Epidemic Diarrhea—PED virus, Severe Acute Respiratory Syndrome—SARS, and the Middle East Respiratory Syndrome-MERS [47]. In 31st of December 2019, the China office of World Health Organization (WHO) has reported cluster of cases with new pneumonia in the city of Wuhan, China. Since then the virus became aggressive and spread quickly to the whole world [48].

The impact of COVID-19 is now linked to every aspect of people's lives such like education, healthcare, jobs and world economy. While we are in August 2020 and the pandemic is still going on, most of the schools and universities are still closed which is causing an interruption in the learning process; with lower prospects of students' growth and development. As well, The delivery of online learning had become big challenge for several educational institutes during this pandemic [49]. In addition, many healthcare services have been reduced or even stopped [3]; including the delivery of surgical services and care for many patients [50]. Moreover, the outbreak and evolution of COVID-19 have created high levels of uncertainty and deactivated the global economy. So, it is hard for decision-makers to formulate a suitable response for the macroeconomic policy [51].

By the mid of August 2020, A total of 21,368,534 cases of Coronavirus have been reported from 213 different countries and territories. Out of the total reported cases, still 6,445,473 cases are active and 14,923,061 were closed due to recovery (95%), or deaths (5%) [52]. These numbers are scaring, and needs a huge global collaboration especially in the scientific research, vaccine industry and the development information technology.

3   Research Methodology

Various digital databases have been explored, by conducting a review for the published studies. The goal was to obtain findings from the studies that empirically have studied the acceptance of technology in healthcare arena. The review of prior related literature is considered key phase through the execution scientific study. In general, reviews are helpful to ease and expand the development of theory, important to fill research gaps when it is mandatory, or lock the research area where a plethora of literature is available [53]. Systematic review is providing the support researchers, to become familiar with their research topic [11] and previous concepts [54–58]. Unlike the traditional or narrative review, Systematic reviews are more rigorous, and suggest well-defined methods to analyze the literature of specific topic [11].

The review within this study was performed based on the guidelines of Preferred Reporting Items for Systematic Reviews and Meta-analysis (PRISMA), as in Fig. 1 [59, 60]. Consequently, the utilized approach to recognize studies and collect the required data has contained various stages: specify the inclusion and exclusion criteria, identify the digital databases to explore, define the strategies to search each digital library, and perform the relevant analysis for the retrieved research papers.

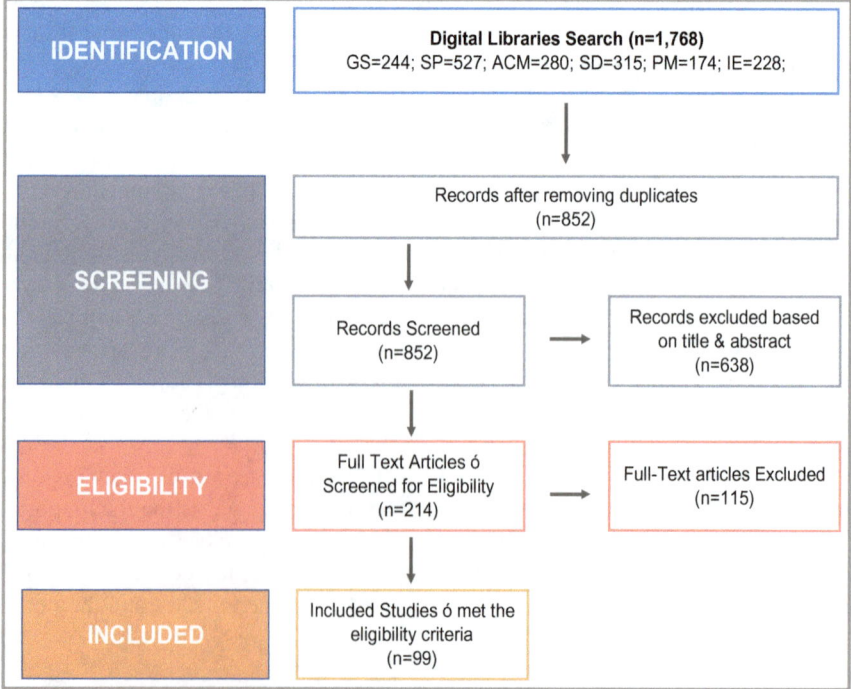

Fig. 1 PRISMA flow diagram

3.1 Search Strategy

Six digital databases were employed to search for the required studies: Google Scholar, Springer, IEEE Xplore, PubMed, Science Direct and ACM digital library. These databases have been explored to retrieve studies, as being published from January 2010 to December 2019 (10 years). The development of search strategy accomplished by defining particular search keywords as in Table 1, and search criteria based on the search features of each digital library. The search keywords and strategy were followed, and the initial results contain 1768 studies as a total as seen in Table 1.

3.2 Selection Criteria

The selection criteria were specified, and included the inclusion and exclusion rules as in Table 2. It is vital to define selection criteria, so the study can easily be classified as valid and guarantee reliability in the collected studies and data analysis. So, the inclusion criteria presented the following rules:

3.3 Data Abstraction and Analysis

The citations of all studies were downloaded to Mendeley reference manager [61]. The rules of inclusion criteria have been applied, a quick screening for titles and abstracts took place to filter the studies. In case of passing those two rounds, the full paper will be retrieved and saved in another folder for final review round. Prior to the data analysis, the eligible studies will be copied to new folder.

Data extraction took place through four phases. The first two stages were to classify the studies as per the type of used model to evaluate the technology acceptance, and to classify the papers as per the studied technology in healthcare.

Table 1 Summary of search keywords

ID	Keywords
1	("Technology Acceptance") AND (Healthcare OR Health OR Medical OR Physician OR Nurse OR Patient)
2	("Technology Adoption") AND (Healthcare OR Health OR Medical OR Physician OR Nurse OR Patient)
3	("Technology Acceptance") AND (Healthcare OR Health OR Medical OR Physician OR Nurse OR Patient) AND ("Intention to use" OR "Actual use")
4	("Technology Adoption") AND (Healthcare OR Health OR Medical OR Physician OR Nurse OR Patient) AND ("Intention to use" OR "Actual use")

Table 2 Inclusion and exclusion criteria

Inclusion criteria	Exclusion criteria
• The study must have been published between January 2010 and December 2019	• The study type is review, position paper, editorials…etc.
• The study should be issued only in English	• Not issued in English
• The study to be within these types: journal article, conference paper, book chapter, Ph.D. Dissertation or Master's thesis	• The subject area is related to animals not human beings
• The study is related to Human beings only	The study's objective is related to the adoption of technology acceptance but not in healthcare domain (e.g. public services, Transportation)
• The studied user groups to include healthcare professionals, medical students and patients	• The discussed model is not related to TAM, not clear, or its constructs are not well-defined
• The study's objective is associated with the utilization of Technology Acceptance Model (TAM), its extensions, modifications or integrated model based on TAM within the healthcare field	• The developed hypotheses are not clear, examined through qualitative approach, or even were not examined at all
• The studied model is clear (Original TAM, Extended, Modified or Integrated)	• The technology in the study is not clearly declared
• The discussed model constructs are clear	• No clear methodology is defined
• The hypotheses are well-defined, and empirically examined.	• The utilized measurement tools, strategy, or used tests are not clearly mentioned.
• A clear declaration for the studied technology	• The findings are not provided or incomplete
• The study includes methodology that is clearly specified	The study type is review, position paper, editorials….etc.
• Well presentation for the used measurement tools and tests	• Not issued in English
• All findings are provided and clear	• The subject area is related to animals not human beings

Third stage to sort the studies as per the year of publication, type, and country where the study has been conducted. The aim of the fourth phase was to explore the extensively used external factors, recognize the confirmed hypotheses among these factors, by analyzing the findings of each study.

4   Results and Discussion

As seen in Fig. 1 and Table 3, the initial search results included 1768 studies as retrieved from the digital databases. After removing the duplicates (916 records), 852 studies were valid to enter the screening stage. Titles and abstracts have been

Table 3 Initial search results

ID	Digital library	Frequency
1	Google Scholar	244
2	IEEE Xplore	228
3	Springer Link	527
4	ACM Digital Library	280
5	Science Direct	315
6	PubMed	174
Total		1768

reviewed for the 852 studies. It was found that 638 studies need to be excluded, due to their unsuitability with the inclusion criteria. At the end, full texts were quickly scanned for 214 publications, in order to confirm the achievement of other inclusion rules. A total of 99 studies were recognized as valid and eligible to be included in the analysis process at this study. Table 4 provide summary for all the eligible publications with integrated models based on TAM.

5 Study Implications

The empirical studies of technology acceptance in healthcare were systematically reviewed, and analyzed, to provide a comprehensive summary the literature in last decade. This summary can be used to shape the direction of research and improve it, where the direction should help to serve the mitigation of the negative impact of COVID-19 on the healthcare service. The mitigation can include the development of new information technologies or enhancements for the current implemented technologies. To achieve these objectives, the acceptance of various technologies in healthcare was reviewed through analyzing the publications between January 2010 and December 2019. The selected theory to review was TAM as proposed by [23], its extensions and modifications since TAM is the prevailing model to understand the user's technology acceptance in healthcare [53].

Perceived ease of use and perceived usefulness are the key constructs of technology acceptance model, and have been extensively used in several studies to measure the level of acceptance for various technologies in healthcare [85–88]. It was confirmed that these two constructs are capable to clarify around 40 percent of user's intention to use and adopt information technologies [41], in different domain plus healthcare [40, 89, 90]. As seen in Fig. 2, the original TAM and its extensions were utilized in 76 studies as per the performed review. While other integrated models based on TAM were proposed in 23 studies, to evaluate the acceptance of technology in healthcare. Such results confirm the suitability and powerful state of TAM, and its constructs to explain the acceptance of different technologies through different types of users. But the result is delivering an implicit message regarding the importance to integrate TAM with other models, and inject other factors to extend the explanatory power of TAM, as proposed by [91].

Table 4 Summary for the studies that discuss integrated acceptance models based on TAM

Source	Year	Type	Technology	Sample size	Sample type	Country	Model
[62]	2010	Conference	Computer assistance Orthopedic surgery system	115	Healthcare Professionals	Taiwan	Integrated model: TAM & TPB
[63]	2010	Journal article	Tele-homecare Technology (Telemedicine)	40	Physicians	USA	Compare two models: TAM & TPB
[64]	2011	Journal article	Healthcare information systems	366	Nurses, Head directors and other related personnel	Taiwan	Integrated model: TAM & IS success model
[65]	2012	Journal article	Health information technology (HIT)	728	Users of online health Information	South Korea	Integrated model-health information technology acceptance model (HITAM): HBM, TPB & TAM
[66]	2012	Conference	Clinic information system	252	Doctors & staff	Malaysia	Integrated model: TAM & TPB
[67]	2013	Journal article	E-learning system	218	Nurses	Taiwan	Integrated model: TAM & flow theory
[68]	2013	Conference	Health information system (HIS)	252	Staff in private healthcare organizations	Malaysia	Integrated model: TAM & TPB
[69]	2013	Journal article	Personal digital assistant (PDA)	222	Physicians	USA	TAM, TPB, and IDT
[70]	2013	Journal article	Clinic information system (CIS)	252	Doctors & staff	Malaysia	Extended hybrid model: TAM & TPB
[71]	2014	Conference	Health cloud services	443	Patients	Taiwan	Integrated model: TAM & SQB

(continued)

Table 4 (continued)

Source	Year	Type	Technology	Sample size	Sample type	Country	Model
[72]	2014	Journal article	Electronic health record (EHR)	150	Physicians	Canada	4 models: TAM, extended TAM, psychosocial model & integrated model
[73]	2014	Journal article	Innovative smartphone	122	Hospital professionals	S. Korea	Integrated model: TRA, TAM & IS success model
[74]	2014	Journal article	Telehealth system	365	Patients	Taiwan	Integrated model: extended TAM & HBM
[75]	2015	Conference	Consumer Health informatics applications	105	Health Consumers	Malaysia	Integrated model: TAM, TRA & UTAUT2
[76]	2015	Journal article	Health-related internet use	293	Female users	Malaysia	Integrated model: HBM & TAM
[77]	2015	Journal article	Mobile electronic medical Records	158	Physicians	Taiwan	Integrated model: TAM and Dual Factor Model
[78]	2016	Journal article	Health information technology: pharmaceutical service systems	1420	Pharmacists/ pharmaceutical assistants	Turkey	Integrated model (P-TAM): TAM, UTAUT & TPB
[79]	2016	Conference	Hospital information systems	100	Hospital staff & doctors	Indonesia	Integrated model: TAM & DeLone and McLean IS success
[80]	2016	Journal article	Computerized clinical practice guidelines	238	Physicians	Taiwan	Integrative model of activity theory and TAM
[81]	2017	Conference	E-health services consumer informatics	91	Citizens	Indonesia	Extended model: TAM & HBM

(continued)

Table 4 (continued)

Source	Year	Type	Technology	Sample size	Sample type	Country	Model
[82]	2017	Journal article	Nursing information system	531	Nurses	Taiwan	Integrated model: TAM & ISSM
[83]	2019	Journal article	Smart wearables	146	60 + years old adults	China	Extended hybrid model: TAM & UTAUT
[84]	2019	Journal article	Telehealth	281	Adults 40+	Taiwan	Integrated model: TAM & SQB

As one of the important objectives in this study, the studied technologies were reviewed. The goal is to recognize what are the prevailing technologies, and to decide whether these technologies are in line with the direction to implement and push for the use and acceptance of E-health services. These services and technologies that can help to mitigate the impact of COVID-19 and reduce its levels of spread, by providing the required treatments or consultations for patients while they stay at home, as a part of the social-distancing precautions. It is essential to study the acceptance of these technologies, to decide how they can be improved, and what is impacting the intention of people to use these applications to support in the battle against the coronavirus.

With a total percentage of 58.6% in term of utilization, it was clear that the research is dominated by three main categories, Telemedicine Solutions, Mobile Health Services, Electronic Records Solutions (e.g. Health information solutions, electronic medical records and electronic health records), as seen in Table 5. Although a plethora exists in the research related to these categories, it is beneficial for the mitigation of COVID-19 impact. Nowadays, telemedicine is experiencing a rapid growth, since it is effective solution to achieve the social-distancing in clinics, and provide the needed health services [1, 4]. Despite the past slow implementations of telemedicine [4], this situation of growth increases the need to study and apply the acceptance of telemedicine, in order to accelerate the adoption, ensure the success of the solution, and facilitate in the mitigation for the impact of COVID-19.

As the virus is quickly spreading and its negative impact is increasing, the active learning curve about it is getting higher. Such daily active learning is helpful to empower the information about its origin, root-cause and the required pre-cautionary measures. Electronic records solutions are playing vital role to collect, store, organize the patient's data, it can increase the production of data in the healthcare field [92]. So, to enhance the data collection and organization

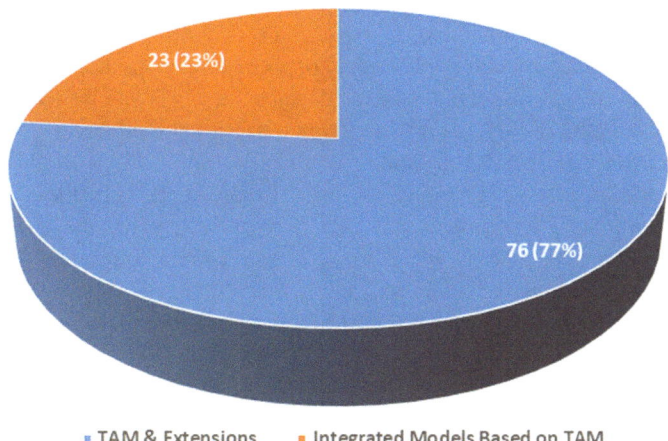

Fig. 2 The utilization of TAM in the literature

Table 5 Statistics of studied technologies

ID	Studied technology	Frequency	(%)
1	Barcode technology	3	3.0
2	Cloud health services	3	3.0
3	E-Learning and education	6	6.1
4	Electronic records solutions	27	27.3
5	Health portals/websites	7	7.1
6	Mobile health services	14	14.1
7	Tele-medicine technology	17	17.2
8	Wearables devices	4	4.0
9	Others	18	18.2
Totals		99	100

processes, it is recommended to review the literature that is related to the adoption and acceptance of electronic records solutions in healthcare. On the other hand, it is mandatory to implement data analytics tools to increase the benefits of electronic records solutions, through proper integration. The review found that there is a lack of studies related to the acceptance of analytics tools. Analytics tools are related to the big data theory, and it is helpful to guide policy-makers in individual country. By giving the chance to design suitable models that can explore and study the activity of the virus, big data analytics solutions can improve the process of preparation for the virus outbreak [3]. Besides, the acceptance of other helpful technologies like robotics can be explored. Robotics can play vital role in the application of E-health services to achieve the precaution measures of social-distancing.

Additionally, the results of mobile health services (14.1%) and health websites (7.1%) studies are promising. Mobile applications and internet-based websites can enable the self-guided data collection on the population level, then the results be swiftly circulated to participants to be apprised about the health emergencies [93]. These analyzed studies can form the foundation to decide what is required to develop and enhance mobile applications, or websites for the healthcare purposes and especially in such pandemic cases. For instance, UAE has launched a mobile digital platform "ALHOSN", as a joint national initiative to protect the community. The application can operate on both: Android [94] and IOS [95], and works by collecting the results of coronavirus tests, from different public and private healthcare entities. These results can help in contact tracing, and determine the geographical hot spots nationwide. The stats of studied technologies as discussed in the analyzed studies can be found in Table 5 and Fig. 3.

Furthermore, we cannot deny where the studies have been conducted which can help to recognize the research gap, in some countries or regions. This study could address the origin of each publication to improve the direction of research, and create additional motivation for researchers. For example, the analysis found that the Central and South America regions have contributed with zero studies to the literature of technology acceptance in healthcare, as highlighted in Fig. 4. These

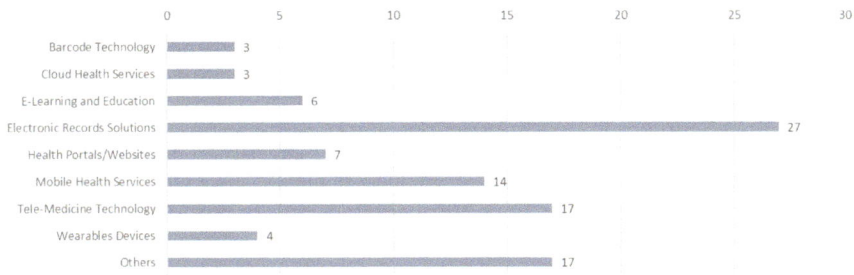

Fig. 3 Number of publications per studied technology

results mean that there is a research gap there, and it needs to be covered. It can be an indication for the rare adoptions and implementations for technology in healthcare field, at these two regions. To prevent any bias, there is a possibility that many technology acceptances studies have been published in Spanish, or Portuguese since these are the commonly used languages there. Additionally, a limited number of studies were achieved in Arab and African as developing regions. Although UAE and Jordan have advanced healthcare systems and successfully implemented a lot of healthcare applications, only three studies were conducted in UAE, and two studies in Jordan. So, it is obvious that these countries are facing a gap in research of technology acceptance in healthcare which may negatively impact the development or improvements of information technologies in healthcare, and consequently affect the mitigation of COVID19 outbreak.

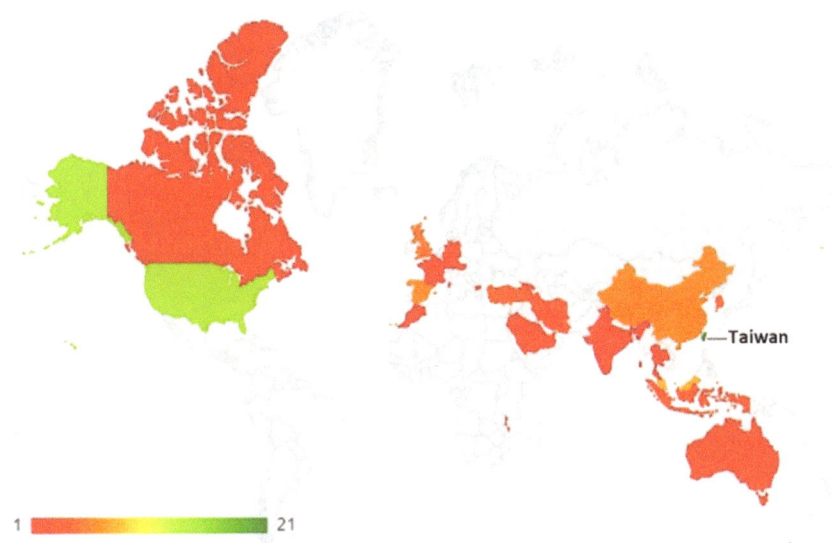

Fig. 4 Geographic chart for the studies as included in this study

On the other hand, Asia was found to have the highest number of published studies (56) in technology acceptance in healthcare. Remarkably, Taiwan has recorded more than 21% of the total analyzed studies, which equals 37.5% of total studies in Asia. This might refer to the well-established healthcare system in Taiwan [96]. Also, the United States as a first runner-up with results of 14 empirical studies to measure the technology acceptance in healthcare. The results of the United states can be considered poor, considering that the united states is the global leader in terms of science and technology research as per the 2018 report of the National Science Foundation's (NSF) Science and Engineering Indicators [97]. More geographical details for the analyzed studies are illustrated in Figs. 5 and 6.

Finally, a categorization per year of publication was performed. An increment can be noticed from 2010 till 2012, with more or less constant frequency till 2016. There was a drop in number of published studies from 2017 and ongoing, which

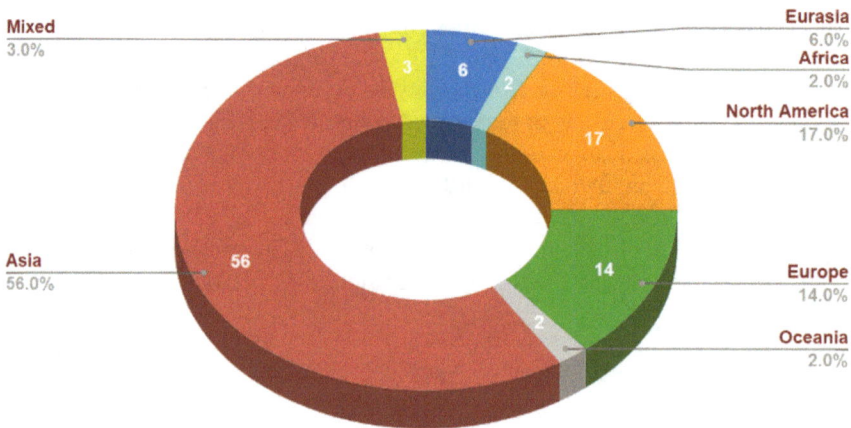

Fig. 5 Publications statistics per region

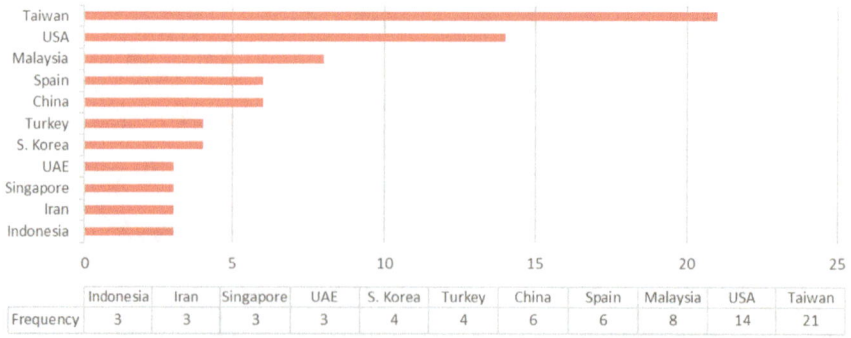

	Indonesia	Iran	Singapore	UAE	S. Korea	Turkey	China	Spain	Malaysia	USA	Taiwan
Frequency	3	3	3	3	4	4	6	6	8	14	21

Fig. 6 Publications statistics per country

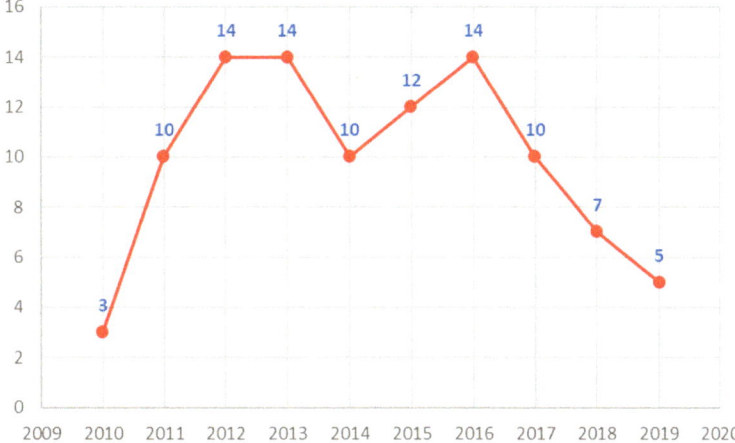

Fig. 7 Frequency of studies per year

can extend the research gap. It seems that the interest to explore the acceptance of technologies in healthcare is getting declined as we can see in Fig. 7. These low numbers, or low interest levels are not helpful in terms of COVID19 mitigation, and huge research work should take place. Government authorities, universities and research centers need to collaborate, formulate appropriate polices, and assign required budgets to motivate the research work of technologies in healthcare, especially those that can facilitate the application of E-health services in such pandemic cases.

6   Study Implications

The study purpose was to provide an overview for the current literature of technology acceptance in healthcare. The studied technologies were reviewed, and assessed to know if they are aligned with the current need to apply the E-health services, due to the situation of COVID19 outbreak. This review considered unique in this crisis's situation, and could provide number of theoretical and practical implications.

First, various technologies were reviewed along with different types of users instead of one type. This added a diversity characteristic to this review, and can be positive to give wider view for researchers, and policy-makers. Second, this review helps to improve the theory of technology acceptance, by recognizing the prevailing technologies in healthcare, the research areas that need to be covered and the location of each research. Third, the review can be helpful to form the direction of technology research, in a way that can serve to mitigate the negative impact of COVID19 on the healthcare services. In contrast, the review can be practical by building the required foundation for other researchers. It gives the required track to follow, to know which technologies are essential in this time of pandemic, and needed currently to be studied in terms of technology acceptance in healthcare. The review is considered supportive, to identify what is impacting the acceptance of these technologies, and how to enhance the user's acceptance to help during the current or future crises. As lessons-learned, information technology providers and healthcare organizations can utilize the findings of this review to improve the current developed solutions, consider optimizations to serve in this pandemic, or avoid mistakes that can lower the levels of user's acceptance for technology.

7 Conclusion and Future Work

The study objective is to review the recent literature technology acceptance in healthcare domain, and build clear perception about the studied technologies and how they can facilitate in the mitigation of COVID19. To achieve these objectives, the study employed a systematic review methodology based on PRISMA guidelines. A total number of 1768 published studies have been reviewed, and 99 studies were identified eligible to be analyzed. The analysis of the review could confirm the availability of promising findings, with a lead for technologies that can be helpful to reduce the impact of COVID19 break, the negative impact on the health services in this case. Those leading technologies include telemedicine solutions, mobile health services, and electronic records solutions (electronic medical records, health information systems…etc.). As well, there is a room to study other technologies like robotics and big data analytics solutions, where these technologies can play vital role in the application of E-health services to comply the precaution measures of social-distancing. In general, it was found that the reviewed studies were mostly performed in Taiwan, and the United States. Arab and African countries as part of developing regions, are still lagging behind in terms of the technology acceptance research. In spite of the advanced healthcare systems, and the successful of adoption for information technology in some Arab countries, i.e. UAE and Jordan, it is obvious that there is a gap of research with relation to technology acceptance in healthcare. This shortage of research can negatively impact the development or improvements of information technologies in healthcare, due to the absence of understanding to the factors that impact user's acceptance, and accordingly affect the mitigation of COVID19 outbreak.

As a clear limitation, the virus is still new and its effects have not been fully disclosed, which caused difficulties to conduct this research due to the scarcity of published literature related to the risks and consequences of COVID19. Also, the available COVID19 literature is still scattered especially from technology perspectives. In the future there should be an obvious direction of how technology and its acceptance can minimize the negative impact of COVID19. The results of this review can be utilized to produce meta-analysis to facilitate in the research direction. Moreover, only one technology model (TAM) was included in this review. To extend this view, the acceptance of technology in healthcare can be reviewed as studied through other theories other than TAM. As well, considering the studied factors and constructed hypotheses in these analyzed studies. This can be helpful to know from where we have to start, and what to include to understand the enablers and barriers of technology adoption in healthcare, with respect to COVID19 outbreak and mitigation.

Acknowledgements This work is a part of a project undertaken at the British University in Dubai.

References

1. Clipper, B.: The influence of the COVID-19 pandemic on technology. Nurse Lead. June (2020)
2. Habes, M., Alghizzawi, M., Ali, S., SalihAlnaser, A., Salloum, S.A.: The Relation among Marketing ads, via Digital Media and mitigate (COVID-19) pandemic in Jordan. Int. J. Adv. Sci. **29**(7), 2326–12348 (2020)
3. Ting, D.S.W., Carin, L., Dzau, V., Wong, T.Y.: Digital technology and COVID-19. Nat. Med. **26**(4), 459–461 (2020)
4. Bashshur, R., Doarn, C.R., Frenk, J.M., Kvedar, J.C., Woolliscroft, J.O.: Telemedicine and the COVID-19 Pandemic, Lessons for the Future. Telemed. e-Health **26**(5), 571–573 (2020)
5. AlShuweihi, M., Salloum, S.A.: Biomedical Corpora and Natural Language Processing on Clinical Text in Languages Other Than English: A Systematic Review. Al-Emran M., Shaalan K., Hassanien A. Recent Adv. Intell. Syst. Smart Appl. Stud. Syst. Decis. Control, vol. 295. Springer, Cham (2021)
6. Alhashmi, M., Alshurideh, B., Al Kurdi, Salloum, S. A.: A systematic review of the factors affecting the ARTIFICIAL intelligence implementation in the health care sector,. In Joint European-US Workshop on Applications of Invariance in Computer Vision, pp. 37–49 (2020)
7. Guo, J., Yuan, X., Cao, Chen, X.: Understanding the acceptance of mobile health services: A service participants analysis. In: 2012 International Conference on Management Science & Engineering 19th Annual Conference Proceedings, pp. 1868–1873 (2012)
8. Ku, W.T., Hsieh, P.J.: Understanding the acceptance of health management mobile services: Integrating theory of planned behavior and health belief model. Int. Conf. Hum.-Comput. Interact. **850**, 247–252 (2018)
9. Lazard, A.J., Watkins, I., Mackert, M.S., Xie, B., Stephens, K.K., Shalev, H.: Design simplicity influences patient portal use: The role of aesthetic evaluations for technology acceptance. J. Am. Med. Informatics Assoc. **23**(e1), e157–e161 (2016)
10. Boon-itt, S.: Quality of health websites and their influence on perceived usefulness, trust and intention to use: an analysis from Thailand. J. Innov. Entrep., **8**(1), December (2019)
11. Almansoori, A., AlShamsi, M., Salloum, S. A., Shaalan, K.: Critical review of knowledge management in healthcare, pp. 99–119 (2021)
12. Arpaci, I., Al-Emran, M., Al-Sharafi, M. A., Shaalan, K.: A novel approach for predicting the adoption of smartwatches using machine learning algorithms. In: Recent Advances in Intelligent Systems and Smart Applications, Springer, pp. 185–195
13. M. Al-Emran, V., Mezhuyev, Kamaludin, A.: Towards a conceptual model for examining the impact of knowledge management factors on mobile learning acceptance. Technol. Soc. (2020)
14. Arpaci, I.: A hybrid modeling approach for predicting the educational use of mobile cloud computing services in higher education. Comput. Human Behav. **90**, 181–187 (2019)

15. Arpaci, I., Karataş, K., Baloğlu, M.: The development and initial tests for the psychometric properties of the COVID-19 Phobia Scale (C19P-S). Pers. Individ. Dif., 110108 (2020)
16. Ketikidis, P., Dimitrovski, T., Lazuras, L., Bath, P.A.: Acceptance of health information technology in health professionals: An application of the revised technology acceptance model. Health Informatics J. **18**(2), 124–134 (2012)
17. Taherdoost, H.: A review of technology acceptance and adoption models and theories. Procedia Manuf. **22**, 960–967 (2018)
18. Taherdoost,H.: Importance of technology acceptance assessment for successful implementation and development of new technologies. Glob. J. Eng. Sci., **1**(3), January (2019)
19. Chau, P.Y.K., Hu, P.J.-H.: Investigating healthcare professionals' decisions to accept telemedicine technology: an empirical test of competing theories. Inf. Manag. **39**(4), 297–311 (2002)
20. Mathieson, K.: Predicting User Intentions: Comparing the Technology Acceptance Model with the Theory of Planned Behavior. Inf. Syst. Res. **2**(3), 173–191 (1991)
21. Blackwell, G.: The future of IT in healthcare. Informatics Heal. Soc. Care **33**(4), 211–326 (2008)
22. Rahimi, B., Nadri, H., Lotfnezhad Afshar, H., Timpka, T.: A systematic review of the technology acceptance model in health informatics. Appl. Clin. Inform., **9**(3):604–634, July (2018)
23. Davis, F. D.: A technology acceptance model for empirically testing new end-user information systems: Theory and results (1985)
24. Davis, F. D.: Perceived usefulness, perceived ease of use, and user acceptance of information technology. MIS Q., 319–340 (1989)
25. Davis, F.D., Bagozzi, R.P., Warshaw, P.R.: User acceptance of computer technology: a comparison of two theoretical models. Manage. Sci. **35**(8), 982–1003 (1989)
26. Al-Qaysi, N., Mohamad-Nordin, N., Al-Emran: Factors affecting the adoption of social media in higher education: a systematic review of the technology acceptance model. Recent Adv. Intell. Syst. Smart Appl. Springer, Cham, pp. 571–584 (2021)
27. Salloum, S.A., Alhamad, A.Q.M., Al-Emran, M., Monem, A.A., Shaalan, K.: Exploring Students' Acceptance of E-Learning Through the Development of a Comprehensive Technology Acceptance Model. IEEE Access **7**, 128445–128462 (2019)
28. Salloum, S. A., Shaalan, K.: Investigating students' acceptance of e-learning system in higher educational environments in the UAE: Applying the extended technology acceptance model (TAM). The British University in Dubai (2018)
29. Alhashmi, S. F. S., Salloum, S. A., Mhamdi, C.: Implementing Artificial Intelligence in the United Arab Emirates healthcare sector: an extended technology acceptance model. Int. J. Inf. Technol. Lang. Stud., **3**(3) (2019)
30. Venkatesh, V., Davis, F.D.: A theoretical extension of the technology acceptance model: Four longitudinal field studies. Manage. Sci. **46**(2), 186–204 (2000)
31. Venkatesh, V., Bala, H.: Technology acceptance model 3 and a research agenda on interventions. Decis. Sci. **39**(2), 273–315 (2008)
32. Bennani, A. E., Oumlil, R.: Do constructs of technology acceptance model predict the ICT appropriation by physicians and nurses in healthcare public centres in Agadir, South of Morocco? In: HEALTHINF 2010—3rd International Conference on Health Informatics, Proceedings, pp. 241–249 (2010)
33. Al-Nassar, B.A.Y., Rababah, K.A., Al-Nsour, S.N.: Impact of computerised physician order entry in Jordanian hospitals by using technology acceptance model. Int. J. Inf. Syst. Change Manag. **8**(3), 191–210 (2016)
34. Gagnon, M. P., Orruño, E., Asua, J., Ben Abdeljelil, A., Emparanza, J.: Using a modified technology acceptance model to evaluate healthcare professionals' adoption of a new telemonitoring system. Telemed. e-Health, **18**(1):54–59, January (2012)
35. Orruño, E., Gagnon, M. P., Asua, J., Ben Abdeljelil, A.: Evaluation of teledermatology adoption by health-care professionals using a modified Technology Acceptance Model. J. Telemed. Telecare, **17**(6):303–307, September (2011)

36. Kowitlawakul, Y.: The technology acceptance model: Predicting nurses' intention to use telemedicine technology (eICU). CIN - Comput. Informatics Nurs. **29**(7), 411–418 (2011)
37. Tubaishat, A.: Perceived usefulness and perceived ease of use of electronic health records among nurses: Application of Technology Acceptance Model. Informatics Heal. Soc. Care **43** (4), 379–389 (2018)
38. McGinn, C.A., et al.: Comparison of user groups' perspectives of barriers and facilitators to implementing electronic health records: a systematic review. BMC Med. **9**(1), 46 (2011)
39. Or, C.K.L., Karsh, B.-T.: A Systematic Review of Patient Acceptance of Consumer Health Information Technology. J. Am. Med. Informatics Assoc. **16**(4), 550–560 (2009)
40. Holden, R.J., Karsh, B.-T.: The Technology Acceptance Model: Its past and its future in health care. J. Biomed. Inform. **43**(1), 159–172 (2010)
41. Peek, S.T.M., Wouters, E.J.M., van Hoof, J., Luijkx, K.G., Boeije, H.R., Vrijhoef, H.J.M.: Factors influencing acceptance of technology for aging in place: A systematic review. Int. J. Med. Inform. **83**(4), 235–248 (2014)
42. Gagnon, M.-P., et al.: Systematic Review of Factors Influencing the Adoption of Information and Communication Technologies by Healthcare Professionals. J. Med. Syst. **36**(1), 241–277 (2012)
43. Mair, F.S., May, C., O'Donnell, C., Finch, T., Sullivan, F., Murray, E.: Factors that promote or inhibit the implementation of e-health systems: an explanatory systematic review. Bull. World Health Organ. **90**(5), 357–364 (2012)
44. Yarbrough, A.K., Smith, T.B.: Technology Acceptance among Physicians. Med. Care Res. Rev. **64**(6), 650–672 (2007)
45. Vaezipour, A., Whelan, B. M., Wall, K., Theodoros, D.: Acceptance of rehabilitation technology in adults with Moderate to severe traumatic brain injury, their caregivers, and healthcare professionals: a systematic review. J. Head Trauma Rehabil. (2019)s
46. Arpaci, I., et al.: Analysis of Twitter Data Using Evolutionary Clustering during the COVID-19 Pandemic. C. Mater. Contin. **65**(1), 193–203 (2020)
47. Lau, S.K.P., Chan, J.F.W.: Coronaviruses: Emerging and re-emerging pathogens in humans and animals. Virol. J. **12**(1), 10–12 (2015)
48. World Health Organisation): Coronavirus (COVID-19) events as they happen
49. Almaiah, M. A., Al-Khasawneh, A., Althunibat, A.: Exploring the critical challenges and factors influencing the E-learning system usage during COVID-19 pandemic. Educ. Inf. Technol., May (2020)
50. Søreide, K., et al.: Immediate and long-term impact of the COVID-19 pandemic on delivery of surgical services. Br. J. Surg., April (2020)
51. McKibbin, W.J., Fernando, R.: The Global Macroeconomic Impacts of COVID-19: Seven Scenarios. SSRN Electron. J. **2**(4), 12–22 (2020)
52. Worldometer: Coronavirus Cases. Worldometer (2020)
53. Marangunić, N., Granić, A.: Technology acceptance model: a literature review from 1986 to 2013. Univers. Access Inf. Soc. **14**(1), 81–95 (2015)
54. Al Mansoori, S., Salloum, S. A.: The impact of Artificial Intelligence and information technologies on the efficiency of knowledge management at modern organizations: a systematic review. Al-Emran M., Shaalan K., Hassanien A. Recent Adv. Intell. Syst. Smart Appl. Stud. Syst. Decis. Control. vol 295. Springer, Cham (2021)
55. Yousuf H., Lahzi, M., Salloum, S. A.: Systematic review on fully homomorphic encryption scheme and its application. Al-Emran M., Shaalan K., Hassanien A. Recent Adv. Intell. Syst. Smart Appl. Stud. Syst. Decis. Control. vol 295. Springer, Cham (2021)
56. Habeh, O., Thekrallah, F., Salloum, S. A.: Knowledge sharing challenges and solutions within software development team: a systematic review. Al-Emran M., Shaalan K., Hassanien A. Recent Adv. Intell. Syst. Smart Appl. Stud. Syst. Decis. Control. vol 295. Springer, Cham (2021)

57. Areed S., Salloum, S. A.: The role of knowledge management processes for enhancing and supporting innovative organizations: a systematic review. Al-Emran M., Shaalan K., Hassanien A. Recent Adv. Intell. Syst. Smart Appl. Stud. Syst. Decis. Control. vol 295. Springer, Cham (2021)

58. Wahdan, K.S.A., Hantoobi, S., Salloum, S.A., Shaalan, K.: A systematic review of text classification research based ondeep learning models in Arabic language. Int. J. Electr. Comput. Eng **10**(6), 6629–6643 (2020)

59. Moher, D., Liberati, A., Tetzlaff, J., Altman, D.G.: Preferred reporting items for systematic reviews and meta-analyses: the PRISMA statement. J. Clin. Epidemiol. **62**(10), 1006–1012 (2009)

60. Moher, D., et al.: Preferred reporting items for systematic review and meta-analysis protocols (PRISMA-P) 2015 statement. Syst. Rev. **4**(1), 1 (2015)

61. Mendeley Ltd.: Mendeley (2020)

62. Lai, D. W., Li, Y. P.: Examining the technology acceptance model of the computer assistance orthopedic surgery system. In: 2010 7th International Conference on Service Systems and Service Management, Proceedings of ICSSSM' 10, pp. 940–945 (2010)

63. Kim, J., DelliFraine, J.L., Dansky, K.H., McCleary, K.J.: Physicians' acceptance of telemedicine technology: An empirical test of competing theories. Int. J. Inf. Syst. Change Manag. **4**(3), 210–225 (2010)

64. Pai, F.Y., Huang, K.I.: Applying the Technology Acceptance Model to the introduction of healthcare information systems. Technol. Forecast. Soc. Change **78**(4), 650–660 (2011)

65. Kim, J., Park, H. A.: Development of a health information technology acceptance model using consumers' health behavior intention. J. Med. Internet Res. **14**(5) (2012)

66. Sarlan, A., Ahmad, R., Wan Ahmad, W. F., Dominic, P. D. D.: Users' behavioral intention to use clinic information system: A survey. In: 2012 International Conference on Computer and Information Science, ICCIS 2012—A Conference of World Engineering, Science and Technology Congress, ESTCON 2012—Conference Proceedings, vol. 1, pp. 37–43 (2012)

67. Cheng, Y.M.: Exploring the roles of interaction and flow in explaining nurses' e-learning acceptance. Nurse Educ. Today **33**(1), 73–80 (2013)

68. Sarlan, A., Ahmad, R., Fatimah, W., Ahmad, W., Dominic, P. D. D., Private Healthcare in Malaysia : Investigation on Technology Profiles and Technology Acceptance Factors. In: Information Systems International Conference (ISICO), 2–4 December 2013, December, pp. 98–103 (2013)

69. Jackson, J.D., Yi, M.Y., Park, J.S.: An empirical test of three mediation models for the relationship between personal innovativeness and user acceptance of technology. Inf. Manag. **50**(4), 154–161 (2013)

70. Sarlan, A., Ahmad, R., Ahmad, W.F.W., Dominic, D.D.: A study of SME private healthcare personnel acceptance of Clinic Information System in Malaysia. Int. J. Bus. Inf. Syst. **14**(2), 238 (2013)

71. Hsieh, P. J., Lai, H. M., Ye, Y. S.: Patients' acceptance and resistance toward the health cloud: An integration of technology acceptance and status quo bias perspectives. In Proceedings—Pacific Asia Conference on Information Systems, PACIS (2014)

72. Gagnon, M.P., et al.: Electronic health record acceptance by physicians: Testing an integrated theoretical model. J. Biomed. Inform. **48**, 17–27 (2014)

73. Moon, B.C., Chang, H.: Technology acceptance and adoption of innovative smartphone uses among hospital employees. Healthc. Inform. Res. **20**(4), 304–312 (2014)

74. Tsai, C.H.: The adoption of a telehealth system: The integration of extended technology acceptance model and health belief model. J. Med. Imaging Heal. Informatics **4**(3), 448–455 (2014)

75. Krishnan, S. B., Dhillon, J. S., Lutteroth, C.: Factors influencing consumer intention to adopt Consumer Health Informatics applications an empirical study in Malaysia. In: 2015 IEEE Student Conference on Research and Development, SCOReD, pp. 653–658 (2015)

76. Ahadzadeh, A. S., Pahlevan Sharif, S., Ong, F. S., Khong, K. W.: Integrating Health Belief Model and Technology Acceptance Model: An investigation of health-related Internet use. J. Med. Internet Res., **17**(2) (2015)
77. Liu, C. F., Cheng, T. J.: Exploring critical factors influencing physicians' acceptance of mobile electronic medical records based on the dual-factor model: A validation in Taiwan. BMC Med. Inform. Decis. Mak., **15**(1) (2015)
78. Sezgin, E., Özkan-Yıldırım, S.: A cross-sectional investigation of acceptance of health information technology: A nationwide survey of community pharmacists in Turkey. Res. Soc. Adm. Pharm. **12**(6), 949–965 (2016)
79. Made Dhanar, I. Y., Reza, M., Meyliana, Widjaja, H. A. E., Hidayanto, A. N.: Acceptance of HIS usage level in hospital with SEM-PLS as analysis methodology: Case study of a private hospital in Indonesia. In: Proceedings of 2016 International Conference on Information Management and Technology, ICIMTech, pp. 112–117 (2016)
80. Hsiao, J. L., Chen, R. F.: Critical factors influencing physicians' intention to use computerized clinical practice guidelines: an integrative model of activity theory and the technology acceptance model. BMC Med. Inform. Decis. Mak., **16**(1), January (2016)
81. Wahyuni, R., Nurbojatmiko: Explaining acceptance of e-health services: An extension of TAM and health belief model approach. In: 2017 5th International Conference on Cyber and IT Service Management, CITSM (2017)
82. Lin, H.C.: Nurses' satisfaction with using nursing information systems from technology acceptance model and information systems success model perspectives. CIN—Comput. Informatics Nurs. **35**(2), 91–99 (2017)
83. Li, J., Ma, Q., Chan, A.H., Man, S.S.: Health monitoring through wearable technologies for older adults: Smart wearables acceptance model. Appl. Ergon. **75**, 162–169 (2019)
84. Tsai, J.M., Cheng, M.J., Tsai, H.H., Hung, S.W., Chen, Y.L.: Acceptance and resistance of telehealth: The perspective of dual-factor concepts in technology adoption. Int. J. Inf. Manage. **49**, 34–44 (2019)
85. Beldad, A.D., Hegner, S.M.: Expanding the Technology Acceptance Model with the Inclusion of Trust, Social Influence, and Health Valuation to Determine the Predictors of German Users' Willingness to Continue using a Fitness App: A Structural Equation Modeling Approach. Int. J. Human-Computer Interact. **34**(9), 882–893 (2018)
86. Lin, W.-Y., Ke, H.-L., Chou, W.-C., Chang, P.-C., Tsai, T.-H., Lee, M.-Y.: Realization and Technology Acceptance Test of a Wearable Cardiac Health Monitoring and Early Warning System with Multi-Channel MCGs and ECG. Sensors **18**(10), 3538 (2018)
87. Liu, M. C., Lee, C. C.: An Investigation of Pharmacists' Acceptance of NHI-PharmaCloud in Taiwan. J. Med. Syst., **42**(11), November (2018)
88. Nadri, H., Rahimi, B., Afshar, H.L., Samadbeik, M., Garavand, A.: Factors affecting acceptance of hospital information systems based on extended technology acceptance model: a case study in three paraclinical departments. Appl. Clin. Inform. **9**(02), 238–247 (2018)
89. King, W.R., He, J.: A meta-analysis of the technology acceptance model. Inf. Manag. **43**(6), 740–755 (2006)
90. Legris, P., Ingham, J., Collerette, P.: Why do people use information technology? A critical review of the technology acceptance model. Inf. Manag. **40**(3), 191–204 (2003)
91. Kim, S., Lee, K.-H., Hwang, H., Yoo, S.: Analysis of the factors influencing healthcare professionals' adoption of mobile electronic medical record (EMR) using the unified theory of acceptance and use of technology (UTAUT) in a tertiary hospital. BMC Med. Inform. Decis. Mak. **16**(1), 12 (2016)
92. Tavakoli, N., Jahanbakhsh, M., Shahin, A., Mokhtari, H., Rafiei, M.: Electronic medical record in central polyclinic of isfahan oil industry: A case study based on technology acceptance model. Acta Inform. Medica **21**(1), 23–25 (2013)
93. Drew, D. A., et al.: Rapid implementation of mobile technology for real-time epidemiology of COVID-19. Science (80-.). **368**(6497):1362–1367, June (2020)

94. Google LLC: Android|The platform pushing what's possible. (2019)
95. Apple. iOS 14 Preview—Features—Apple (2020)
96. Wu, T.-Y., Majeed, A., Kuo, K.N.: An overview of the healthcare system in Taiwan. London J. Prim. Care (Abingdon) **3**(2), 115–119 (2010)
97. National Science Board: S&E Indicators 2018|NSF—National Science Foundation. National Science Board Science and Engineering Indicators (2018)

Psychological and Socio-Economic Effects of the COVID-19 Pandemic on Turkish Population

Mustafa Baloglu⑩, **Kasım Karatas**⑩, **and Ibrahim Arpaci**⑩

Abstract The negative effects of the COVID-19 pandemic are not limited to psychological, but also include social and economic effects. This study investigated psychological, social, and economic effects of COVID-19 pandemic on the Turkish population. COVID-19 Phobia Scale (C19P-S) was used to collect data from 2143 participants. Results indicated that women showed the highest phobic reactions on the economic subscale whereas men showed the highest phobic reactions on psycho-somatic subscale. Patterns of differences varied among geographical regions but in general, eastern regions scored higher than western regions. Significant differences were also observed based on educational attainment; lower-middle class showed the highest scores on all the subscales.

Keywords Coronavirus · COVID-19 · Phobia · Psychological · Social · Economic

1 Introduction

After its first appearance in China at the end of 2019, corona virus disease (COVID-19) quickly became a worldwide pandemic. COVID-19 currently effects all continents and 213 countries. A total of confirmed cases as of May 2020 are 5,370,375 and confirmed death tolls are 344,454 [1]. Many countries have to take

M. Baloglu
School of Education, Hacettepe University, Ankara, Turkey
e-mail: baloglu@hotmail.com

K. Karatas
Department of Educational Sciences, Karamanoglu Mehmetbey University, Karaman, Turkey
e-mail: kasimkaratas@kmu.edu.tr

I. Arpaci (✉)
Department of Computer Education and Instructional Technology, Tokat Gaziosmanpasa University, Tokat, Turkey
e-mail: ibrahim.arpaci@gop.edu.tr

© The Author(s), under exclusive license to Springer Nature Switzerland AG 2021   245
I. Arpaci et al. (eds.), *Emerging Technologies During the Era of COVID-19 Pandemic*, Studies in Systems, Decision and Control 348,
https://doi.org/10.1007/978-3-030-67716-9_15

serious measures to slow it down or control the spread rate of COVID-10. Some of these measures include restricting travel abroad, quarantining people, vacationing schools, closing down places of worship, introducing strict curfews, and temporarily closing down offices and shopping centers [2].

Because of large number of infected and death cases and no cure available, COVID-19 cause serious fear and anxiety among public. Similar epidemics such as H1N1, SARS, MERS, Ebola, and Zika viruses have also had similar negative physiological and psychological effects [3]. Anecdotal reports also confirm that COVID-19 leads to increases in the level of anxiety worldwide. Similarly, COVID-19 associated mental difficulties are already being reported [4] and its negative effects are expected to continue for an unknown period of time.

The COVID-19 pandemic have social, somatic, economic, and particularly, psychological effects in Turkey. Studies on the psychological effects of COVID-19 in Turkey is yet limited but available studies already confirm the negative effects of COVID-19 [5]. The current study investigated the psychological social, somatic, and economic effects of the COVID-19 from a heterogeneous sample. The study is expected to contribute to existing literature and pave the way for future research. More specifically, the study aimed to answer the several research questions: Is there a significant difference in COVID-19 phobia between men and women? (R1), Is there a significant difference in COVID-19 phobia among different geographical regions of the country? (R2), Is there a significant difference in COVID-19 based on the participants' educational attainment? (R3), and is there a significant difference in COVID-19 based on the participants' socioeconomic statuses? (R4).

2 Literature Review

The review of the studies on the COVID-19 and mental health can be categorized into three predominant themes: Scale development, psychological problems (i.e. anxiety, stress, and depression) associated with COVID-19, and mental-health problems faced by healthcare workers. Studies in the first category mostly focused on the development of assessment instruments related to COVID-19. In this context, the first published study, Ahorsu et al. developed the Fear of COVID-19 Scale (FC-19S) and tested its reliability and validity properties [6]. They reported that 7-item scale has adequate internal reliability and validity. This scale is already adapted to several languages. In addition, Arpaci et al. developed and conducted initial tests for the psychometric properties of the COVID-19 Phobia Scale (C19P-S) [7]. They confirmed reliability and validity of the 20-item and 4-factor scale. The four factors of the C19P-S were psychological, psycho-somatic, economic, and social. This is the first research in the literature that showed multi-factor structure of COVID-19 phobia. In another study, Lee developed and evaluated the properties of the Coronavirus Anxiety Scale (CAS) to identify dysfunctional anxiety associated with the COVID-19 [8]. The results indicated that the 5-item scale has an adequate reliability and validity. Finally, Taylor et al. developed a 36-item

COVID Stress Scales (CSS) to measure the COVID-related anxiety and stress [9]. They proposed a 5-factor scale: (1) Danger and contamination fears, (2) Fears about economic consequences, (3) Xenophobia, (4) Compulsive checking and reassurance seeking, and (5) Traumatic stress symptoms.

In the second group of COVID-19 studies, Rajkumar [10] conducted a literature review on the COVID-19 and mental health [10]. The findings indicated that anxiety and depression symptoms and self-reported stress were common psychological reactions to the COVID-19 pandemic. For example, Ahmed et al. [11] investigated the psychological problems associated with COVID-19 outbreak in China [11]. Their results indicated a much higher rate of depression, anxiety, alcohol consumption, and a lower mental well-being among Chinese people due to the COVID-19 outbreak. Similarly, Huang and Zhao [12] investigated the Chinese participants' anxiety, depression, and poor sleep quality during the COVID-19 pandemic [12]. They reported that younger people had higher anxiety and depression symptoms compared to older people and healthcare workers had the highest level of poor sleep quality. McKay et al. [13] investigated the relationships among anxiety sensitivity, disgust propensity and sensitivity, and fear of contracting COVID-19 and supported the moderating roles of disgust propensity and sensitivity in the relationship between anxiety sensitivity and fear of contracting COVID-19 [13].

Moghanibashi-Mansourieh [14] investigated the anxiety levels of Iranians during the COVID-19 outbreak [14]. They showed that anxiety levels were higher among women, individuals with COVID-19 contact, people who followed the news more frequently and the age group between 21 to 40 year olds. Wang et al. [15] investigated levels of anxiety, psychological impact, stress, and depression during the initial stages of the COVID-19 outbreak in China [15]. They found that anxiety, depression, and stress were reported as psychological responses to the COVID-19. Similarly, Roy et al. [16] investigated the attitude, anxiety, and perceived-mental-healthcare need among Indian adults during the COVID-19 pandemic and reported higher levels of anxiety [16]. Xiao et al. [17] investigated the effect of social capital on sleep quality, anxiety, and stress among self-isolated people during the COVID-19 pandemic in China. Their results indicated that anxiety positively predicted stress, sleep quality, social capital [17].

In the third group of COVID-19 studies, Spoorthy [18] conducted a literature review on the COVID-19 and mental-health problems faced by healthcare workers [18]. They found that self-efficacy and poor social support were related to anxiety, stress, insomnia, and depressive symptoms. Similarly, Xiao et al. [19] investigated the levels of anxiety, stress, self-efficacy, sleep quality, and social support among Chinese medical staff during the COVID-19 pandemic [19]. Their results showed that anxiety predicted stress, sleep quality, social support, and self-efficiency. Zhang et al. [20] investigated mental health and psycho-social problems of the medical health workers during the COVID-19 pandemic in China [20]. Their also reported that medical health workers had higher prevalence of insomnia, anxiety, depression, somatization, and obsessive-compulsive symptoms. Lai et al. [21] investigated the degree of symptoms of anxiety, depression, distress, and insomnia among health workers during the COVID-19 pandemic in China [21]. Their results indicated that health care workers experienced anxiety, depression, distress, and insomnia symptoms, especially women and nurses. Liang et al. [22] investigated Chinese medical staffs' mental health during the COVID-19 outbreak and reported that several medical staffs experienced clinically significant depressive symptoms. Table 1 summarizes the review of these studies [22].

Table 1 The review of the studies on the COVID-19 and mental health

Study	Sample	Instrument(s)	Analysis	Major findings
[6]	717 Iranian participants	Hospital Anxiety and Depression Scale (HADS), Perceived Vulnerability to Disease Scale (PVDS)	EFA, CTT and Rasch analysis	Development of the Fear of COVID-19 Scale (FC-19S)
[7]	3393 Turkish participants	C19P-S	EFA, CFA	Development of the COVID-19 Phobia Scale (C19P-S)
[8]	775 adults	Work and Social Adjustment Scale (WSAS)	EFA, CFA	Development of the Coronavirus Anxiety Scale (CAS)
[9]	3479 Canadian and 3375 American participants	Patient Health Questionnaire-4 (PHQ-4), Short Health Anxiety Inventory (SHAI), Obsessive Compulsive Inventory-Revised (OCI-R), Xenophobia Scale (XS), Marlowe Crowne Social Desirability Scale Short Form (MCSD-SF)	EFA, CFA	Development of the 36-item COVID Stress Scales (CSS).
[10]	28 articles	Literature review	Thematic analysis	Anxiety and depression symptoms and self-reported stress were common psychological reactions to the COVID-19 pandemic
[11]	1074 Chinese participants	Beck Anxiety Inventory (BAI), BDI-II, Alcohol Use Disorder Identification Test (AUDIT), Warwick Edin- burgh Mental Wellbeing Scale (WEMWBS)	Chi-square test	COVID-19 outbreak resulted in a much higher rate of depression, anxiety, alcohol consumption, and a lower mental well-being among Chinese people

(continued)

Table 1 (continued)

Study	Sample	Instrument(s)	Analysis	Major findings
[12]	7236 Chinese participants	Generalized anxiety disorder (GAD-7)	Chi-square test	Younger people have a higher anxiety and depression symptoms than older people and healthcare workers have the highest level of poor sleep quality
[13]	908 Chinese adults	Depression Anxiety Stress Scale-21 (DASS-21), Generalized Anxiety Disorder Scale-7 for COVID-19 Anxiety (CoVGAD-7), Anxiety Sensitivity Index-3rd edition (ASI-3), Disgust Propensity and Sensitivity Scale-Revised (DPSS-R)	Moderation analysis	Disgust propensity and sensitivity have a moderating role in the relationship between anxiety sensitivity and fear of contracting COVID-19
[14]	10754 Iranian individuals	DASS-21	Chi-square, t-test, ANOVA	Anxiety level was higher among women, individuals having a positive contact, people who more follow the news related to COVID-19, and the age group between 21 to 40 years
[15]	1210 Chinese participants	DASS-21, Impact of Event Scale-Revised (IES-R)	Linear regressions	Anxiety, depression, and stress were reported as psychological responses to the COVID-19
[16]	662 Indian adults	Self-reported questionnaire	Descriptive	A high level of anxiety was reported
[17]	170 self-isolated Chinese individuals	SAS, SASR, PSQI, PSCI-16	Path analysis	Anxiety positively predict the stress and negatively predict the sleep quality and social capital, which positively predict the sleep quality

(continued)

Table 1 (continued)

Study	Sample	Instrument(s)	Analysis	Major findings
[18]	6 articles	Literature review	Thematic analysis	Self-efficacy and poor social support were positively related to the anxiety, stress, insomnia, and depressive symptoms among health care workers
[19]	180 Chinese medical staff	General Self-Efficiency Scale (SES), Self-Rating Anxiety Scale (SAS), Stanford Acute Stress Reaction Questionnaire (SASR), Social Support Rate Scale (SSRS), Pittsburgh Sleep Quality Index (PSQI)	Structural equation model (SEM)	Anxiety positively predicted the stress and negatively predicted the sleep quality, social support, and self-efficiency
[20]	2182 Chinese subjects	Insomnia Severity Index (ISI), Symptom Check List-revised (SCL-90-R), Patient Health Questionnaire-4 (PHQ-4), 2-item depression scale (PHQ-2)	Multivariate logistic regression	Medical health workers had a higher prevalence of insomnia, anxiety, depression, somatization, and obsessive-compulsive symptoms
[21]	1257 Chinese health care workers	9-item Patient Health Questionnaire, 7-item Generalized Anxiety Disorder scale, 7-item Insomnia Severity Index, 22-item Impact of Event Scale–Revised	Multivariate logistic regression	Health care workers experienced anxiety, depression, distress, and insomnia symptoms, especially women and nurses
[22]	23 doctors and 36 nurses in China	Self-rating depression scale (SDS), Self-rating anxiety scale (SAS)	T-test	Medical staffs experienced clinically significant depressive symptoms

3 Method

3.1 Population and Sample

Population of the current study is individuals who currently reside in Turkey. Convenience sampling from the population obtained a total of 2143 participants (60.3% women). The sample consisted of a wide range of age groups (12–92 years old, M = 39.66, SD=16.87) who willingly completed the online survey. Among the group, majority were married (59%) and middle class (60.1%). Participants came from 72 out of 81 different cities of Turkey [mostly from Konya (27.7%), Karaman (9.3%), Ankara (9.2%), Mersin (6.6%), Istanbul (3.4%), Izmir (3.4%), Adana (3.3%), and Antalya (3.1%)].

More than one fifth of the sample (20.9%) had chronic health problems and most common problems included hypertension (3.9%), diabetics (3.7%), heart problems (1.5%), and asthma (1.4%). A total of 10 participants indicated that they recently diagnosed with COVID 19 (0.5%); 8.9% knew someone close to them diagnosed with COVID 19; and 2.1% knew someone close to them died recently due to COVID 19. Further descriptive information on the participants is reported in Table 2.

Table 2 Descriptive data of the participants

		Frequency	Percent
Marital Status	Married	1264	59.0
	Single	815	30.0
	Other	64	3.0
Educational attainment	No formal schooling	119	5.6
	Primary school	440	20.5
	Secondary school	119	9.3
	High school	264	12.3
	Undergraduate student	531	24.8
	College degree	435	20.3
	MSc/PhD student	54	5.5
	Graduate degree	62	2.9
	Missing	19	1.5
Social/Economic Status	Upper	32	1.5
	Upper middle	432	20.2
	Middle	1288	60.1
	Lower-middle	272	12.7
	Low	118	5.5
Chronic Disease	Yes	488	20.9
	No	1694	79.0
COVID-19	Positive	10	0.5
	Negative	2133	99.5

4 Instruments

The Coronavirus Phobia 19 Scale (CP19-S) and a set of demographic questions were used to collect the data. The CP19-S is a self-report instrument with a five-point Likert-type scale to assess the levels of coronavirus (COVID-19) phobia [7]. The scores on the scale range between 20 and 100 and higher scores indicate greater phobia symptoms in the respected subscales and total scale. The validity and reliability properties of the CP19-S are evidenced in a recent study [7].

5 Procedure

The IRB had been received and an online portal that included the demographic questions and the items of the COVID-19 was made publicly available. Data were collected in a period of two weeks and analyzed using parametric statistics. Before the analysis, the assumptions of parametric analyses were screened and found satisfactory.

6 Results

Means, standard deviations, reliability coefficients and normality data on the CP19-S are computed and reported in Table 3. On a five-point Likert scale, overall CP19S score was 3.03 in which both women and men showed the highest phobic reactions on the psychological subscale and the lowest on the psycho-somatic subscale.

A one-way multivariate analysis of variance (MANOVA) showed that there was statistically significant difference between men and women, Wilks' $\lambda = 0.97$, $F_{(3.2138)} = 16.58$, $p < 0.0001$, $\eta^2 = 0.03$, power $= 1.00$. On all subscales, women scored significantly higher than men ($p < 0.0001$) but effect sizes of the differences were small (ranged from 0.01 to 0.03). In terms of marital status, there was no multivariate difference, Wilks' $\lambda = 0.99$, $F_{(8.4274)} = 1.91$, $p > 0.05$, power $= 1.00$. One-way MANOVA also tested and found significant differences among the seven regions of Turkey (Fig. 1), Wilks' $\lambda = 0.96$, $F_{(28.7508)} = 2.82$, $p < 0.0001$, $\eta^2 = 0.01$, power $= 1.00$. Patterns of differences varied among regions but in general, eastern regions scored significantly higher than western regions ($p < 0.05$).

Significant differences were observed based on the educational attainment of the participants, Wilks' $\lambda = 0.96$, $F_{(28.7548)} = 3.02$, $p < 0.0001$, $\eta^2 = 0.01$, power $= 1.00$ and SES, Wilks' $\lambda = 0.98$, $F_{(16.6520)} = 2.58$, $p < 0.0001$, $\eta^2 = 0.01$, power $= 0.97$. On all subscales, phobia scores decreased with increasing educational attainment (Fig. 2). For example, on the psychological subscale, those who did not

Table 3 Descriptive statistics, intercorrelations and internal consistency coefficients on the CP19-S total and subscale scores for men and women

Variables	Correlations				
	1.	2.	3.	4.	5.
1. Psychological subscale					
2. Psycho-somatic subscale	0.36 (0.33)				
3. Social subscale	0.63 (0.61)	0.52 (0.46)			
4. Economic subscale	0.47 (0.46)	0.53 (0.58)	0.49 (0.50)		
5. Total CP19-S	0.82 (0.82)	0.73 (0.71)	0.85 (0.83)	0.76 (0.77)	
Means	20.70 (18.89)	10.23 (9.41)	15.67 (14.35)	9.80 (9.17)	56.40 (51.81)
Standard deviations	5.22 (5.71)	3.90 (3.84)	4.52 (4.63)	3.79 (3.64)	13.83 (14.04)
Internal consistency coefficients	0.87 (0.88)	0.89 (0.91)	0.84 (0.86)	0.88 (0.88)	0.92 (0.93)

Note All Pearson product-moment correlation coefficients are significant ($p < 0.01$)
Note Values outside of parentheses belong to women whereas values enclosed in parentheses belong to men

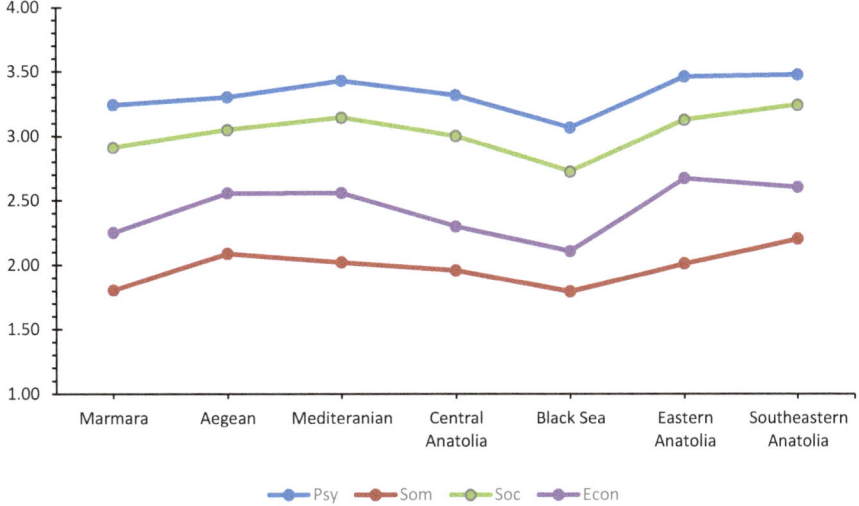

Fig. 1 Regional differences in COVID 19 effects

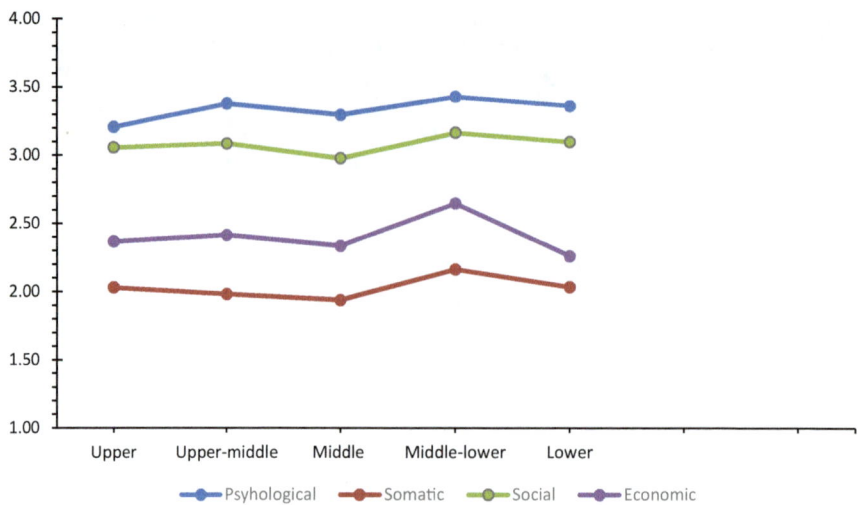

Fig. 2 Differences in COVID 19 based on socioeconomic status

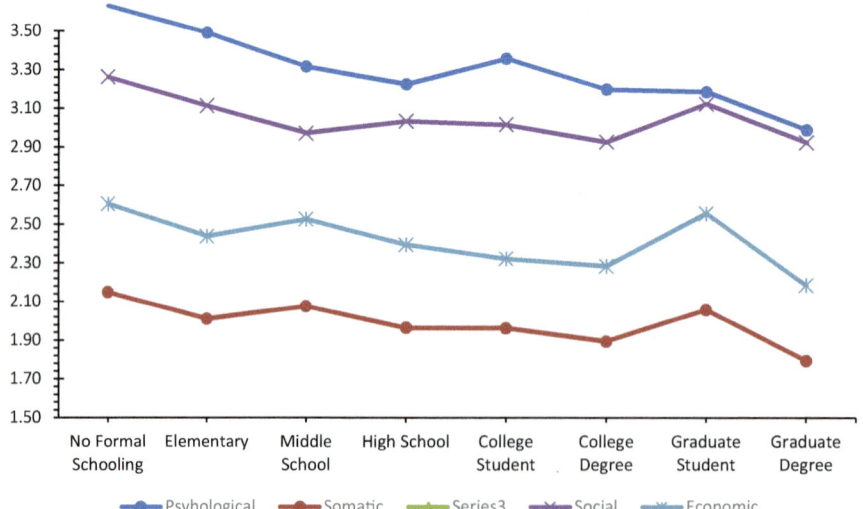

Fig. 3 Differences in COVID 19 based on educational attainment

have any formal education scored the highest and those with graduate degrees scored the lowest. However, in terms of SES, lower-middle class showed the highest scores on all the subscales and (Fig. 3).

7 Discussion and Conclusion

Even though pandemics cause negative psychological effects such as depression, anxiety, fear of death, phobias, and psychotic symptoms [23], people's reactions in the face of such stressful events vary greatly among different cultures as well as within the same culture. Similarly, the COVID-19 pandemic is currently affecting a large number of world's population directly or indirectly. Problems caused by COVID-19 usually lead to fear, anxiety, and phobias in somatic, social, economic, and psychological dimensions, which are already being reported by researchers in China [24]. Similarly, studies conducted in different parts of the world also show that individuals experience similar psychological problems due to COVID-19 [11]. This study was an attempt to show the effects of COVID-19 in the Turkish population.

According to our results Turkish women had higher levels of COVID-19 phobia. This finding supports the finding of recent research conducted in Turkey [5] as well as in other countries [14]. All together these results are in agreement with previous finding in epidemics such as EBOLA, SARS and Zika viruses where women were affected more negatively than men in terms of psychological, economic and general health [25].

COVID-19 restricts access to resources and so people are afraid of not meeting their needs. One of the findings of this research is that women are more afraid of running out of food due to COVID-19, so they tend to accumulate food (This was an item under the economical subscale of the CP19-S). Similarly, Balkhi et al. 26 reported that women were more likely to purchase additional amounts of groceries in fear of them running out [26]. Fear of famine, loss of control over the environment, and insecurity may be the primary reasons responsible for the panic buying phenomenon [27]. According to Ryan et al. [28] COVID-19's economic restraint and negative effects on the ability of individuals to meet their basic human needs causes stress and fear [28].

One of the risk factors for specific phobias such as corona phobia is educational attainment. Bjelland et al. [29] assert that people with higher educational levels are protected from stress and depression throughout life [29]. As the level of education increases people show stronger emotional responses and are more willing to use problem-focused coping strategies [24]. We also found that individuals with higher educational attainment had lower levels of COVID-19 phobia, which is case among Chinese [30] and Australian samples [31]. On the contrary, Balkhi et al. [26, 32] with Pakistani and Moccia et al. [32] with Italian samples found no significant effect of education in terms of psychological and behavioral problems caused by COVID-19 [26, 32]. Therefore, we conclude that there are cultural differences in terms of the effects of educational attainment on psychological problems caused by COVID-19. In the case of Turkey, we argue that individuals with higher educational levels are more conscious about how to protect themselves from COVID-19 and their coping capacities are better; as a result, they experience lower levels of COVID-19 phobia.

Groups with lower SES in general are more prone to phobias, have weaker coping styles, poorer social support systems, and less access to health services [33]. Similarly, Bitan et al. [34] reported that socioeconomic status (SES) was significantly associated with the fear of COVID-19, in which lower SES reported higher rates of fear [34]. However, Haktanir et al. [5] have found that people with middle SES were more afraid of coronavirus compared to those with higher SES [5]. Yet, our study with more participants from a wider geographic distribution showed no effect of SES on corona phobia. Therefore, we call for more research on the subject.

Based on the results we recommend that psychosocial support systems be provided especially for women who are at higher risk in developing phobic reactions against COVID-19. There are a number of limitations in the current study that need to be addressed at this point. The fact that phobic reactions did not differ significantly in terms of SES in COVID-19 does not match with some research findings in the literature. The reason(s) for this finding can be examined in depth using qualitative research methodologies. Second, the effects of COVID-19 phobia can be demonstrated not only by cross-sectional studies such as the current one but also by longitudinal studies. Correlational studies can be conducted with various psychological structures that are predicted to be related to COVID-19 phobia. The mediators and regulatory psychological structures that lead to COVID-19 phobia can be identified. Finally, the variables investigated in this study do not imply causality in any shape or form because the nature of the current research was descriptive. Experimental research is essential to identify the *cause(s)* of corona phobia.

References

1. World Health Organization (WHO), Coronavirus disease (COVID-19) pandemic, April 2020 (Online). Available: https://www.who.int/emergencies/diseases/novel-coronavirus-2019. Accessed on May. 15, 2020
2. Cable News Network (CNN), Coronavirus: Which countries have travel bans? April 2020 (Online). Available: https://edition.cnn.com/travel/article/coronavirus-travel-bans/index.html Accessed on May. 15, 2020
3. Tausczik, Y., Faasse, K., Pennebaker, J., Petrie, K.: Public anxiety and information seeking following the H1N1 outbreak: blogs, newspaper articles, and wikipedia visits. Health Commun. 27(2), 179–185 (2011). https://doi.org/10.1080/10410236.2011.571759
4. Bhuiyan, A., Sakib, N., Pakpour, A., Griffiths, M., Mamun, M.: COVID-19-related suicides in Bangladesh due to lockdown and economic Factors: case study evidence from media reports. Int. J. Ment. Health Addict. (2020). https://doi.org/10.1007/s11469-020-00307-y
5. Haktanir A., Seki, T., Dilmaç, B.: Adaptation and evaluation of Turkish version of the fear of COVID-19 Scale. Death Stud., 1–9 (2020). https://doi.org/10.1080/07481187.2020.1773026
6. Ahorsu, D.K., Lin, C.Y., Imani, V., Saffari, M., Griffiths, M.D., Pakpour, A.H.: The fear of COVID-19 scale: development and initial validation. Int. J. Ment. Health Addict. (2020). https://doi.org/10.1007/s11469-020-00270-8
7. Arpaci, I., Karataş, K., Baloğlu, M.: The development and initial tests for the psychometric properties of the COVID-19 Phobia Scale (C19P-S). Pers. Individ. Dif., 164:110108 (2020). https://doi.org/10.1016/j.paid.2020.110108

8. Lee, S.A.: Coronavirus anxiety scale: A brief mental health screener for COVID-19 related anxiety. Death Stud. **44**(7), 393–401 (2020). https://doi.org/10.1080/07481187.2020.1748481

9. Taylor, S., Landry, C. A., Paluszek, M. M., Fergus, T. A., McKay, D., Asmundson, G. J. G.: Development and initial validation of the COVID Stress Scales. J. Anxiety Disord., **72**:102232 (2020)s https://doi.org/10.1016/j.janxdis.2020.102232

10. Rajkumar, R. P.: COVID-19 and mental health: A review of the existing literature. Asian J. Psychiatrvol. **52**(March): 102066 (2020) https://doi.org/10.1016/j.ajp.2020.102066

11. Ahmed, M. Z., Ahmed, O Aibao, Z., Hanbin, S., Siyu, L., Ahmad, A.: Epidemic of COVID-19 in China and associated Psychological Problems. Asian J. Psychiatr., **51** (April):102092 (2020) https://doi.org/10.1016/j.ajp.2020.102092

12. Huang, Y., Zhao, N., Chinese mental health burden during the COVID-19 pandemic. Asian J. Psychiatr., **51**(March):102052 (2020) https://doi.org/10.1016/j.ajp.2020.102052

13. McKay, D., Yang, H., Elhai, J., Asmundson, G. J. G.: Anxiety regarding contracting COVID-19 related to interoceptive anxiety sensations: The moderating role of disgust propensity and sensitivity. J. Anxiety Disord., **73**(April):102233 (2020) https://doi.org/10.1016/j.janxdis.2020.102233

14. Moghanibashi-Mansourieh, A.: Assessing the anxiety level of Iranian general population during COVID-19 outbreak. Asian J. Psychiatr., **51**(April):102076 (2020) https://doi.org/10.1016/j.ajp.2020.102076

15. Wang et al., C.: Immediate psychological responses and associated factors during the initial stage of the 2019 coronavirus disease (COVID-19) epidemic among the general population in China. Int. J. Environ. Res. Public Health, **17**(5) (2020). https://doi.org/10.3390/ijerph17051729

16. Roy D., Tripathy, S., Kar, S. K., Sharma, N., Verma, S. K., Kaushal, V.: Study of knowledge, attitude, anxiety & perceived mental healthcare need in Indian population during COVID-19 pandemic. Asian J. Psychiatr., **51**(April):102083 (2020). https://doi.org/10.1016/j.ajp.2020.102083

17. Xiao, H., Zhang, Y., Kong, D., Li, S., Yang, N.: Social capital and sleep quality in individuals who self-isolated for 14 days during the coronavirus disease 2019 (COVID-19) outbreak in January 2020 in China. Med. Sci. Monit. **26**, 1–8 (2020). https://doi.org/10.12659/MSM.923921

18. Spoorthy, M.S.: Mental health problems faced by healthcare workers due to the COVID-19 pandemic–A review. Asian J. Psychiatr. **51**(April), 2018–2021 (2020). https://doi.org/10.1016/j.ajp.2020.102119

19. Xiao, H., Zhang, Y., Kong, D., Li, S., Yang, N.: The Effects of Social Support on Sleep Quality of Medical Staff Treating Patients with Coronavirus Disease 2019 (COVID-19) in January and February 2020 in China. Med. Sci. Monit. **26**, e923549 (2020). https://doi.org/10.12659/MSM.923549

20. Zhang et al., W. R.: Mental health and psychosocial problems of medical health workers during the COVID-19 epidemic in China. Psychother. Psychosom., **100053**(45) (2020) https://doi.org/10.1159/000507639

21. Lai, J., et al.: Factors associated with mental health outcomes among health care workers exposed to coronavirus disease 2019. JAMA Netw. open **3**(3), e203976 (2020). https://doi.org/10.1001/jamanetworkopen.2020.3976

22. Liang, Y., Chen, M., Zheng, X., Liu, J.: Screening for Chinese medical staff mental health by SDS and SAS during the outbreak of COVID-19. J. Psychosom. Res. **133**(March), 16–18 (2020). https://doi.org/10.1016/j.jpsychores.2020.110102

23. Taylor, S.: The psychology of pandemics. Cambridge Scholars Publishing (2019)

24. Huang, L., Lei, W., Xu, F., Liu, H., Yu, L.: Emotional responses and coping strategies in nurses and nursing students during Covid-19 outbreak: A comparative study. PLoS ONE **15** (8), e0237303 (2020). https://doi.org/10.1371/journal.pone.0237303

25. Wenham, C., Smith, J., Morgan, R.: COVID-19: the gendered impacts of the outbreak. The Lancet **395**(10227), 846–848 (2020). https://doi.org/10.1016/s0140-6736(20)30526-2

26. Balkhi, F., Nasir, A., Zehra, A., Riaz R., Psychological and behavioral response to the coronavirus (COVID-19) pandemic. Pandemic Cureus **12**(5) (2020) https://doi.org/10.7759/cureus.7923
27. Arafat, S., Kar, S., Marthoenis, M., Sharma, P., Hoque Apu, E., Kabir, R.: Psychological underpinning of panic buying during pandemic (COVID-19). Psychiatry Res. **289**:113061 (2020) https://doi.org/10.1016/j.psychres.2020.113061
28. Ryan, B. J., Coppola, D., Canyon, D. V., Brickhouse, M., Swienton, R.: COVID-19 community stabilization and sustainability framework: an integration of the Maslow hierarchy of needs and social determinants of health. Disaster Med. Public Health Prep., 1–7 (2020)
29. Bjelland, I., Krokstad, S., Mykletun, A., Dahl, A.A., Tell, G.S., Tambs, K.: Does a higher educational level protect against anxiety and depression? The HUNT study. Soc. Sci. Med. **66** (6), 1334–1345 (2008)
30. Tian, F., Li, H., Tian, S., Yang, J., Shao, J., Tian, C.: Psychological symptoms of ordinary Chinese citizens based on SCL-90 during the level I emergency response to COVID-19. Psychiatry Res. **288**, 112992 (2020)
31. Coelho, C.M., Gonçalves-Bradley, D., Zsido, A.N.: Who worries about specific phobias?—A population-based study of risk factors. J. Psychiatr. Res. **126**, 67–72 (2020)
32. Moccia, L., et al.: Affective temperament, attachment style, and the psychological impact of the COVID-19 outbreak: an early report on the Italian general population. Brain Behav. Immun. **87**, 75–79 (2020). https://doi.org/10.1016/j.bbi.2020.04.048
33. Song, A., Kim, W.: The association between relative income and depressive symptoms in adults: Findings from a nationwide survey in Korea. J. Affect. Disord. **263**, 236–240 (2020). https://doi.org/10.1016/j.jad.2019.11.149
34. Tzur Bitan, D., Grossman-Giron, A., Bloch, Y., Mayer, Y., Shiffman, N., Mendlovic, S.: Fear of COVID-19 scale: Psychometric characteristics, reliability and validity in the Israeli population. Psychiatry Res., **289**:113100 (2020). https://doi.org/10.1016/j.psychres.2020.113100

Behavioral Intention of Students in Higher Education Institutions Towards Online Learning During COVID-19

Qasim AlAjmi, Mohammed A. Al-Sharafi, and Amr Abdullatif Yassin

Abstract Across the world, Higher Education Institutions (HEIs) were mandatory to transform their teaching model from face-to-face interaction to online learning. Neither the faculty nor the students were ready for this sudden transformation, most of the HEIs were practicing online learning for the sake of changing or technology adoption. COVID-19 changes the HEIs standpoint towards online delivery, as of mid of March 2020, the only online learning come to be the reasonable solution for teaching in schools and HEIs. Therefore, this study aimed to develop a research model to explore the effect of perceived enjoinment, perceived ease of use, perceived usefulness, social influence on the students' intention and behavioral towards online learning during COVID-19 in Oman. The developed model was validated by employing Structural Equation Modeling (SEM) and the data collected via online questionnaire from 191 students whom have just completed the "Educational Technology" course on Spring 2020 at A'Sharqiyah University-Oman (ASU). The main finding of this study indicating that the identified factors; PEOU and SI significantly influenced students' Behavioral Intention (BI) towards online learning, where others; PE and PU were not supported in this study.

Keywords Higher Education Institutions · Covid-19 pandemic · Online learning

Q. AlAjmi (✉)
Department of Education, College of Arts and Humanities, A'Sharqiyah University, Ibra, Oman
e-mail: alajmi.qasim@gmail.com

M. A. Al-Sharafi
Institute for Artificial Intelligence and Big Data, Universiti Malaysia Kelantan, City Campus, Pengkalan Chepa, 16100 Kota Bharu, Kelantan, Malaysia
e-mail: alsharafi@ieee.org

A. A. Yassin
Center of Languages and Translation, Ibb University, Ibb, Yemen
e-mail: amryassin84@gmail.com

© The Author(s), under exclusive license to Springer Nature Switzerland AG 2021
I. Arpaci et al. (eds.), *Emerging Technologies During the Era of COVID-19 Pandemic*, Studies in Systems, Decision and Control 348,
https://doi.org/10.1007/978-3-030-67716-9_16

1 Introduction

Online learning defined as the process of transferring knowledge to the learner on a site residency or work rather than the learner's transfer to the educational institution, which is built on a foundation to Communicate knowledge, skills and educational materials to the learner through various technical means and techniques [1]. So, when the learner is away from the teacher, technology can be used to fill this gap between both parties, simulating face-to-face connection to improve knowledge reception among students [2]. Caleb Philips was the first one who delivered online learning through Correspondence Class newspaper in 1729, then in 1922 the University of Pennsylvania delivered some of the courses throgh the Radio, and then via TV by Standford University in 1968. Thenafter, in 1999 Learning management systems(LMS) cane to us like Blackboard and Canvas. In 2002, Massachusetts Institute of Technology (MIT) released 2000 courses online for free which were used by 65 million users from 215 country, followed by Khan Academy in 2008 with 71 million users [3].

On 12th of January (2020), World Health Organization(WHO) named the new Virus which spread in Wuhan(China) as "coronavirus", then on 30th of January, the new coronavirus was listed under a public health emergency of international concerns, and on February 12th, the new coronavirus infection was named as "COVID-19" [4]. As UNESCO report, 61 countries from Middle East, Africa, Asia, North America and south America have closed their schools and universities [4]. One of the worlds' leading education business corporations "Pearson" announced that more than 300 million students' education across the world interrupted by Coronavirus outbreak [5]. After One month, UNESCO officially said that 1.37 billion students and youths across 138 countries were affected by the pandemic. The sultanate of Oman government has taken a decision to suspend the study across the country duo to widespread of coronavirus diseases (COVID-19) [6]. Consequently, many countries decreed public and private higher education institution in different parts of the world to suspend face-face teaching and to find an alternative method for online learning.

Students worldwide, pursuing there higher education, found themselves in a sudden change from face-to-face to online learning. A few higher education institutions suspended teaching and some of the institutions continued course offering through online platforms [6]. With a short notice, whole higher education institution had to start online delivery across the world, millions of students had to stay at homes and attending their classes online. The student and teachers have encountered many challenges with using online teaching platforms such as; teaching flow, technical issues, material designing, ... etc. Teaching materials had been redesigned to suite online learning, and the assessment concerns were the biggest challenge [7, 8]. In response to these immediate changing, the teachers shifted their teaching applying "flipped classroom" strategy, as it was the case in Oman, in order to increase the students' involvement and interaction [9].

Despite the wide spread of online learning, no country level statistics is available regarding the students and teachers who experienced e-learning prior to Coronavirus outbreak [10]. For example, a team of researchers in Sultan Qaboos University (SQU), Oman, conducted a research to study the effect of online learning on the SQU students performance, and the main finding was that the students from practical college were not happy with online learning compared to the students from theoretical colleges [11]. However, this result cant assure that the students are ready for online learning. This study, therefore, will measure the students' perception and behavioral intention towards the online learning in terms of four domains, based on previous literature, namely Perceived Enjoyment, Perceived Usefulness, Perceived Ease of Use and Social Influence. These five domains were extracted from the literatures [12–14], and this study will propose a Fit research model for online learning measurement. Hence, this study aims to address the following research objectives:

1. To identify the factors that influence and motivate students' behavioral intention towards online learning.
2. To develop a research model to analyze the students' behavioral intention towards online learning.
3. To validate the proposed model using Structural Equation Modeling (SEM) approach.

So, the currents study measures the students' behavioral intention and perception towards online delivery of courses. The remainder of the chapter is organized into five sections: Model development (Sect. 2), Research Methodology (Sect. 3), Finding and Discussion (Sect. 4). Finally, Sect. 5 discusses the conclusion, limitations of the study and future research directions.

2 Model and Hypothesis Development

In this study, the previous literature has been reviewed in order to accomplish to the first objective of this study. The factors of this study have been clustered as follows: Perceived Enjoyment (PE), Perceived Ease of Use (PEOU), Perceived Usefulness (PU), and Social Influence (SI), which are hypothesized to have an effect on students' behavioral intention (BI) as showed in Fig. 1. This model is based on past studies that have used various information technology adoption theories, including Technology Acceptance Model (TAM) and the Unified Theory of Acceptance and Use of Technology (UTAUT) for assessing e-learning users' behavioral intention [15–18].

Based on the previous literature, we have concluded the factors that hypothesized to influence students' behavioral intention on online delivery, these factors are listed in Table 1 and detailed in the below subsections:

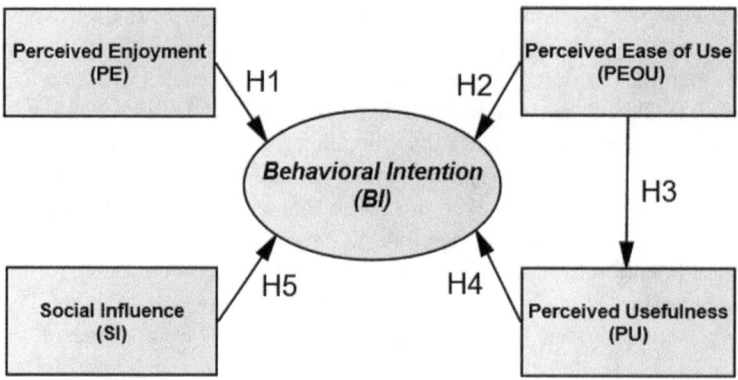

Fig. 1 Proposed Research Model

Table 1 Review of Previous literature

Literature	Model applied	I.V	D.V
[19]	TAM SEM Path Analysis	PEU, PU, Educational Quality, Content and information quality, Service Quality	Satisfaction and intention of using e-learning
[20]	TAM PCA SEM	PU, PEU, SN, PL	Adoption and usages behavior of Social media
[13]	TAM SEM	PE, PU, PEU, SI	Intention to use Twitter as a source of information
[21]	TAM SEM	PU, PEU	Intention to use Social networks
[22]	TAM and DOI SEM	PU, PEU, SI, etc.	Intention to uses Facebook as a virtual classroom
[23]	TAM SEM	PU, PEU, BI, User attitude	YouTube Adoption
[24]	TAM Regression T-Test	PU, PEU, Collaborative learning	Intention to use social media to improve academic performance

2.1 Perceived Enjoyment (PE)

The PE known as playfulness [20], and for this study we can define it as the degree of the students' enjoyment while studying online. Hence, the following hypothesis is proposed;

H1: PE has a positive significant influencing on students' behavioral intention towards online learning.

2.2 Perceived Ease of Use (PEOU) and Perceived Usefulness (PU)

Perceived ease of use is defined as the degree to which a person believes that using a particular system would be free of effort [25]. The PEOU and PU were examined and showed that PEOU has a positive significant effect on PU [12, 21]. It means that, as far as the students finds the online learning useful, they will feel significant ease in using it. For this, the researchers proposed the following hypothesis;

H2: PEOU has a positive significant influence on students' behavioral intention towards online learning.
H3: PEOUU has a positive significant influence on students' behavioral intention towards online learning.
H4: PU has a positive significant influence on students' behavioral intention towards online learning.

2.3 Social Influence (SI)

SI refers to measurements of the perception of how individuals are affected by people around them. It is defined as an important component of the behavioral intention [12, 26]. Consequently, for this study, SI can be defined as the influence of the students' behavioral intention towards online delivery. Hence, it is postulated that:

H5: SI has a positive significant influence on students' behavioral intention towards online learning.

For the above factors and hypotheses, the research model is developed as Fig. 1 to achieve the first and second objectives of the study.

3 Research Methodology

3.1 Context of Study

Oman like any other country was affected by the outbreak of COVID-19, which has increased the use of internet and online learning as a solution to the issue of education, resulted from COVID-19 outbreak. With reference to the global digital report statistics which was released in February 2020; 4.66 million people using internet in Oman with an internet infiltration rate of 92% in January 2020, and this number has increased by 18% compared to a year before of the same month. Currently, 2.80 million social media users reside in Oman, the infiltration rate of

Social media in Oman is 56%. This was accompanied by a decree to HEIs in Oman to move to online learning instead of face-to-face learning.

The study was carried out in A'Sharqiyah University-Oman (ASU), which is a private university working under Omani ministry of higher education umbrella accommodating 3650 students. ASU was the first HEIs who started the online learning from the second day of the study suspension, and ASU technological infrastructure was the enabler for this quick response to shifting the teaching model from face-face interaction to virtual communication using various e-learning platforms, such as Moodle, MS Teams, Zoom, YouTube channels, … etc.

In order to validate the proposed research model of this study, a self-administrated online survey has been employed. The link of an online survey was sent to 196 students of ASU who had just completed the course "Educational Technology" in the Spring semester of 2020.

These students have studied first-half of the semester through face-to-face interaction and due to COVID-19, they studied online for the second-half of the semester. For this reason, the authors selected them as a sample for the study, as they can compare between two models of learning (Face-To-Face and Online learning). The survey consists of 18 questions based on Table 1 of previous studies. The selected sample studied "Educational Technology" course beside other courses through the online mode for the second half of Spring 2020 using Moodle, Microsoft Teams, Zoom, YouTube channels and WhatsApp groups as channels of communications. In response to the online survey, 191 students participated out of 196 students, registered for the course, to whom the invitation link was sent via the WhatsApp group which was created previously for the purpose of communication. The sample of this study was restricted to the students who had just completed the Educational Technology course.

In total, 191 students from Educational course subject. The participants (191 females) there were no male registered for this subject in this semester, with 99% of (18–24) age group, were the majority of them with "Good level" in computer skills level (49.2%) and (80%) of the respondents were using both (Computer and Mobile) as a channel of communication while studding online. Further details shown in Table 2.

3.2 Measurement Development and Pilot Study

The dependent variable of this study covered the behavioral intention towards online learning. Table 2, showed the independent variables and dependent variable which represent the model factors, with each factor consisting of numerous items that are measured on a five-point Likert scale (ranging from 1 = 'strongly disagree' to 5 = 'strongly agree'). The utilized items were all extracted and adopted from previously validated scales as shown in Table 3. A minimum of three items were adopted to measure each factor to ensure proper reliability as suggested by [27].

Table 2 Respondents' profile

Variable	Data	Respondents	Percent
Age-Group	18–24	189	99
	25–30	2	1
Computer Skills Level	Beginner	21	11.0
	Average	62	32.5
	Good	94	49.2
	Advanced	14	7.3
Channel of Communication	Computer	12	6.3
	Mobile	17	8.9
	Both	162	84.8
Residence area	Semi-city	128	67.0
	City	44	23.0
	Remote Village	19	9.9
Internet Connection Type	WiFi	113	59.2
	4G	78	40.8
Total		**191**	**100%**

Table 3 Measurement variables and representative items

Variables	Code	Items	Sources
PE	PE1	I experience great pleasure when using Online learning more than traditional learning	[13, 28]
	PE2	I feel satisfied when I use Online learning more than traditional learning	
	PE3	I feel pleasure when I use Online learning more than traditional learning	
PU	PU1	Using Online learning helps me to complete my assignments quickly more than traditional learning	[14]
	PU2	Using Online learning increases my learning achievement more than traditional learning	
	PU3	Using Online learning makes it easier for me to better understand	
PEOU	PEOU1	I find it easy to use online learning	[14]
	PEOU2	It is easy for me to become skillful in using Online learning	
	PEOU3	Overall, Online learning is easy for me to use	
SI	SI1	People who are important to me think that I should use Online learning rather than traditional learning	[13, 28]
	SI2	Students who use online learning enjoy greater recognition from teachers than those who use traditional learning	
	SI3	Students who use online learning at university enjoy better grades than those who use traditional learning	
	SI4	People nearby me recommend to use online learning rather than traditional learning	

(continued)

Table 3 (continued)

Variables	Code	Items	Sources
	SI5	Overall, use of online learning will improve my social image among the college	
BI	BI1	I intend to begin/continue use Online learning to interact with colleagues and lecturers	[14]
	BI2	I will continue to use Online learning as a learning tool	
	BI3	I will strongly recommend use of Online learning to others	
	BI4	Overall, I intend to continue using Online learning rather than traditional learning	

Table 4 Reliability Result

Variables	No. Items	Cronbach's Alpha
Perceived Enjoyment while studying Educational Technology Course (PE)	3	0.919
Perceived Usefulness while studying Educational Technology Course (PEOU)	3	0.836
Perceived Ease of Use while studying Educational Technology Course (PEOU)	3	0.825
Social Influence while studying Educational Technology Course (SI)	5	0.849
Behavioral Intention while studying Educational Technology Course (BI)	4	0.802

The Survey consisted of two divisions; an introduction and demographic section, and the questions section. The first section introduced the research information about the objectives and the educational terminologies with their consent request and authors obligation to ensure them with their privacy right. The demographic part contains gender, age group, computer level skills, and the used devices. The questions section was divided into Five-Dimensions (PE, PEOU, PU, SI, and BI) based on a 5-point Likert scale.

In order to verify our collected data at earlier phases of this study, we wanted to confirm the collected data reality, to do so, we have checked the reality of first 30 respondents out of 191. With SPSS V.23, we have checked the data reliability. According to [29], Cronbach's Alpha cut-off is 0.7. Table 4 below shows that all constructs had a satisfactory value (above 0.7) of Cronbach's Alpha, indicating that the used measurement is reliable and this confirms the reliability of the survey.

4 Finding

4.1 Common Method Bias (CBM)

In order to make sure that the collected data were free of any nature of biasness, we have used Common Method Bias (CBM) through Harman's single test factor using SPSS V.23. In this study, the first component of "Total Variance Explained" recorded less than 50% of all variables in the instrument. Podsakoff et al. [30] has recommended this for behavioral research, as the first factor accounted for the collected data of this study is about 43% of the overall variance, which confirms that CBM does not affect the result, as showed in Table 5.

4.2 Measurement Model Assessment

The study used SEM method to validate the proposed research model in order to achieve the third objective of this research, and SMART PLS 3 used for data analysis. Composite Reliability (CR) was used to assess the reliability value of the model variables, and the threshold for CR should not be less than 0.06 in order to be accepted, and the values above 0.70 are considered satisfactory [31]. Table 6 showed that the CR values for the variables were above 0.07, thereby this indicates good internal consistency reliability. Furthermore, the factor loading and Average Variance Extracted (AVE) were used to assess the convergent validity of the measurement model. As we can notice from Table 6, the factor loading in the rage of 0.712–0.899, while the benchmark value is 0.7 as per Hair et al. [29]. Thereby, our measurement validity for the proposed model has strong validity and acceptable reliability as shown in Fig. 2. The authors also checked the discriminate validity by Cross-loading for the items-level and Fornell-Larker criterion for the constructs level. The loading values must be higher against their respective construct compared to other constructs, and our research model loading values are higher against their respective constructs as showed in Table 7.

Table 5 Total Variance Explained

Component	Initial Eigenvalues			Extraction Sums of Squared Loadings		
	Total	% of Variance	Cumulative %	Total	% of Variance	Cumulative %
1	9.146	43.555	43.555	9.146	43.555	43.555

Extraction Method: Principal Component Analysis

Table 6 Result of measurement model-convergent validity

Constructs	Items	Loading*	AVE**	CR***
Perceived Enjoyment	PE1	0.844	0.748	0.899
	PE2	0.883		
	PE3	0.867		
Perceived Usefulness	PU1	0.829	0.674	0.861
	PU2	0.852		
	PU3	0.780		
Perceived Ease of Use	PEOU1	0.868	0.776	0.912
	PEOU2	0.895		
	PEOU3	0.880		
Social Influence	SI1	0.727	0.541	0.855
	SI2	0.758		
	SI3	0.712		
	SI4	0.723		
	SI5	0.757		
Behavioral Intention	BI1	0.746	0.7	0.903
	BI2	0.827		
	BI3	0.899		
	BI4	0.868		

*Individual Item reliability (>0.70), **Composite reliability (>0.70), ***Average Variance Extracted (>0.50)

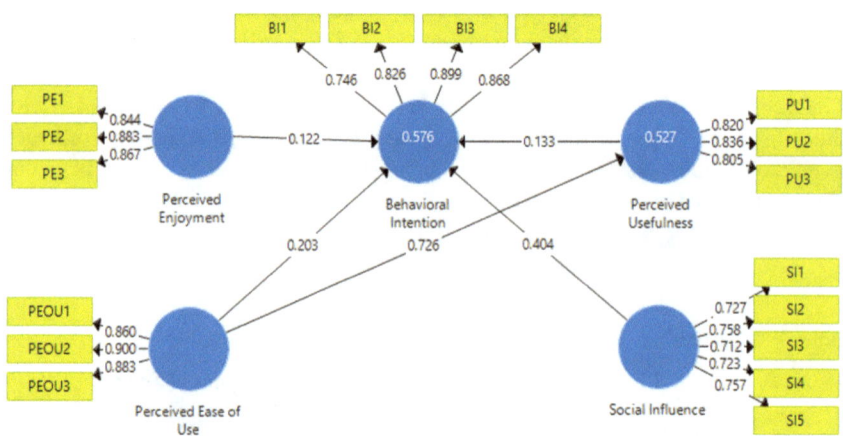

Fig. 2 Measurement model result

Table 7 Discriminant Validity by Fornell-Larker Criterion

	Perceived Ease of Use	Social Influence	Behavioral Intention	Perceived Enjoyment	Perceived Usefulness
Behavioral Intention	**0.837**				
Perceived Ease of Use	0.643	**0.881**			
Perceived Enjoyment	0.636	0.685	**0.865**		
Perceived Usefulness	0.63	0.718	0.703	**0.821**	
Social Influence	0.705	0.639	0.697	0.653	**0.735**

The value in the **BOLDFACE** are the square root of AVE

4.3 Structural Model Assessment

As long as the research model passed the first stage of assessment (Measurement Model), then the study moved to the structural model assessment to test the hypothesis of the proposed research model. (i.e., H1–H4; as showed in Fig. 1. Table 8, which shows the result of validation, and the results show that H2, H3 and H5 are supported as t-values are above 1.96 and p-values were below 0.05, and H1 and H4 have t-values below 1.96 and p-values above 0.05, which are not supported as recommended by Hair et al. [29].

The above result showed that BI is affected strongly by (SI) and (PEOU); ($\beta = 0.404$) and ($\beta = 0.203$) respectively, and PEOU has strong effect on PU, with less effect by (PU) and (PE); ($\beta = 0.133$) and ($\beta = 0.122$) respectively on BI.

Cohen (1988) [32], suggests that if the value of R^2 above 0.67 is high, value ranging from 0.33 to 0.67 are moderate, whereas the values between 0.19 and 0.33 are weak, and values less than 0.19 are unacceptable, thereby our research model is

Table 8 Path Coefficient of the research hypothesis

Hypo	Relationship	β	Std. Error	T-Value	P-Value	Result
H1	PE— >BI	0.122	0.089	1.371	**0.171**	Not Supported
H2	PEOU— >BI	0.203	0.091	2.226	**0.026**	**Supported***
H3	PEOU— >PU	0.726	0.040	18.125	**0.000**	**Supported****
H4	PU— >BI	0.133	0.092	1.440	**0.151**	Not Supported
H5	Social Influence — >BI	0.404	0.077	5.259	**0.000**	**Supported****

Significant at P** = <0.01, P* < 0.05

Table 9 Effect size f^2

Construct relation	f-Square	Result
Perceived Ease of Use	0.040	Small
Perceived Enjoyment	0.013	Small
Perceived Usefulness	0.020	Small
Social Influence	0.170	Medium

moderate as the square value was 0.577 [32]. This result suggests that the proposed research model variable (PE, PEOU, PU, SI) explained 0.56% of the variance of behavioral intention. Therefore, above 0.35 of the effect sizes is considered to be large, ranging 0.15–0.35 is considered as medium effect size, between 0.02 and 0.15 is considered to have small effect size, and less than 0.02 is considered to have no effect size. Table 9 elaborates the result of this research model. Predictive Relevance Q^2 in SMART PLS3, according to [33], the Q^2 should be more than 0. In the current study, the result is 0.393, which is an accepted value.

5 Discussion

This study aimed to investigate the behavioral intention of students in HEIs towards online learning. The first hypothesis investigates the influence of PE on students' behavioral intention towards online learning, and the findings showed this hypothesis was not supported. This finding is different from past studies [13, 28]. This finding is also not on line with past studies that confirms that technology and e-learning are preferred by students because it changes the method of teaching from teacher-centered approach to student-centered approach [34, 35]. This might be attributed to COVID-19 outbreak, which has affected the students' learning, especially that online learning might be the first experience of the majority of the students. This first experience might explain the novelty effect of the use of technology on the students' behavioral intention towards online learning [36]. So, they did not find online learning of that difference from traditional learning.

Besides, the study also investigated the influence of PEOU on students' behavioral intention towards online learning, and the findings supported this hypothesis, which is in line with past studies [14, 37]. This finding supports that students' behavioral intention towards online learning increases when they the find the use of online learning easy. This might be explained by that the student of the current generation might not need any ICT training since technology has become a part of our everyday life. So, students, especially university students, might not need training to use technology for online learning, since they have a good experience in terms of technology use [34]. Hence, easiness is not significant in its influence on students' behavioral intention towards online learning.

The third hypothesis of the study investigated the influence of PEOU on PU, and the findings supported this hypothesis, which is in line with past studies [12, 21].

That is, the aim of the students is to gain knowledge and to improve their learning, so as long as they achieve such learning goals, the use of technology will be easy. This is in line with previous studies which showed that technology might provide students with different learning materials and tools, which help them to improve their learning [38].

The fourth hypothesis investigated the influence of PU on students' behavioral intention towards online learning, which was not supported in this study, which is different from the findings of past studies [14]. Hence, PU does not increase the intention of students to use online learning. This might be explained by that the outbreak of COVID-19 has a negative impact on students' psychology and their restfulness, and this might affect their learning achievement. Such specific situation of COVID-19 might explain why this hypothesis was not supported.

The fifth hypothesis investigated the influence of SI on students' behavioral intention towards online learning. This hypothesis was supported, and it is in line with past studies [13, 28]. This finding shows that social image of students among their classmates has a significant influence on their intention towards online learning, and their self-image has an impact on online learning. That is, students learn in a small community, and technology might be viewed as a part of the prestige of the students. This might explain the influence of IS on students' intention to study through the online mode.

To conclude, the study has five hypotheses, from which two hypotheses were not supported. The unsupported hypothesis might be due that students of the current generation do not find any difficulty in using technology, so ease of use is not significant in its influence on their behavioral intention towards online learning. Also, COVID-19 outbreak has a major impact on students' psychology such as anxiety, motivation and restfulness. Such psychological aspects might be the explanation why PU did not influence students' behavioral intention towards online learning. Besides, the proposed model of the study showed that it is effective in investigating students' behavioral intention towards online learning, so it can be used in future studies or replicated in other learning context.

6 Conclusion, Limitation, and Future Research

The main finding of this study was a developed model to analyze the students' behavioral intention of online delivery based on the data collected from "Educational Technology" students from ASU. SPSS was used for descriptive statistics and Smart PLS used for measurement and structural model assessment. Theoretically, this study contributes to the current concerns on students' intention towards online learning. (a) The study confirms the students' behavioral intention towards online learning. (b) It confirms the relationship between perceived usefulness and ease of use. The research result has a practical value in terms of understanding students' intention of online learning using various technologies.

This study, with a special focus on A'Sharqiyah university students, identified a group of significant factors affecting their intention to continue learning in an online.

Few limitations were encountered, collected data was collected from students with basic skills of educational technology tool, so the result can't be generalized for other students and university. Therefore, the future researchers advised to collect data from different institutions and with different methodologies.

References

1. Al-Emran, M., Al-Maroof, R., Al-Sharafi, M.A., Arpaci, I.: What impacts learning with wearables? An integrated theoretical model. Interact. Learn. Environ., 1–21 (2020)
2. Alajmi, Q.A., Kamaludin, A., Arshah, R.A., Al-Sharafi, M.A.: The effectiveness of cloud-based e-learning towards quality of academic services: an Omanis' expert view. E-learn. **9**(4) (2018)
3. UNESCO: Online learning: definitions, tools and strategies. Manual. Decis. Mak., 14–38
4. UNESCO: COVID-19 educational disruption and response. https://en.unesco.org/themes/education-emergencies/coronavirus-school-closure. Accessed July 2020
5. Pearson: Disruption of education across the world. https://www.pearson.com/uk/educators/higher-education-educators/training-and-support/training-support-digital-products/supporting-academic-continuity-during-covid-19.html. Accessed March 2020
6. Observer, O.: Oman suspends schools and universities from Sunday. In: Oman Observer (ed.) (2020)
7. Al-Emran, M., Teo, T.: Do knowledge acquisition and knowledge sharing really affect e-learning adoption? An empirical study. Educ. Inf. Technol. (2019). https://doi.org/10.1007/s10639-019-10062-w
8. Arpaci, I., Al-Emran, M., Al-Sharafi, M.A.: The impact of knowledge management practices on the acceptance of Massive Open Online Courses (MOOCs) by engineering students: a cross-cultural comparison. Telemat. Inform., 101468 (2020). https://doi.org/10.1016/j.tele.2020.101468
9. Chen, F., Lui, A.M., Martinelli, S.M.: A systematic review of the effectiveness of flipped classrooms in medical education. Med. Educ. **51**(6), 585–597 (2017)
10. DATAREPORTAL: DIGITAL 2020: OMAN. https://datareportal.com/reports/digital-2020-oman. Accessed July 2020
11. Reporter O.N.: Study of the effect on online learning on SQU students' performance, Oman. In: Oman (ed.) (2020)
12. Al-Sharafi, M.A., Mufadhal, M.E., Arshah, R.A., Sahabudin, N.A.: Acceptance of online social networks as technology-based education tools among higher institution students: structural equation modeling approach. Sci. Iran. **26**, 136–144 (2019)
13. Al-Daihani, S.M.: Students' adoption of Twitter as an information source: an exploratory study using the technology acceptance model. Malays. J. Libr. Inf. Sci. **21**(3), 57–69 (2016)
14. Davis, F.D.: Perceived usefulness, perceived ease of use, and user acceptance of information technology. MIS Quart., 319–340 (1989)
15. Gangwar, H., Date, H., Ramaswamy, R.: Understanding determinants of cloud computing adoption using an integrated TAM-TOE model (in English). J. Enterp. Inf. Manag. **28**(1), 107–130 (2015). https://doi.org/10.1108/Jeim-08-2013-0065
16. Shiau, W.L., Chau, P.Y.K.: Understanding behavioral intention to use a cloud computing classroom: a multiple model comparison approach (in English). Inf. Manag. **53**(3), 355–365 (2016). https://doi.org/10.1016/j.im.2015.10.004

17. AlAjmi, Q., Arshah, R.A., Kamaludin, A., Sadiq, A.S., Al-Sharafi, M.A.: A conceptual model of e-learning based on cloud computing adoption in higher education institutions. In: 2017 International Conference on Electrical and Computing Technologies and Applications (ICECTA), Ras Al Khaimah, United Arab Emirates, pp. 1–7 (2017). https://doi.org/10.1109/icecta.2017.8252013
18. Alajmi, Q., Sadiq, A.: Opportunities and challenges in management information systems. Int. J. Waljat Coll. **3**(2), 76–77 (2016)
19. Mohammadi, H.: Investigating users' perspectives on e-learning: an integration of TAM and IS success model. Comput. Hum. Behav. **45**, 359–374 (2015)
20. Dumpit, D.Z., Fernandez, C.J.: Analysis of the use of social media in Higher Education Institutions (HEIs) using the technology acceptance model. Int. J. Educ. Technol. Higher Educ. **14**(1), 5 (2017)
21. Elkaseh, A.M., Wong, K.W., Fung, C.C.: Perceived ease of use and perceived usefulness of social media for e-learning in Libyan higher education: a structural equation modeling analysis. Int. J. Inf. Educ. Technol. **6**(3), 192 (2016)
22. Milošević, I., Živković, D., Arsić, S., Manasijević, D.: Facebook as virtual classroom–social networking in learning and teaching among Serbian students. Telemat. Inform. **32**(4), 576–585 (2015)
23. Chintalapati, N., Daruri, V.S.K.: Examining the use of YouTube as a learning resource in higher education: scale development and validation of TAM model. Telemat. Inform. **34**(6), 853–860 (2017)
24. Al-rahmi, W.M., Othman, M.S., Yusuf, L.M.: Social media for collaborative learning and engagement: adoption framework in higher education institutions in Malaysia. Mediterr. J. Soc. Sci. **6**(3 S1), 246 (2015)
25. Davis, F.D.: Perceived usefulness, perceived ease of use, and user acceptance of information technology (in English). MIS Quart. **13**(3), 319–340 (1989). https://doi.org/10.2307/249008
26. Choi, G., Chung, H.: Elaborating the technology acceptance model with social pressure and social benefits for social networking sites (SNSs). In: Proceedings of the American Society for Information Science and Technology, vol. 49, no. 1, pp. 1–3 (2012)
27. Al-Emran, M., Mezhuyev, V., Kamaludin, A.: PLS-SEM in information systems research: a comprehensive methodological reference. In: Hassanien, A.E., Tolba, M.F., Shaalan, K., Azar, A.T. (eds.) Proceedings of the International Conference on Advanced Intelligent Systems and Informatics 2018, Cham, pp. 644–653. Springer International Publishing, Heidelberg (2019)
28. Al-Sharafi, M.A., Mufadhal, M.E., Arshah, R.A., Sahabudin, N.A.: Acceptance of online social networks as technology-based education tools among higher institution students: structural equation modeling approach. Sci. Iran. **26**(Special Issue on: Socio-Cognitive Engineering), 136–144 (2019)
29. Hair Jr, J.F., Hult, G.T.M., Ringle, C., Sarstedt, M.: A Primer on Partial Least Squares Structural Equation Modeling (PLS-SEM), 2 ed. Sage Publications, 2016
30. Podsakoff, P.M., MacKenzie, S.B., Lee, J.-Y., Podsakoff, N.P.: Common method biases in behavioral research: a critical review of the literature and recommended remedies. J. Appl. Psychol. **88**(5), 879 (2003)
31. Nunnally, J.C.: Psychometric Theory 3E. Tata McGraw-hill education (1994)
32. Cohen, J. (ed.): Statistical Power Analysis for the Behavioral Sciences, 2nd edn. Á/L. Erbaum Press, Hillsdale, NJ, USA (1988)
33. Wold, H.: Model construction and evaluation when theoretical knowledge is scarce: theory and application of partial least squares. In: Evaluation of Econometric Models, pp. 47–74. Elsevier (1980)
34. Razak, N.A., Yassin, A.A., Maasum, T.N.R.T.M.: Formalizing informal CALL in learning english language skills. In: Enhancements and Limitations to ICT-Based Informal Language Learning: Emerging Research and Opportunities, pp. 161–182. IGI Global (2020)

35. Valverde-Berrocoso, J., Garrido-Arroyo, M.D.C., Burgos-Videla, C., Morales-Cevallos, M. B.: Trends in educational research about e-learning: a systematic literature review (2009–2018). Sustainability **12**(12), 5153 (2020)
36. Yassin, A.A., Razak, N.A.: Investigating the relationship between foreign language anxiety in the four skills and year of study among Yemeni University EFL learners. 3L: Lang. Linguist. Lit. **23**(3) (2017)
37. Al-Qaysi, N., Mohamad-Nordin, N., Al-Emran, M., Al-Sharafi, M.A.: Understanding the differences in students' attitudes towards social media use: a case study from Oman. In: 2019 IEEE Student Conference on Research and Development (SCOReD), IEEE, pp. 176–179 (2019)
38. Yassin, A.A., Razak, N.A., Maasum, N.R.M.: Investigating the need for computer assisted cooperative learning to improve reading skills among Yemeni University EFL students: a needs analysis study. Int. J. Virtual Pers. Learn. Environ. (IJVPLE) **9**(2), 15–31 (2019)

Exploring the Main Determinants of Mobile Learning Application Usage During Covid-19 Pandemic in Jordanian Universities

Mohammed Amin Almaiah, Ahmad Al-Khasawneh, Ahmad Althunibat, and Omar Almomani

Abstract Recently, higher education sector has been affected by Covid-19 pandemic significantly. Where, several universities have started to adopt online distance learning tools such as mobile learning applications. However, in order to success mobile learning applications during this pandemic, it is important to understand the necessary factors that ensure the actual use among students in post implementation. The findings showed that factors of technology, awareness, training and experience had a significant and positive influence on the actual use of mobile learning applications. While, the results indicated that psychological factors had a negative effect on the actual use. Furthermore, the results also revealed that technological and individual factors play a crucial role in solving the psychological issues among students. The findings of this research will offer useful recommendations for educational institutions in order to encourage the use of mobile learning applications effectively during Covid-19 pandemic.

Keywords Mobile learning applications · Covid-19 pandemic · Actual use · Jordanian universities

M. A. Almaiah (✉)
Department of Computer Networks and Communications, College of Computer Sciences and Information Technology, King Faisal University, Al-Ahsa 31982 Saudi Arabia
e-mail: malmaiah@kfu.edu.sa

A. Al-Khasawneh
CIS Department, President of Irbid National University, Hashemite University, Zarqa, Jordan

A. Althunibat
Department of Software Engineering, Al-Zaytoonah University of Jordan, Amman, Jordan

O. Almomani
Computer Network and Information Systems Department, The World Islamic Sciences and Education University, Amman 11947, Jordan

© The Author(s), under exclusive license to Springer Nature Switzerland AG 2021   275
I. Arpaci et al. (eds.), *Emerging Technologies During the Era of COVID-19 Pandemic*, Studies in Systems, Decision and Control 348,
https://doi.org/10.1007/978-3-030-67716-9_17

1 Introduction

According to a recent report issued by UNESCO, Corona Virus (Covid-19) pandemic has enforced more than 145 countries to close universities and schools, which impacting on 80% of students across the world [1]. In addition, most of universities and colleges have cancelled all teaching and learning activities such as traditional lectures, exams and workshops inside the universities [1]. Covid-19 pandemic has caused many universities to switch from face-to-face to online distance learning. Therefore, several universities across the world have started to resume their lectures and exams through online learning tools such as e-learning and mobile learning applications [2–4].

Mobile learning applications are playing a crucial role during this pandemic because it has several features such as portability, where mobile learning applications can be taken in different locations at home, office and others by using mobile devices [5], instant connectivity, where mobile learning applications can be used to access a variety of information and learning activities anytime and any-where with instant connectivity facility between students and instructors [6], context sensitivity, where mobile devices can be used to find and gather real or simulated data [7], interactivity and mobility [8].

In Jordanian universities, the implementation of mobile learning applications is still in early stages [9], where there are several Jordanian universities that already have been developed mobile learning information systems in their settings such as mobile student information system [6]. In most of mobile learning studies, an improvement in learning is noted because of using mobile learning applications among students, since the mobility of these applications allows for learning anywhere and anytime without having to come the university. In this case, the introduction of mobile learning applications will reduce the spread of Covid-19 among students as well as will ensure the continuous of learning in active way with flexible manner and without any constraints.

However, mobile learning in Covid-19 is still a pending topic [1], where some university students and instructors are still reluctant to use mobile learning applications due to several reasons such as lack of technological resources and infrastructure, lack of training and knowledge and psychological aspects [10, 11]. These aspects are paramount in exceptional and emergency situations where the use of mobile learning is not optional, but it is mandatory to maintain student learning with the use of mobile learning applications. Therefore, first, improvement of technological resources and infrastructure is a primary purpose for universities in order to use mobile learning benefits effectively. Second, training of university students and instructors of mobile learning applications is a priority task for universities, which should be compulsory.

Therefore, taking into above critical issues as well as consideration the current situation of Covid-19 and the importance of the use of mobile technologies to conduct the learning remotely, this research addressed the following objectives:

- To explore the factors that could influence on the actual use of mobile learning applications during Covid-19 pandemic in Jordanian universities, specifically,
- To check the importance of technological, individual and psychological factors on the actual use of mobile learning applications during Covid-19 pandemic in Jordanian universities.

2   Literature Review

As we see now in the world, the global spread of Covid-19 has motivated many universities to adopt online distance learning systems. However, some challenges like technological issues, awareness, literacy and economic aspects might act as a hindrance to the usage of online learning systems, specifically in developing countries like Jordan and Saudi Arabia [12–14]. According to a global survey from the International Association of Universities (IAU) showed that two third of the universities replaced their classroom teaching to online learning [1]. A recent study conducted by Almaiah, Al-Khasawneh and Althunibat [1], showed that Availability of technological infrastructure, access to internet network, competence and pedagogies of online distance learning were the major challenges faced the universities in Jordan and Saudi Arabia.

Another study in UK by Eyles, Gibbons and Montebruno [15], indicated that university students who have less access to the resources like Wi-Fi, internet, smart devices will be the most disadvantaged students from continuous of learning than any other group of students due to sudden closure, and thus, will affect on students' achievement negatively. In Jordan, public education have been transferred for school students through collaboration of national TV. Similarly, for universities have used several types of online social applications like Microsoft Zoom, Microsoft teams, Youtube, Skype and WhatsApp [1]. In Nigeria, Jegede [16], indicated that online learning is the best alternative tools of classroom teaching and learning in future pandemic occurrence by 89%. A study conducted by Radha et al., [17], during Covid-19 pandemic indicated that majority of students preferred smart phones for conducting the online learning than other devices. They faced number of challenges during online learning due to technical failure like internet connectivity, data limit, device problem and less face to face interaction. Furthermore, more than half of students found online learning as convenience learning in comfortable environment [17].

In fact, Covid-19 pandemic has significantly changed educational system in the world in terms of how to teach, learn and communicate with students. Many countries like Saudi Arabia, Jordan, China and others have started to adopt online virtual classroom to engage the students in teaching learning activities [1]. While, some other countries like Pakistan, India and Bangladesh have completely shut down their universities and schools due to poor access to online learning tools and applications during Covid-19 lockdown [18]. However, these countries need to identify the essential requirements for ensuring the successful implementation of the online learning systems during Covid-19 pandemic [1]. Therefore, this study mainly aims to propose a conceptual model for exploring the essential factors for mobile learning applications usage during Covid-19 pandemic.

2.1 *Online Learning and Covid-19 Pandemic in Jordanian Universities*

In Jordan, more than one 100,000 university students from 30 universities have been deprived of education and learning due to the COVID-19 lockdown. Where Ministry of Higher Education of Jordan has allocated budget to start the preparations for online learning in public universities. In fact, many educators have started to use online learning tools such as video conferencing to conduct online classrooms. In addition, they have used pre-recorded online videos were shared to facilitate the teaching and learning process for Engineering, Education, Arts, management and science students of Bachelor and post graduate level.

Before COVID-19, most Jordanian universities looked to online learning as a supplementary way to support the learning process for students [6]. Few of students in these universities may have taken few online classes voluntarily, or professors incorporated online lessons into in-person classes. In fact, before few years Ministry of Higher Education of Jordan closed many online universities in Jordan. Now with Covid-19, the form of education has changes and online learning has become the main rescuer for many universities over the world. In order to help universities to maintain the continuity of learning process for their students during Covid-19 pandemic, there is need to investigate the necessary factors that ensure the actual use of mobile learning applications among students in post implementation [19–21].

Despite many universities already used online learning tools, now Covid-19 made online learning the only feasible way to maintain continuity of education in this pandemic. However, this transition to online learning requires great efforts from universities, teachers, students and governments in order to ensure the successful usage of online learning systems among university students. Unfortunately, there is little research on investigating the success factors of mobile learning adoption during Covid-19, and the past literature on mobile learning research cannot be generalized to the experiences of students during the pandemic. Therefore, this study aims to fill this gap by exploring the main factors of mobile learning system usage among university students during COVID-19.

3   Hypotheses and Research Model

The proposed framework aims to investigate the main determinants of mobile learning adoption during Covid-19 based on examining the impact of technological, individual and psychological factors on mobile learning usage among university students. These factors were adopted from previous studies [22–25].

3.1   Technological Factors

Several previous literature [26–28] have indicated that technological aspects are considered one of the associated factors of mobile learning implementation. Technological aspects in this study include five factors related to internet speed, easy access, availability of technological resources and equipments, availability of necessary applications and software, technical maintenance and other mobile devices and network facilities [29–34]. In fact, lack of technological resources (software, hardware, internet network and technical maintenance) could result to unsuccessful usage of mobile learning applications, and this leading to increase the challenges of both students and university lecturers during Covid-19. This is the main reason of why prior studies focused on the important role of technological aspects on impacting students' usage of online learning technologies such as e-learning and mobile learning [22]. This result is supported by Aldowah et al., [22], they indicated that there is a positive relationship between technological aspects and students' usage of e-learning, which leads to successful implementation of e-learning system. Based on these recommendations, this study assumed that investigating the impact of technological aspects on students' usage of mobile learning could help universities and service providers to provide the necessary technological resources to effectively increase the use of mobile learning applications among university students during Covid-19, and thus, this will ensure the continuity of learning process by using mobile learning applications. Therefore, this study hypothesis the following:

H1: There is a relationship between technological aspects and students' usage of mobile learning applications during Covid-19.
H2: There is a relationship between technological aspects and psychological factors for actual use of mobile applications.

3.2   Individual Factors

Individual factors also have been considered as one of the primary determinants of success or failure of mobile learning applications [35]. Chavoshi and Hamidi [35], found that individual factors are the most effective factors to motivate students' to accept mobile learning. In fact, understanding individual factors and their impact on mobile learning usage during this pandemic could help in supporting the successful implementation of mobile learning applications in universities. In this study, individual factors focus on three dimensions including knowledge, awareness and training for both students and instructors. A study conducted by Aldowah et al., [22], found that availability of knowledge and necessary skills for both students and instructors have a high impact on e-learning system usage. In addition, Fayyoumi et al., [36], indicated that the primary key to accept new technologies by individuals is continuously offer the required training sessions on how to use new technologies in order to enhance skills of both students and instructors. These results motivated Almaiah and Alyoussef [4], to investigate the main reasons about students' resistance to accept e-learning system in their learning. The study found that students and instructors' limited knowledge about new technologies and how to use it minimized their opportunity to use e-learning systems effectively. Based on that, there is strong evidences supporting the impact of individual factors on students' use of new technologies. Thereby, understanding the impact of individual factors could play an a key role in increasing the use of mobile learning applications among university students and lecturers during Covid-19, and thus, this will ensure the continuity of learning process by using mobile learning applications. Based on that:

H3: There is a relationship between individual factors and students' usage of mobile learning applications during Covid-19.

H4: There is a relationship between individual factors and psychological factors for actual use of mobile applications.

3.3 Psychological Factors

Psychological factors such as stress, worries about privacy and anxiety can cause failure of using online learning technologies, which may impede the learning [37]. Several studies in the literature have indicated that psychological factors can affect in negative way on students' technological skills, low participate in online lectures and poor communication with lecturers via online sessions [38]. Taat and Francis [39], found that psychological factors were correlated significantly with students' actual use of e-learning system in Malaysia during Covid-19 pandemic. In addition, they found that students' anxiety had negative effect on students' ability to use online learning tools and their awareness about the benefits of online learning systems. Another studies revealed that students' stress and anxiety of online learning systems could be reduced by providing the necessary technical infrastructure, high speed of internet and training sessions of how to use new educational technologies such as mobile learning applications. Based on the above discussion, this study investigates the impact of psychological factors on students' usage of mobile learning applications under Covid-19 pandemic circumstances. Therefore:

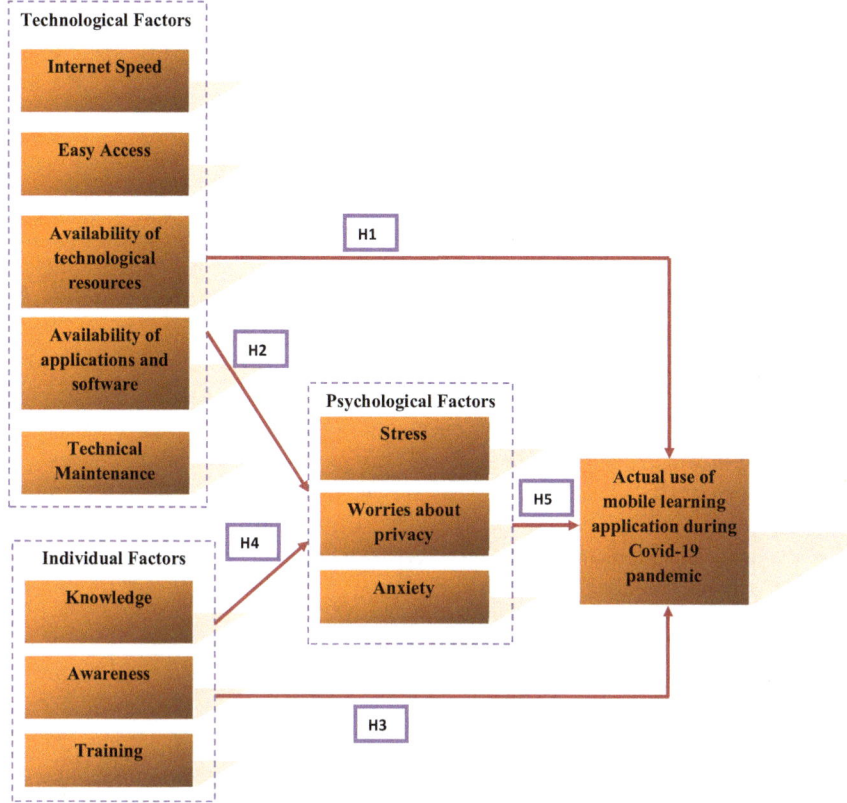

Fig. 1 The proposed model

H5: There is a relationship between psychological factors and students' usage of mobile learning applications during Covid-19 (Fig. 1).

4 Research Method

4.1 Data Collection

In this study, online questionnaire survey was used for data collection due to the Covid-19 pandemic lockdown situation. The researchers of this study had distributed the online questionnaire via email and Google sheet for students at five universities in Jordan. These universities have already developed mobile learning systems in their settings. Using online survey questionnaire, students were invited to participate in this study through online classes, during the summer semester

2020. All participants in this study were briefed on the academic research purpose and confidentiality of the survey, in which they voluntarily agreed to participate.

4.2 Participants

Due to Covid-19 and its prevention, face-to-face data collection was not possible. The questionnaire in the form of Google Sheet has been sent to the students through university lecturers. They were asked to fill the form and submit their responses if they were willing to participate in the research. The participants had not forced to participate in this research and it was clearly mentioned that if they did not like to continue then they were not forced to submit the form. In total, 487 online questionnaires were distributed, with 397 questionnaires being returned, indicating an 81.52% response rate. Most of responses had incomplete or invalid answers and therefore were excluded. Hence, 397 responses were considered valid for further analysis. Among 397 valid responses, 60.7% of respondents were female, while 39.3% were male. Moreover, 52.6% of respondents who responded were undergraduate; 47.4% were postgraduate students.

4.3 Research Instrument

The items and measurements for testing the constructs in the proposed model were adopted from existing previous research. The online questionnaire included four constructs (technological factors, individual factors, psychological factors and actual use) and included demographic information (e.g., gender and age). The items for measuring technological factors and individual factors were developed from the measurements used by Almaiah and Al Mulhem [2]. The measurement items for psychological factors was derived from Taat and Francis [39]. Actual use were adapted from Delone and McLean [40]. In this study, a 5 point scale similar to Likert model was utilized for measuring every item, ranging from "strongly disagree = 1" to "strongly agree = 5". In order to examine the appropriateness and clarify of the questionnaire, we invited five university professors, each of them holding significant expertise in the mobile learning felid. After that, pre-tested was carried out with 25 post-graduate students from University of Jordan, with the results indicating that the instructions and questions were completely understood.

4.4 Data Analysis Methods

In this study, we have applied two main techniques in order to analyze the data as well as evaluate the proposed hypotheses in the research model. The first method is the confirmatory factor analysis (CFA) in order to evaluate the measurement model in terms of reliability, convergent validity, and discriminant validity. In the second method, structural equation modelling (SEM) method was applied to test the hypotheses in the proposed model.

5 Results

5.1 Results of Cronbach's Alpha

The Cronbach's alpha analysis was employed to evaluate the reliability of items for each construct in the proposed research model. As the results summarized in Table 1, the value of this coefficient ranged between 0.773 and 0.894, exceeding the critical value of 0.7 as suggested by Kannan and Tan [41], and indicating satisfactory reliability for all variables in the proposed research model.

5.2 Results of Convergent and Discriminant Validity

In this study, all variables in the proposed model were evaluated using two types of validity analysis: convergent and discriminant validity. For convergent validity analysis, Table 1 shows that the average variance extracted (AVE) was above (0.5). According to Hair et al. [42], specify that a variance greater than 0.5 is acceptable. Therefore, the convergent validity values for the research constructs are acceptable. Concerning the discriminant validity analysis, the square root of AVE was obtained to correlate the latent constructs. Table 2 highlights that the square root of the AVE for each construct is greater than the pairwise correlations. This result means that the psychometric characteristics of the instrument are also deemed acceptable in terms of their discriminant validity [43].

Table 1 Results of Cronbach's Alpha and AVE

Variables	Cronbach's alpha	Average Variance Extracted (AVE > 0.5)
Technological Factors	0.894	0.773
Individual Factors	0.773	0.741
Psychological Factors	0.887	0.796
Actual Use	0.865	0.801

Table 2 Results of Discriminant validity analysis

Variables	Technological	Individual	Psychological	Actual Use
Technological Factors	**0.925**			
Individual Factors	0.797	**0.971**		
Psychological Factors	0.630	0.758	**0.910**	
Actual Use	0.646	0.684	0.545	**0.953**

5.3 *Results of the Structural Equation Modelling (SEM)*

The proposed model was examined using SEM method. The findings of hypotheses testing, presented in Table 3, indicated that all hypotheses were supported. Based on the results, technological factors had significant and positive effects on students for motivating them towards using mobile learning application effectively during Covid-19 pandemic. This result supports hypothesis H1, (H1: β-value = 0.527, p < 0.001). In addition, the results indicated that there is significant positive relationship between individual factors and students' actual use of mobile learning applications, which supported hypothesis H3 (H3: β-value = 0.498, p < 0.001). This implies that individual factors had significant and positive effect on students for motivating them towards using mobile learning application effectively during Covid-19 pandemic. The results also indicated that psychological factors had significant negative effect on actual use of mobile learning applications among university students, which supported hypothesis H5 (H5: β-value = −0.532, p < 0.001). This implies that psychological factors could prevent students to use mobile learning applications effectively during this pandemic, and thus, this will cause failure of mobile learning applications usage. In addition, the results showed that psychological factors influenced positively by both of technological and individual factors (H2: β-value = 0.570, p < 0.001) and (H4: β-value = 0.502, p < 0.001). Based on these results, hypotheses H2 and H4 were supported.

6 Discussions

In fact, the global spread of Covid-19 has motivated many universities to adopt mobile learning applications in order to ensure the continuous of learning among students. But, some issues related to technological challenges, individual characteristics and psychological problems may act as hindrances to the usage of these applications among university students. Therefore, there is need to investigate the main determinants that could affect on students' acceptance or rejection of mobile learning applications during Covid-19 pandemic. In order to success mobile learning during this pandemic, it very is important to understand the necessary factors that ensure the actual use among students in post implementation. Based on that, this study proposed a model to examine the effect of technological, individual

Table 3 Results of SEM analysis

Hypotheses	Standardized coefficient (β)	SE (P)	T-value	Supported
H1: There is relationship between technological aspects and students' usage of mobile learning applications during Covid-19	0.527*	0.008	4.542	Positive Effect
H2: There is a relationship between technological aspects and psychological factors for actual use of mobile applications	0.570*	0.008	4.921	Positive Effect
H3: There is relationship between individual factors and students' usage of mobile learning applications during Covid-19	0.498*	0.005	4.130	Positive Effect
H4: There is a relationship between individual factors and psychological factors for actual use of mobile applications	0.502*	0.005	4.443	Positive Effect
H5: There is a relationship between psychological factors and students' usage of mobile learning applications during Covid-19	−0.532*	0.010	4.712	Negative Effect

and psychological factors on actual use of mobile learning applications among university students during Covid-19 pandemic. The proposed model was examined using structural equation modelling (SEM). The findings of this study will offer useful recommendations for educational institutions to ensure the success of mobile learning applications usage during Covid-19 pandemic. The findings of this study will be discussed in details below.

First, the findings of this study found that actual use of mobile learning applications during Covid-19 influenced significantly by technological factors in terms of internet speed, easy access, availability of technological resources, availability of software and applications and technical maintenance. These results imply that availability of technological resources such as of laptops and mobile devices with high speed of internet network are considered the most prerequisite requirements for both university lecturers and students to participate in online teaching and learning through mobile learning applications. On the other hand, challenges increase when technical support is not consistently available or there is not a plan to periodically maintenance of technical challenges. In some developing countries, many students have not been able to complete their learning due to poor internet access, poor data service, or even lack of laptop or mobile devices. Therefore, it is important that universities should introduce and implement mobile learning applications to ease the online teaching and learning activities during the Covid-19 lockdown for both lecturers and students.

Second, the study findings found that individual factors (knowledge, awareness and training) had significant and positive effect on students for motivating them towards using mobile learning applications effectively during Covid-19 pandemic. This implies that there is significant positive relationship between individual factors and students' actual use of mobile learning applications. The findings indicates that successful online learning through using mobile learning applications during Covid-19 is feasible only when university students and instructors have the adequate skills, knowledge and awareness on how to exploit the advantages of mobile learning applications. Although the many advantages of mobile learning applications, which making teaching and learning accessible to students regardless of geography or culture, it requires awareness, training, and knowledge of these applications. Universities should not assume that all students and instructors have the same knowledge, awareness and willingness to engage in online learning technologies like mobile learning applications. On the other hand, inadequate technical skills of students and instructors will cause stress about them and thus, the result is failure of mobile learning applications usage. Therefore, the findings of this study recommends that university administration and decision makers need to provide adequate online training sessions for their students and instructors on how to employ mobile learning applications in learning and teaching process to make online learning successful during this pandemic.

Third, the findings indicated that psychological factors affected on actual use of mobile learning applications in a negative way. In addition, the results showed that psychological factors influenced positively by both of technological and individual factors. This means that psychological issues such as stress, phobia of working independently, worries about privacy, and anxiety could impede the learning and teaching via mobile learning applications. Therefore, universities should enhance students' comfort through easily access to mobile learning classes, enhance their ability to enjoy through using mobile learning applications and increase their awareness about the benefits of mobile learning applications.

Finally, the model proposed in this research determined the main determinants of mobile learning applications actual use that could be beneficial for Jordanian universities to ensure the effective usage of mobile learning applications during Covid-19. In addition, this study added to the current literature by investigating the effect of technological, individual and psychological factors on mobile learning usage among university students. Therefore, the results of this study will serve as an important reference for university policy makers, managers and IT technicians regarding the effective factors that contribute in enhancing the usage of mobile learning applications among students during Covid-19, and thus, ensure the continuous of learning process during this pandemic.

7 Research Implications

The findings of this study will serve as an important reference for university policy makers, managers and IT technicians regarding the effective factors that contribute in enhancing the usage of mobile learning applications among students during Covid-19. The findings suggest that technological infrastructure and resources play a vital role in increasing the use of mobile learning applications effectively during Covid-19. Therefore, in order to success the use of mobile learning in universities, technical department in universities should provide the necessary technological resources such as fast, affordable or free Internet access, providing laptops, iPads, or tablets and providing periodically maintenance of technical problems and solve it. This study also recommends for university decision makers that students' stress and anxiety of mobile learning applications can be reduced by providing the necessary technical infrastructure, high speed of internet and training sessions of how to use mobile learning applications.

This study also provides a useful recommendation for university decisions makers on a deeper understanding on the importance of individual factors in accepting and using mobile learning applications effectively, through preparing university students and instructors and equipping them with the necessary technical skills, knowledge and awareness of the several advantages of mobile learning applications are vital to increase the use of mobile learning applications effectively and thus, this will reduce the bad consequences and challenges of online learning. Therefore, universities should not ignore like these factors that could contribute in enhancing the use of mobile learning applications during Covid-19.

In fact, technological resources such as laptop and mobile devices with high speed of internet network are considered the most prerequisite requirements for both university lecturers and students to participate in online teaching and learning process. In many developing countries, many students have not been able to complete their learning due to poor internet access, poor data service, or even lack of laptop or mobile devices. Therefore, it is important that universities should introduce and implement mobile learning applications to ease the online teaching and learning activities during the COVID-19 lockdown for both lecturers and students.

8 Conclusions

The use of mobile learning applications to continuous the learning process during Covid-19 pandemic became very useful. Critical situations such as Covid-19 pandemic shed light on the importance of mobile learning applications and their benefits for both university students and instructors. Faced with this exceptional situation, mobile learning applications are useful tools to ensure the continuous of learning in an active way with flexible manner and without any constraints. In order to success these applications during this pandemic, it very is important to understand the necessary factors that ensure the actual use among students in post implementation.

Therefore, in this paper we responded to these objectives, we empirically investigated the factors that could influence on the actual use of mobile learning applications during Covid-19 pandemic in Jordanian universities. Where, we identified some important aspects that lead to success implementation of mobile learning applications and the main factors that motivate to use mobile learning among students. In addition, the technological, individual and psychological factors that influenced on the actual use of mobile learning applications were verified, with highlighting eleven aspects: technological resources (internet speed, easy access, availability of technological resources and equipments, availability of necessary applications and software, technical maintenance), individual factors (knowledge, awareness and training) and psychological factors (stress, worries about privacy and anxiety).

The findings found that availability of technological resources, awareness among students and lecturers, training on the use of mobile learning apps and experience of mobile learning before the pandemic were positively influence on the actual use of mobile learning applications among university students. While, the effect of psychological factors was negative the actual use of mobile learning applications. Furthermore, the results showed that psychological factors influenced positively by both of technological and individual factors.

Although this research has several contributions, it does have some limitations. In the first place, the data were collected from a limited set of universities. Thus, further studies in more universities or in more countries are required to improve the generalizability of the findings. Secondly, besides technological, individual and psychological factors, there may be other important factors that could affect on mobile learning actual use, such as organizational, culture, quality, and others. Future work could examine their effects.

References

1. Almaiah, M.A., Al-Khasawneh, A., Althunibat, A.: Exploring the critical challenges and factors influencing the e-learning system usage during COVID-19 pandemic. Educ. Inf. Technol., 1 (2020)
2. Almaiah, M.A., Al Mulhem, A.: Analysis of the essential factors affecting of intention to use of mobile learning applications: a comparison between universities adopters and non-adopters. Educ. Inf. Technol. 24(2), 1433–1468
3. Al-Emran, M., Salloum, S.A.: Students' attitudes towards the use of mobile technologies in e-evaluation. Int. J. Interact. Mobile Technol. (IJIM) 11(5), 195–202 (2017)
4. Almaiah, M.A., Alyoussef, I.Y.: Analysis of the effect of course design, course content support, course assessment and instructor characteristics on the actual use of E-learning system. IEEE Access 7, 171907–171922 (2019)
5. Almaiah, M.A., Jalil, M.A., Man, M.: Extending the TAM to examine the effects of quality features on mobile learning acceptance. J. Comput. Educ. 3(4), 453–485 (2016)

6. Almaiah, M.A.: Acceptance and usage of a mobile information system services in University of Jordan. Educ. Inf. Technol. **23**(5), 1873–1895 (2018)
7. Al-Emran, M., Mezhuyev, V., Kamaludin, A., ALSinani, M.: Development of M-learning application based on knowledge management processes. In: Proceedings of the 2018 7th International Conference on Software and Computer Applications, pp. 248–253 (2018)
8. Almaiah, M.A., Jalil, M.A.: Investigating students' perceptions on mobile learning services. Int. J. Interact. Mobile Technol. (IJIM) **8**(4), 31–36 (2014)
9. Almaiah, M.A., Man, M.: Empirical investigation to explore factors that achieve high quality of mobile learning system based on students' perspectives. Eng. Sci. Technol. Int. J. **19**(3), 1314–1320 (2016)
10. Al-Emran, M., Shaalan, K.: Academics' awareness towards mobile learning in Oman. Int. J. Comput. Digital Syst. **6**(01), 45–50 (2017)
11. Almaiah, M.A., Jalil, M.A., Man, M.: Preliminary study for exploring the major problems and activities of mobile learning system: a case study of Jordan (2016)
12. Almaiah, M.A., Alismaiel, O.: A Examination of factors influencing the use of mobile learning system: an empirical study. Educ. Inf. Technol. **24**(1), 885–909 (2019)
13. Al-Emran, M., Mezhuyev, V.: Examining the effect of knowledge management factors on mobile learning adoption through the use of importance-performance map analysis (IPMA). In: International Conference on Advanced Intelligent Systems and Informatics, pp. 449–458. Springer, Cham (2019)
14. Arpaci, I., Al-Emran, M., Al-Sharafi, M.A., Shaalan, K.A.: Novel approach for predicting the adoption of smartwatches using machine learning algorithms. In: Recent Advances in Intelligent Systems and Smart Applications, pp. 185–195. Springer, Cham (2021)
15. Eyles, A., Gibbons, S., Montebruno Bondi, P.: Covid-19 school shutdowns: what will they do to our children's education? (2020)
16. Jegede, D.: Perception of undergraduate students on the impact of COVID-19 pandemic on higher institutions development in federal capital territory Abuja, Nigeria (2020)
17. Radha, R., Mahalakshmi, K., Kumar, V.S., Saravanakumar, A.R.: E-Learning during lockdown of Covid-19 pandemic: a global perspective. Int. J. Control Autom. **13**(4), 1088–1099 (2020)
18. Mhlanga, D., Moloi, T.: COVID-19 and the digital transformation of education: what are we learning on 4IR in South Africa? Educ. Sci. **10**(7), 180 (2020)
19. Al-Emran, M., Arpaci, I., Salloum, S.A.: An empirical examination of continuous intention to use m-learning: an integrated model. Educ. Inf. Technol., 1–20 (2020)
20. Almaiah, M.A, Almulhem, A.: A conceptual framework for determining the success factors of e-learning system implementation using Delphi technique. J. Theor. Appl. Inf. Technol. **96** (17) (2018)
21. Al-Emran, M., Mezhuyev, V., Kamaludin, A.: Towards a conceptual model for examining the impact of knowledge management factors on mobile learning acceptance. Technol. Soc., 101247 (2020)
22. Aldowah, H., Al-Samarraie, H., Ghazal, S.: How course, contextual, and technological challenges are associated with instructors' individual challenges to successfully implement e-learning: a developing country perspective. IEEE Access **7**, 48792–48806 (2019)
23. Arpaci, I.: A hybrid modeling approach for predicting the educational use of mobile cloud computing services in higher education. Comput. Hum. Behav. **90**, 181–187 (2019)
24. Arpaci, I.: Understanding and predicting students' intention to use mobile cloud storage services. Comput. Hum. Behav. **58**, 150–157 (2016)
25. Almaiah, M.A., Al-Khasawneh, A.: Investigating the main determinants of mobile cloud computing adoption in university campus. Educ. Inf. Technol., 1–21 (2020)
26. Arpaci, I., Al-Emran, M., Al-Sharafi, M.A.: The impact of knowledge management practices on the acceptance of Massive Open Online Courses (MOOCs) by engineering students: a cross-cultural comparison. Telemat. Inform., 101468 (2020)

27. Almaiah, M.A., Alamri, M.M., Al-Rahmi, W.: Applying the UTAUT model to explain the students' acceptance of Mobile learning system in higher education. IEEE Access **7**, 174673–174686 (2019)
28. Almaiah, M.A., Alamri, M.M.: Proposing a new technical quality requirements for mobile learning applications. J. Theor. Appl. Inf. Technol. **96**, 19 (2018)
29. Almaiah, M.A., Alamri, M.M., Al-Rahmi, W.M.: Analysis the effect of different factors on the development of mobile learning applications at different stages of usage. IEEE Access **8**, 16139–16154 (2019)
30. Almaiah, M., Al-Khasawneh, A., Althunibat, A., Khawatreh, S.: Mobile government adoption model based on combining GAM and UTAUT to explain factors according to adoption of mobile government services (2020)
31. Shawai, Y.G., Almaiah, M.A.: Malay language mobile learning system (MLMLS) using NFC technology. Int. J. Educ. Manage. Eng. **8**(2), 1 (2018)
32. Alamri, M.M., Almaiah, M.A., Al-Rahmi, W.M.: Social media applications affecting students' academic performance: a model developed for sustainability in higher education. Sustainability **12**(16), 6471 (2020)
33. Alksasbeh, M., Abuhelaleh, M., Almaiah, M.: Towards a model of quality features for mobile social networks apps in learning environments: an extended information system success model (2019)
34. Almaiah, M.A., Nasereddin, Y.: Factors influencing the adoption of e-government services among Jordanian citizens. Int. J. Electron. Gov. **16**(3), 236–259 (2020)
35. Chavoshi, A., Hamidi, H.: Social, individual, technological and pedagogical factors influencing mobile learning acceptance in higher education: a case from Iran. Telemat. Inform. **38**, 133–165 (2019)
36. Fayyoumi, E., Idwan, S., AL-Sarayreh, K., Obeidallah, R.: E-learning: challenges and ambitions at Hashemite University. Int. J. Innov. Learn. **17**(4), 470–485 (2015)
37. San Martín, H., Herrero, Á.: Influence of the user's psychological factors on the online purchase intention in rural tourism: integratinsg innovativeness to the UTAUT framework. Tour. Manag. **33**(2), 341–350 (2012)
38. Yu, T.K., Lin, M.L., Liao, Y.K.: Understanding factors influencing information communication technology adoption behavior: the moderators of information literacy and digital skills. Comput. Hum. Behav. **71**, 196–208 (2017)
39. Taat, M.S., Francis, A.: Factors influencing the students' acceptance of e-learning at teacher education institute: an exploratory study in Malaysia. Int. J. High. Educ. **9**(1), 133–141 (2020)
40. Delone, W.H., McLean, E.R.: The DeLone and McLean model of information systems success: a ten-year update. J. Manage. Inf. Syst. **19**(4), 9–30 (2003)
41. Kannan, V.R., Tan, K.C.: Just in time, total quality management, and supply chain management: understanding their linkages and impact on business performance. Omega **33**(2), 153–162 (2005)
42. Hair, J.F., Black, J.W., Babin, B.J., Anderson, R.E.: Multivariate data analysis, 7th edn. Prentice-Hall, Englewood Cliffs, NJ, USA (2010)
43. Fornell, C., Larcker, D.F.: Evaluating structural equation models with unobservable variables and measurement error. J. Marketing Res. **18**(1), 39–50 (1981)

Digitizing Learning During the Outbreak of COVID-19 Pandemic: Lessons Learned from the Most Infected Countries

Maryam N. Al-Nuaimi, Mohammed N. Al-Kabi, and Mostafa Al-Emran

Abstract In late December 2019, a novel coronavirus (COVID-19) was determined in Wuhan, China. Within a short period, more than 100 countries were infected with this epidemic. To mitigate the development of this virus, many countries have announced the closure of their educational institutions. The closure decision has left many institutions unable to select the appropriate technology for delivering the learning materials to their students. To assist those institutions attempting to digitize their learning during this pandemic, the main aim of this research is to review the leading technologies used for delivering the learning materials by considering the experiences of the most infected countries at the time of conducting this study. Through a mind map, we have concluded that the most leading technologies used for online learning are video conferencing, Massive Open Online Courses (MOOCs), educational national portals, recorded video lectures, cloud computing platforms, Microsoft Teams, and educational TV. We have also discussed the main challenges that might encounter the delivery of online learning. It is believed that the conclusions drawn from this study would assist those institutions trying to digitize their learning materials during the outbreak of COVID-19.

Keywords Novel coronavirus · COVID-19 · Educational technologies · Online learning

M. N. Al-Nuaimi
Department of English Language, Al Buraimi University College, Al Buraimi, Oman
e-mail: maryam@buc.edu.om

M. N. Al-Kabi
Department of Information Technology, Al Buraimi University College, Al Buraimi, Oman
e-mail: mohammed@buc.edu.om

M. Al-Emran (✉)
Faculty of Engineering & IT, The British University in Dubai, Dubai, UAE
e-mail: mustafa.n.alemran@gmail.com

© The Author(s), under exclusive license to Springer Nature Switzerland AG 2021
I. Arpaci et al. (eds.), *Emerging Technologies During the Era of COVID-19 Pandemic*, Studies in Systems, Decision and Control 348,
https://doi.org/10.1007/978-3-030-67716-9_18

1 Introduction

In late December 2019, a novel coronavirus (COVID-19) was discovered in Wuhan, China [1]. In less than two months, this epidemic has spread from China to more than 100 countries worldwide [2]. As of today (28-03-2020), the number of confirmed cases has increased to 622,316, and more than 28,800 death cases were reported across the infected countries [3]. This urgent threat has forced universities and schools across numerous countries to close their doors and digitize their learning activities.

Due to this pandemic, the educational systems across many countries have gradually dismantled. The spread of COVID-19 has affected students' ability to attend classes and acquire their learning activities. This has changed the way on how millions of students would be educated globally. These changes would lead to a paradigm shift in the educational systems, whether for the better or the worse situations in the long run [4]. In light of the rapid spread of COVID-19 across Asia, Europe, the USA, and the middle east, many universities and schools have decided to deliver the learning materials to students through online technologies in an attempt to mitigate the development of the epidemic. These urgent decisions have forced students to take a home-schooling environment, particularly in the most infected countries at the time of conducting this study, like China, Italy, the USA, Spain, Germany, Iran, and South Korea.

The call for online teaching during the wake of this epidemic has hurriedly encouraged educational institutions to turn their physical classes into virtual ones [5]. While this might be the best immediate reaction from the pedagogic and epidemiologic perspectives, it has affected the instructors in reformatting and retrofitting the learning materials in a way to be acceptable by the students [5]. A large number of research studies were conducted in the past concerning the issues related to the delivery of online teaching [6, 7]. However, none of these studies have discussed the issues related to the use of information technologies in promoting educational activities during the outbreaks of such an epidemic. As a means of supporting those universities trying to digitize their learning materials during the outbreak of COVID-19, this research aims to provide a holistic view of the leading technologies used to deliver the learning materials from the lenses of the most infected countries. Due to the lack of scientific articles concerning the use of educational technologies during the spread of this epidemic and specifically at the time of conducting this study, it is imperative to report that the sources through which this study relied on, were derived from newsletters, magazines, and other relevant websites.

2 Delivery of Learning in China

In response to the deadly coronavirus pandemic, many higher education institutions have made dramatic paradigm shifts to adapting pedagogy to online tuition [8]. To conserve equality in education and enable students to sustain their studies during the novel coronavirus outbreak, Chinese universities have launched digital instructional platforms to deliver a variety of courses utilizing ultra-modern digital tools [9]. Although it is a contingent delivery method, this inclusive model of distance teaching and learning targets all students, including international students and worldwide learners [8]. Since the crisis has provided the impetus for transforming campus-based degrees into fully-fledged online degrees, it has become imperative for online education to go viral [10].

Because COVID-19 is highly contagious, the real-time interaction between instructors and students has been jeopardized [11]. That being the case, the government-mandated closure of campuses has oriented tertiary education in China towards the endorsement of Massive Open Online Courses (MOOCs), aided by live-streaming apps and synchronized video-conferencing systems to fulfill practical teaching needs, such as Zoom, Skype, Alibaba's DingTalk, and Google Hangouts [8, 12]. A national portal called "National Cloud-Platform for Educational Resources and Public Service" was also used to provide millions of students with continued educational services [13]. Within a matter of weeks, the COVID-19 has become a catalyst for the trajectory of higher education to be digitally expediting learning innovation. In mainland China, 120 million students have accessed and shared learning materials via live broadcasts, virtual reality tuition, and learning cloud platforms. Such a trend has been consolidated by the government initiative of launching national internet cloud educational services [4, 12]. Additionally, more than 20 online curriculum platforms, 24,000 courses for higher education institutions, and library digital resources have been offered for free within a short period of time since the initial cases have been diagnosed [11]. In Chinese territories like Hong Kong, the vast majority of students have commenced using interactive apps while learning at home [4]. The flourishing of 5G technology in China has brought along the momentum for the prevalence of new modalities of digital education. Nonetheless, it has made the pedagogical efficacy of digital modalities of education vulnerable to pitfalls in terms of the infrastructure of the online education system and the impact of several aspects of the digital divide on students' and instructors' engagement with distance education [14]. In essence, the COVID-19 public health crisis has risen red flags of other hidden emergencies; viz-to-viz, "digital divide", or "digital exclusion". That is to say, in more impoverished or more rural areas with underdeveloped digital technology infrastructures, the construction of the medium for online learning to take place would be an enormously overwhelming task. Simultaneously, the access to digital information sources does not necessarily warrant that learning does occur, which makes the credibility of online-pedagogy questionable.

The decisive measures were undertaken by Chinese authorities to revoke the hazards of face-to-face instruction by setting into motion state-of-the-art digital learning programs that persist in being haphazard and short-term for two reasons. First, rather than being holistic and persistent in the long run, the transfer from offline to online education remains contingent on the fade or otherwise the outbreak of the epidemic. Second, for digital education to succeed substantially, an extraordinarily high level of systematic collaboration and cooperation among higher education institutions is considerably in demand to develop a new cloud-based online learning and broadcasting platform as well as to upgrade a suite of education infrastructure [4, 9]. This state of affairs vindicates the argument that on more profound levels of analysis, there comes to the surface an acute deficiency in the sustainable endeavors to bridge the digital divide on the levels of technology infrastructures, legislations, logistic readiness, inclusiveness, digital equity, bandwidth connectivity, and digital literacy among various hierarchical levels of stakeholders in the educational enterprise. These pillars are essentially vital in accelerating and facilitating the transition from offline to online education in virtual contexts for students, parents, instructors, administrators, and digital curriculum developers, while taking caution to mitigate the skeptical remarks on the quality of electronic tuition, especially in the era following the fourth industrial revolution, in which the world has become wireless.

The unprecedented distance learning model that has been emerging and evolving under the exceptional epidemic circumstances resembles, to a noticeable extent, the MOOCs paradigm. Thus, shortcomings that are inherent in MOOCs such as the low rates of completion, as opposed to small private online courses, and the difficulty to maintain attentiveness and engagement into the online format of education are amongst the salient challenges that hinder fully reaping the pedagogical gains of online instruction. Furthermore, securing clear and concise communication channels to manage learners' anxiety, the adaptation of assessment tools to ensure robust learning, and nurturing individualized cognitive and metacognitive development of learners constitute major milestones to cultivate the growth of remote online learning [15].

3 Delivery of Learning in Italy

Italy has the second top COVID-19 confirmed cases after China, with an increasing number of patients suffering from coronavirus [16]. The growing number of COVID-19 infections leads to paralyzing different aspects of Italian lives [17]. Many schools, universities, and companies rely on e-learning and smart working as alternatives to conventional education across different levels. Smart working cannot solve the production processes problems and workflows in factories [17]. The digital transformation can solve some of the COVID-19 issues, but it is not a magic solution to the COVID-19 problem.

On March 4th, 2020, Italy's education minister, Lucia Azzolina, announced the closure of all academic establishments, including schools and universities, starting from March 5th, 2020 to March 15th, 2020. The closure later is extended to April 3rd, 2020 [17–19]. To handle the closure situation, Italy has used two national portals. The first one is called "La scuola continua", which was dedicated to school students. The other portal called "Nuovo Coronavirus", which was devoted to show how to deal with COVID-19 in emergency cases [13].

Matthew Loveless is an associate professor at The University of Bologna in Italy who decided to use Microsoft Teams as a platform to deliver lectures to his students [20]. Microsoft Teams is a platform for unified communication and collaboration. It is a hub for teamwork in Microsoft Office 365 Suite. Therefore, it can be used for group chat, video meetings, file storage, and hosting online meetings. Microsoft Teams is compatible with almost all the available operating systems and can be used on desktop and mobile devices. To solve the problem of academic establishments closure, the use of online learning is a good option since it lets the students to pursue their studies from homes regardless of how far these homes are from their schools or universities. The University of Bologna decides to convert all its traditional convertible courses to online courses, whereas some cannot be converted like lab courses [21]. While the library at the University of Bologna is also closed, Antonino Rotolo, the vice-rector for research at the University of Bologna, said that this crisis would help academic establishments to accelerate the process of adopting digital technologies for teaching [21].

Furthermore, online learning is a solution to the problem of timetables for students who have mutually contradictory time courses, as well as avoiding the chances of infections. Online learning is an economical option since we get rid of the transportation and residential costs. The present circumstances enforce the necessity to consider online learning and home-schooling as a viable option for education at different levels for the present and the future. The Italian universities have adopted digital technologies to minimize the harmful effects of placing the whole country under quarantine. Therefore, all courses, exams, and research were digitized to enable these universities to deliver online courses and hold thesis defenses through videoconferences. The lab-based courses could be postponed until the closure and quarantine are lifted, and life returns to normal [21].

4   Delivery of Learning in the USA

The transformation from face-to-face education to online and distance learning is going on all around the world, due to the increasing infections of COVID-19. In the USA, Harvard University stated that it would move for online and distance learning for graduate and undergraduate classes starting from March 23, 2020 [22]. The transformation to online and distance learning is not only restricted to Harvard University, but other universities also adopted it in the USA, such as "American University", "Amherst College", "Barnard College", "Columbia University",

"University of California San Diego and Berkeley", "Hofstra University in New York", "Ohio State University", "Princeton University", "Seattle University", "University of Southern California", "Stanford University", and "University of Washington", according to multiple news outlets.

Prior to the COVID-19 crisis, Johns Hopkins University (JHU) has handled the problem of shortages of dead human bodies that were used to teach autopsies through the use of multimedia platforms as a solution [23]. During the spread of the COVID-19 epidemic, the JHU published a practical guide for faculty members through its website to explain the process of moving their online courses as an alternative to traditional face-to-face courses [24]. The JHU launched a website called "Keep Teaching @ JHU", which is a hub of hyperlinks and frequently asked questions to provide the faculty members with all necessary guidance to prepare their online materials.

It is urgent under the current circumstances to move into virtual classes. This transition requires creative learning activities and to think deeply about how to conduct assessments. In this sense, the faculty members can use Zoom and Panopto, which are web-based video conferencing tools, to communicate with their students and deliver their lectures. Transitioning to online learning is not easy due to the differences between face-to-face classes and virtual ones [24]. Many colleges and universities in the USA have closed their face-to-face classes, and the transition to online learning is coming ahead [25, 26]. The transition to online learning is not an easy option since there is a large portion of faculty members who are not familiar with educational technologies. Furthermore, there is a large portion of students who do not have access to reliable and fast Internet services [26].

5 Delivery of Learning in Spain

Concerning the delivery of learning in schools during the spread of COVID-19, the Ministry of Education in Spain offers three national learning platforms. First, INTEF as a pedagogical resource to support distance learning. Second, Procomún collection with around 100,000 educational resources and learning objects. Third, an online channel called Educlan [13].

In terms of higher education, the University of Deusto in Spain offers several teaching support resources such as Moodle, a free and open-source learning management system, and Google Meet to make video conferencing between the students and their instructors [27]. The University of Deusto advised the faculty members and students to use emails, Google Calendar, Google Hangouts, and Google Drive to communicate virtually. In Spain, there are few distinguished online universities like the Open University of Catalonia (UOC) and the National Distance Education University (UNED) [23]. These two Spanish universities offer online learning, while others offer traditional (face-to-face) learning. The traditional learning provided by the vast majority of Spanish universities can change face-to-face learning into distance and online learning. Still, the question is, would this transition be good enough for the students? Many think that distance/online learning means the use of different technological tools only, such as recorded lectures, educational platforms, PowerPoint slides, PDFs, simulations, virtual reality (VR), etc. This concept of distance/online learning that depends on technical tools ignores the quality through which the interaction between the mentor and his/her students is affected [23]. This means that the most essential factor is the adopted pedagogical model.

The Spanish Ministry of Universities and the Conference of Rectors have launched an educational platform to enable the instructors and their students to attend classes virtually. This educational platform was mainly designed by both UNED and UOC. This platform mainly aims to provide guidance and training resources for instructors to enable them to convert their face-to-face classes into virtual ones [28].

6   Delivery of Learning in Germany

Germany, as other European countries, forced restrictions on schools and universities to slow down the spread of COVID-19. The Technical University of Munich (TUM), as one of the German Universities which forced these restrictions, had tried to use online learning to overcome this crisis [29]. The TUM has informed the instructors to teach online instead of teaching on its campus [30]. A booklet entitled "Flexible Solutions for Digital Teaching" was published on the web by the TUM in order to enable the instructors to convert their traditional classes into virtual ones easily [31]. The booklet shows the instructors the appropriate tools used for conducting an online lecture, seminar, or meeting. Furthermore, this booklet shows how to record different online activities and create educational videos [31].

7   Delivery of Learning in Iran

The Iranian students have already experienced the distance learning approach for the first time during the eighties in the first gulf war using their state TV to broadcast different lectures to their students. This means that the distance learning method has already been experienced around 35 years ago in Iran. It is, therefore, that Iran has adopted the same approach in the COVID-19 era. Currently, the Iranian used different techniques in their second distance learning experience, which enables various parties to interact with each other [32]. From their first experience with distance learning through TV, the Iranians realized the difficulties of including all school classes and courses and conducting different exams, and that was due to the one-way communication method [32]. To overcome these limitations and to deal with the current urgent situation, the University of Tehran has published on one of its webpages about the availability of many online classes prepared by the university instructors under the guidance of the e-learning center in the university [33].

Concerning the delivery of learning at Iranian schools, it has been argued that the Iranian school students may have to continue their studies next summer, and this means there will be no summer holiday for school students [34]. Furthermore, the idea of adopting distance education, offering videoconferencing classes, and State educational TV is highly encouraged [34]. It is hoped that adopting these technologies may help to improve the quality of the educational system in Iran. In line with the increasing number of national portals to help students in pursuing their education across several countries, Iran has also launched a national portal called "Dedicated TV programming" that was created by the Ministry of Education to deliver the learning for different grades during this crisis [13].

8 Delivery of Learning in South Korea

Unlike the other highly infected countries, South Korea has resorted to its functional and consistent standard operation procedure to hold sway over risk factors in public health crises. With the eruption of the fatal disease, the closure of higher educational institutions and retrieving their students from study abroad programs, have been compulsory approaches to restrain public gatherings, and hence, to contain the prevalence of the novel coronavirus [22, 35, 36]. Since South Korea is one of the most technologically equipped countries in the world, mobile mass indiscriminate awareness-raising and smartphone apps providing GPS maps have been set into action to observe the infection spread carefully [36].

South Korea has demonstrated a role model for the world in the employment of smartphone apps to constrict the contagion, and consequently to reduce mortality. On the other hand, the South Korean vigorous information technology infrastructure has not been manifested comparably and robustly in the conversion to online education. In a nation where education is highly competitive on global levels, South Korean universities have not managed to confer their students, including nationals and international students, convenient and pathological distance online learning experiences in compensation for face-to-face instruction. As such, the crisis has left higher educational institutions in South Korea to be incapable of alleviating the negative feedback of frustrated students [37].

Apart from online education initiatives, Seoul has devoted the vast technological resources it enjoys at its disposal to primarily save lives by creating and implementing smartphone apps with two main aims; (a) mass testing and tracking of suspected cases, and (b) educating the population of the preventive mechanisms. To that end, the first app is mandatory for people arriving in South Korea from immensely impacted areas. The app forwards users to the teleworking executive to report any suspicious symptoms on a regular basis. The other app is not compulsory, but it alerts public health officials whenever someone leaves the isolation zone. Concurrently, the Centers for Disease Control and Prevention in South Korea have published a full amount of transparent information punctually every day to educate minutely the public, including experts and citizens to improve their understanding of how the virus evolves and functions in addition to the conveyance of relevant data that relieve the population concerns. Such valuable information has been communicated to residents using a national mobile phone alert system [38].

South Korea has a penetration rate of over 90%, which is anticipated to narrow the digital divide in the country to a considerable extent. Despite that, university students in South Korea might perceive the recorded video lectures and real-time telelectures as a poor substitute for face-to-face instruction. This stems from the design of universities online platforms, which simulate MOOCs and minimizing lively interactive learning experiences [10]. South Korea has used the educational broadcasting system as a national portal to provide learners with advanced educational services based on the use of multimedia [13]. Bridging the digital divide in South Korea has not been adequately associated with sustainable strategic planning for crisis management in education, which seems to be far underdeveloped as compared to risk management in healthcare sectors. Instead, the core of South Korean educative endeavors under the spread of the epidemic lies in the utilization of smart apps and high-speed connectivity as agencies to the quick containment of the epidemic through the constant disclosure of crystal-clear and intensive up-to-date information to the public, swift widespread testing, social distancing, and digital monitoring of those under quarantine [39].

The lessons gleaned from the South Korean case revolve around the aggressive employment of smart technologies in plotting the trajectory of the pandemic by securing information necessary for effecting extreme proactive measures to break the exponential growth cycle of the infection. However, the country has compromised the quality of online alternatives of conventional face-to-face tuition during lockdowns of campuses for the sake of striking down its epidemic curve [40]. In this respect, the coronavirus catastrophe has been tackled by integrating and readjusting multiple forms of artificial intelligence to share, transmit, process, and reciprocate timely high-quality digital data between the public and governmental organizations, which is vital to make detrimental decisions to eradicate the epidemic [41]. The South Korean experience poses the question as to whether governments should place the information technology capital investment into the healthcare industry or education under pervading disease quandaries. Such a question leads to a more insisting debate as to whether the educational sectors demand well-established benchmarks for an effective standard operation procedure that prognosticates and correspondingly applies a comprehensive scheme of digital solutions to remunerate for emergency shutdowns of educational institutions.

9 Discussion and Conclusion

Due to the outbreaks of COVID-19, many countries have announced the closure of their educational institutions, including schools and universities [42]. The closure of these institutions has led to a paradigm shift in the educational process in a way to transform the face-to-face classrooms into virtual ones. This transformation has left many institutions unable to decide to select the appropriate technology for delivering the learning materials to their students. In order to help those institutions attempting to digitize their learning during this pandemic, the main aim of this research is to review the leading technologies used for delivering the learning materials with a particular focus on the most infected countries. In that, we have reviewed the educational technologies used in China, Italy, the USA, Spain, Germany, Iran, and South Korea.

It has been observed that a large number of technologies were used to deliver online learning as an alternative to traditional learning among the selected countries. To draw a comprehensive picture of those technologies, we have summarized the leading delivery technologies used for online learning through a mind map, as shown in Fig. 1. These educational technologies include video conferencing, MOOCs, educational national portals, recorded video lectures, cloud computing platforms, Microsoft Teams, and educational TV. It is believed that the conclusions drawn from this study would assist those institutions trying to move into online learning by enabling them to select the appropriate technology that suits their infrastructure capabilities.

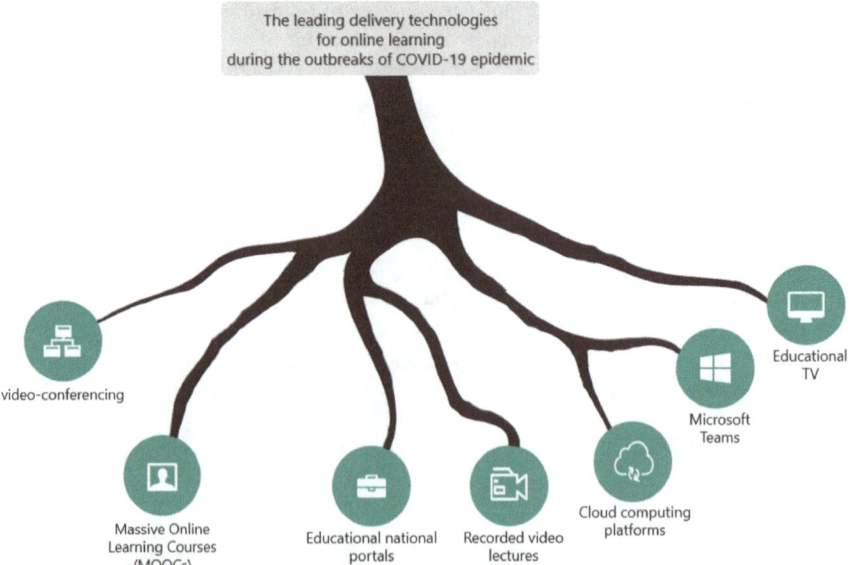

Fig. 1 Leading delivery technologies for online learning

While several technologies were suggested, some challenges might still exist through the transition to online learning across several universities. First, many faculty members and students are not familiar with such technologies. Second, many faculty members and students do not have access to the internet, especially those living in remote areas. Third, the differences in quality between traditional classrooms and online learning might be a substantial factor that hinders the transition to online classes.

References

1. Zhu, N., et al.: A novel coronavirus from patients with pneumonia in China, 2019. N. Engl. J. Med. (2020). https://doi.org/10.1056/NEJMoa2001017
2. Ceukelaire, W.D., Bodini, C.: We need strong public health care to contain the global corona pandemic. Int. J. Heal. Serv. (2020)
3. Worldometers: COVID-19 coronavirus outbreak. Worldometers. https://www.worldometers.info/coronavirus/ (2020)
4. Tam, G., El-Azar, D.: 3 ways the coronavirus pandemic could reshape education. World Economic Forum. https://www.weforum.org/agenda/2020/03/3-ways-coronavirus-is-reshaping-education-and-what-changes-might-be-here-to-stay/ (2020)
5. Sage: 16 Answers to your questions about teaching online. Social Science Space. https://www.socialsciencespace.com/2020/03/16-answers-to-your-questions-about-teaching-online/ (2020)
6. Kebritchi, M., Lipschuetz, A., Santiague, L.: Issues and challenges for teaching successful online courses in higher education: a literature review. J. Educ. Technol. Syst. **46**(1), 4–29 (2017)
7. Arpaci, I., Al-Emran, M., Al-Sharafi, M.A.: The impact of knowledge management practices on the acceptance of Massive Open Online Courses (MOOCs) by engineering students: a cross-cultural comparison. Telemat. Inf. (2020)
8. Zhaohui, W.: How a top Chinese university is responding to coronavirus. World Economic Forum. https://www.weforum.org/agenda/2020/03/coronavirus-china-the-challenges-of-online-learning-for-universities/ (2020)
9. Kennedy, K.: Online learning shift will bring positives, but adapted not replicated content is key. The Pie News. https://thepienews.com/news/sector-shares-mixed-views-on-the-surge-in-online-learning-due-to-travel-bans/ (2020)
10. Lau, J., Yang, B., Dasgupta, R.: Will the coronavirus make online education go viral?. The World University Rankings. https://www.timeshighereducation.com/features/will-coronavirus-make-online-education-go-viral (2020)
11. Winthrop, R.: How has the coronavirus impacted the classroom? On the frontlines with Dr. Jin Chi of Beijing Normal University. Brookings. https://www.brookings.edu/blog/education-plus-development/2020/02/27/how-has-the-coronavirus-impacted-the-classroom-on-the-frontlines-with-dr-jin-chi-of-beijing-normal-university/ (2020)
12. Fabienne, L.: Schools in China switching to online education amid coronavirus outbreak. Interesting Engineering. https://interestingengineering.com/schools-in-china-switching-to-online-education-amid-coronavirus-outbreak (2020)
13. UNESCO: National learning platforms and tools. UNESCO. https://en.unesco.org/themes/education-emergencies/coronavirus-school-closures/nationalresponses (2020)
14. Zinan, C.: Online teaching, distance learning discussed at symposium. China Daily.com.cn. https://global.chinadaily.com.cn/a/202003/17/WS5e70c2f5a31012821727fd7e_2.html (2020)

15. Snelling, J., Fingal, D.: 10 strategies for online learning during a coronavirus outbreak. Deas, content and resources for leading-edge educators. https://www.iste.org/explore/10-strategies-online-learning-during-coronavirus-outbreak (2020)
16. Microsoft: COVID-19 tracker. Microsoft (2020)
17. Saini, V.: Coronavirus: lessons from Italy. EUobserver. https://euobserver.com/coronavirus/147753 (2020)
18. Giuffrida, A., Tondo, L., Beaumont, P.: Italy orders closure of all schools and universities due to coronavirus. The Guardian. https://www.theguardian.com/world/2020/mar/04/italy-orders-closure-of-schools-and-universities-due-to-coronavirus (2020)
19. Feuer, W.: Italy expands its quarantine to the entire country as coronavirus cases and deaths surge. CNBC LLC. https://flipboard.com/topic/intensivecare/italy-expands-its-quarantine-to-the-entire-country-as-coronavirus-cases-and-deat/f-adee285409%2Fcnbc.com (2020)
20. Loveless, M.: Academe in the red zone-Matthew loveless describes life as a professor in Italy during the coronavirus outbreak. Inside Higher Ed. https://www.insidehighered.com/views/2020/03/11/professor-italy-describes-life-there-during-coronavirus-opinion (2020)
21. Zubaşcu, F.: Italian universities scramble to move teaching and research online during coronavirus lockdown. Science|Business. https://sciencebusiness.net/news/italian-universities-scramble-move-teaching-and-research-online-during-coronavirus-lockdown (2020)
22. VOA: Colleges, universities move classes online amid coronavirus outbreak. VOA. https://www.voanews.com/science-health/coronavirus-outbreak/colleges-universities-move-classes-online-amid-coronavirus (2020)
23. Sangra, A.: Study online in times of coronavirus. Sociedad Vascongada de Publicaciones (2020)
24. Cruickshank, S.: How to adapt courses for online learning: a practical guide for faculty. Johns Hopkins University. https://hub.jhu.edu/2020/03/12/how-to-teach-online-courses-coronavirus-response/ (2020)
25. Abdalla, J.: US universities switch to online courses due to coronavirus. Al Jazeera Media Network. https://www.aljazeera.com/news/2020/03/universities-switch-online-courses-due-coronavirus-200310202804023.html (2020)
26. Pfleger, P.: The coronavirus outbreak and the challenges of online-only classes. NPR. https://www.npr.org/2020/03/13/814974088/the-coronavirus-outbreak-and-the-challenges-of-online-only-classes (2020)
27. Deusto: Action at the university of Deusto in the face of the new coronavirus. University of Deusto. https://www.deusto.es/cs/Satellite/deusto/en/university-deusto/action-at-the-university-of-deusto-in-the-face-of-the-new-coronavirus (2020)
28. Elperiodico: Universities and associations launch free distance education initiatives|coronavirus. Elperiodico (2020)
29. TUM: Coronavirus: university operating under restrictions since 18 March 2020. The Technical University of Munich (TUM). https://www.tum.de/en/about-tum/news/coronavirus/ (2020)
30. TUM: Information for instructors. Technical University of Munich (2020)
31. TUM: Flexible solutions for digital teaching. Technical University of Munich (2020)
32. Nassir, B.: Coronavirus and Iran's experience on distance learning. https://www.tehrantimes.com/news/446207/Coronavirus-and-Iran-s-experience-on-distance-learning (2020)
33. University of Tehran: Announcement: holding classes online during coronavirus disease 2019 (COVID-19). University of Tehran. https://ut.ac.ir/en/news/11099/announcement-holding-classes-online-during-coronavirus-disease-2019-covid-19- (2020)
34. Agency, T.N.: Fallout of coronavirus: Iran schools may run make-up classes in summer. Tasnim News Agency. https://www.tasnimnews.com/en/news/2020/03/12/2221856/fallout-of-coronavirus-iran-schools-may-run-make-up-classes-in-summer (2020)
35. Fendos, J.: Lessons from South Korea's COVID-19 outbreak: the good, bad, and ugly. The Diplomat. https://thediplomat.com/2020/03/lessons-from-south-koreas-covid-19-outbreak-the-good-bad-and-ugly/ (2020)

36. Hou, C.: How South Korea is handling the coronavirus outbreak better than other countries. Changing America. https://thehill.com/changing-america/well-being/prevention-cures/487465-how-south-korea-is-handling-the-coronavirus (2020)
37. Joung, M.: In COVID locked-down South Korea, students long for the classroom. Voice of America. https://www.voanews.com/science-health/coronavirus-outbreak/covid-locked-down-south-korea-students-long-classroom (2020)
38. Braun, A.S.: Commitment, transparency pay off as South Korea limits COVID-19 spread. EURACTIV. https://www.euractiv.com/section/coronavirus/news/commitment-transparency-pay-off-as-south-korea-limits-covid-19-spread/ (2020)
39. Kasulis, K.: South Korea's coronavirus lessons: quick, easy tests; monitoring. ALJAZEERA. https://www.aljazeera.com/news/2020/03/south-korea-coronavirus-lessons-quick-easy-tests-monitoring-200319011438619.html (2020)
40. Branswell, H.: Understanding what works: how some countries are beating back the coronavirus. STAT. https://www.statnews.com/2020/03/20/understanding-what-works-how-some-countries-are-beating-back-the-coronavirus/ (2020)
41. Xie, B., et al.: Global health crises are also information crises: a call to action. J. Assoc. Inf. Sci. Technol., 1–5 (2020)
42. UNESCO: COVID-19 educational disruption and response. UNESCO. https://uil.unesco.org/covid-19-educational-disruption-and-response (2020)

Using mHealth Apps in Health Education of Schoolchildren with Chronic Disease During COVID-19 Pandemic Era

Abdulaziz Mansoor Al Raimi, Chan Mei Chong, Li Yoong Tang, Yan Piaw Chua, and Latifa Yahya Al Ajeel

Abstract COVID-19 significantly affects all our normal life daily especially health care services, so it's important to find and implement innovative approaches to help individuals at a high risk to resume normal life daily. The usage of digital technologies and social networking has grown rapidly over the last decades, and these technologies are increasingly being incorporated into health education. In this study, we discussed the importance of using the mHelath technology for schoolchildren with chronic disease during the COVID-19 era, and we have used Social Learning Theory and Technology Acceptance Model from the Theory of Reasoned Action (TRA) as the theoretical framework for the present study. The previous study concluded the mobile device being studied is a reliable way of helping schoolchildren increase awareness their disease, but further research efforts should assess the impact of application usage on disease outcomes over a more extended follow-up period as compared to traditional care.

Keywords mHelath · Mobile apps · Health education · COVID-19

A. M. Al Raimi (✉) · C. M. Chong · L. Y. Tang · L. Y. Al Ajeel
Department of Nursing, Faculty of Medicine, University Malaya, 50603 Kuala Lumpur, Malaysia
e-mail: aziz-mansoor@hotmail.com

C. M. Chong
e-mail: mcchong@um.edu.my

L. Y. Tang
e-mail: liliantang@um.edu.my

L. Y. Al Ajeel
e-mail: latifaagel@yahoo.com

A. M. Al Raimi · L. Y. Al Ajeel
Department of Health Sciences, Seiyun Community College, Seiyun, Yemen

Y. P. Chua · L. Y. Al Ajeel
Institute of Educational Leadership, University Malaya, 50603 Kuala Lumpur, Malaysia
e-mail: chuayp@um.edu.my

I. Arpaci et al. (eds.), *Emerging Technologies During the Era of COVID-19 Pandemic*, Studies in Systems, Decision and Control 348,
https://doi.org/10.1007/978-3-030-67716-9_19

1 Introduction

Electronic Health (eHealth) refers to "health services and information delivered or enhanced through the internet and related technologies" [1], while mobile health (mHealth) is subbranch of eHealth and can be defined as "the use of mobile computing and communication technologies in health care and public health" [2]. So Mobile Apps "It's software Apps specifically designed for and available on smartphones and tablets, have been cheerfully adopted by users of smartphones and tablets and proposed as a delivery mechanism for self-management health experimental" [3].

The mobile phone is one of the most rapid developing sectors in the information technology and its resulting can impact in medicine, it's the one of the most of communication system all over the place and it has dynamic trends, it also can be utilized for communication, surfing internet, and utilizing concrete Apps [4].

Digital methods use technology to enhance adherence have been rapidly created and validated over the past decade, including instant messages, smart wellness devices, and immersive websites. Digital approaches are successful in growing awareness of different disease, minimizing limitations of movement, enhancing self-management such use of action plans to improving quality of life and maximizing drug usage [5].

Smartphones are used and used by about 50% of ages 12–17 and 75% of adults aged 30–49 years of age, so there is a strong need for an intervention in mHealth directed at supportive family assistance as early adolescents with chronic disease move to take more responsibility for their treatment [6].

Mobile health (mHealth) apps have the ability to promote self-management of patients by incorporating prescription reminders, facilitating symptom self-monitoring, enhancing access to and accuracy of knowledge shared with physicians, and supplying patients and parents with educational tools [7].

The use of mobile apps for prevention and improvement of health care is also known as mHealth. These resources include text messaging, exercise machines, and smartphone apps—the most common features used in smartphone mHealth. Conscious of the promise of mHealth in disease prevention, the number of health-related apps has gradually grown [8]. Mobile health Apps can promote disease prevention and can be a highly useful method for providing successful mental health services to encourage supportive family support as early teens develop and learn self-management habits [6].

Smartphones have been commonly deployed around the world over the past few decades but have recently seen several therapeutic uses during the COVID-19 pandemic. These new technologies help to avoid face-to-face interactions with the patients by the healthcare provider, thereby maintaining social distance and preventing virus transmission. Such phones are useful for clinical assessment, diagnosis, prompt referral, prescription and also for tracking patients from home and distant areas [9].

The increasing importance of mobile Health globally has led to a significant effort by official health organizations such as World Health Organization (WHO) which has been publishing every year since 2009 on a reports which covering initiatives in electronic and mobile health [10].

In particular, few studies are available evaluating the effectiveness of smartphone use in children's health education and how health education can improving patient specially schoolchildren with chronic disease [11], So Schneider et al. [8] recommended in their study future studies should be focused to examine the impact of device usage on chronic disease results over a longer follow-up time and relative to standard treatment.

The remainder of the paper is organized as follows: Sect. 2: Health Education via mobile app describes the using of mHelath apps as new methods for delivering health education, Sect. 3: Role of mHelath apps during COVID-19 era in health education of schoolchildren which concluded the recent articles and research emerging during the COVID-19 regarding using mHelath apps for heath delivering education and Sect. 4: Theoretical Framework, which de the theorical framework used in this study.

2   Health Education via Mobile App

Education is a main key to improving health knowledge, skill and compliance especially for the patient with poor knowledge and compliance [12]. Therefore, we need to develop educational strategies that will motivate patients and increase health knowledge and skills for self-management after setting individual education needs [11].

Health education vis mobile app can potentially mitigate the adverse effects of diseases by helping users detect and treat early symptoms prior to moving to a poorly controlled state [13]. These education methods will make users aware of symptoms and causes and allow them to take prompt and effective action to resolve symptoms or avoid further worsening of symptoms [13].

Mobile Apps to provide teens agers with knowledge about how well their disease is managed, though, will be focused about proven evidence-based guidance because many of the existing mHealth Apps not align with the evidence-based guidelines [14].

More and more mobile health apps have been created in the past few years to help empower the patient to properly track and treat their disease. There is no question that patients who take control of their disease and stay committed to treatment will have decreased complications, increased quality of life and lowered cost of health care [15].

Face-to-face, health education was designed to facilitate commitment and self-management for chronic disease affected young people. Such measures showed effectiveness by increasing adherence to treatment, lung capacity, self-efficacy and attendance at school as well as reducing movement restrictions and visits to emergency departments.

However, the implementation of these face-to-face interventions often presents significant barriers. Several barriers to adherence to patient medication are also barriers to patient and family involvement, and involvement in disease management interventions in the individual. Transportation hurdles, for example, may prohibit a family from receiving refills, which can also find it impossible to attend appointments [5].

Mobile phones are one of E-learning tools used recently, it's unlike desktop computers or even laptops, which are nearly always with the person. Many of people are rarely leave away from his mobile phones for a while, and it in our hand or in our pocket or purse [16]. Mobile applications and other social networking services like Facebook, goggle plus are becoming more widespread in educational environments, with educators exploring how such tools can be used for teaching and learning [17].

A systematic analysis research conducted to assess digital interventions for pediatric treatment found that all results indicate that electronic interventions aimed at improving adherence are positive, as well as enhancing safety outcomes in addition to adherence to medicine, suggested that future studies explore specific digital intervention platforms (apps, social networking, text messages) to examine the efficacy and participation of young people with chronic disease in evaluating the degree of care and human activity necessary and appropriate for better long-term adherence [5].

Also Schneider et al. [8] found in their study results that the mobile device being studied is a reliable way of helping teenagers control their disease. They recommend further study actions would further evaluate the influence of technology usage on patient effects over a longer follow-up period compared to conventional treatment.

3   Role of mHelath During COVID-19 Era in Health Education of Schoolchildren

On 30 January 2020 the spread of a new coronavirus strain was identified by the World Health Organization (WHO) as a "public health emergency of international significance." It was declared a "pandemic" on 11 March. COVID-19 spread to 187 countries according to World Health Organisation (WHO) statistics. The virus continues to influence large populations from many different ways including psychological, emotional, political, and cultural [18].

According to WHO last survey released on 1st June 2020 found in many countries the public health systems have been partly or entirely compromised. More than half (53%) of the countries surveyed have partially or completely disrupted hypertension treatment services; 49% for diabetes and complications due to diabetes; 42% for cancer diagnosis and 31% for cardiovascular disease emergencies, rehabilitation programs have also been affected for about two-thirds (63%) of countries, while rehabilitation is essential to a successful recovery following a severe COVID-19 illness [19].

The health condition of schoolchildren in the same manner generally effected, generally the teenagers are less severe COVID-19 cases than those of adult patients, but some research found both ages tended to be vulnerable to COVID-19, although there was no major sex disparity [18, 20].

Parents of children with some chronic disease may have raised worries regarding disease management during a respiratory disease pandemic, and may improve attention to their children health status [21]. A such cases a child or youth with COVID-19 infected may experience serious morbidity due to combined effects on the respiratory tract [22], so poor of disease control is a contributing factor for enhanced virus-induced exacerbation of severity [23]. Since all methods of disease control optimization, have been shown to significantly reduce the risk of complications, most of which are virus-induced [24].

Using mHealth care can reach to the places where poor or no healthcare is available and can also allow people to access some healthcare services where is difficult to reach them [25]. One of the main benefits of wireless networking technologies is progress through cost savings, more productive procedures and fulfilling some of healthcare professionals' workload needs. The main challenges facing any wireless networking technologies can also be caused by both the lack of network coverage and system malfunctions and infrastructure part failures [26].

School-based disease prevention programs provide an opportunity to meet children at risk for inadequate management of disease while they are well [27]. Influenza epidemic results show that school suspensions have some advantages in high-transmissibility outbreaks, as with COVID-19 [28].

During school closing due to COVID-19 outbreak there is less people crowded and less vehicles on the street and children spending more time indoors, they should be less vulnerable to air pollution in outdoors with a resulting increase in air quality, both combined with a reduction in childhood referrals to emergency departments (ED). From the onset of the COVID-19 pandemic, air quality has already changed, and this change will lead to better health in a good way [21]. At the same time, being indoors that's make children become more vulnerable to indoor conditions that may worsen of some allergies disease, including secondary sensitivity to cigarette smoke and occupational allergens such as molds, rodents and roaches [21].

Social media is seen as an important channel for promoting risk communication during previous epidemics such as Zika and Ebola. Similarly, in the COVID-19 outbreak, students who use social media effectively to acquire health information [29].

The current data is consistent with a wide variety of impacts of school closing, ranging from no benefit to transmission declines to more significant consequences, school closing has very high economic costs and possible damages [28]. Despite the lack of evidence, the experts strongly agree that all children and adolescents with chronic disease will stay on their treatment schedule throughout without any change during the COVID-19 outbreak to be safe [30].

Mobile learning has traditionally been developed more strongly within the field of informal education, this trend is attributed partly to the fact that mHealth requires learning in a more subjective, individualized and placed manner. Integrating informal education systems and services into formal education may lead to providing educational alternatives to help schools respond to the demands of an ever-changing environment that calls for more responsive and individualized instruction, and to place students at the center of the teaching-learning process and provide them with more power. As a result, there is a increasing number of observations about the use of mHealth as a bridge between informal and formal schooling [31].

Due to the broad using of mobile phone technology, health applications may be able to reach a large people, particularly in settings where are poor infrastructure and lack to access printed materials or face-to-face consultations [32].

4   Theoretical Framework

In this study we used two learning theories as the theoretical framework for the present study, which is Social Learning Theory [33–35] and Technology Acceptance Model from the Theory of Reasoned Action (TRA) (Vygotsky 1978).

4.1   Social Learning Theory

Social cognitive theory indicates that people are not only responding to external stimuli but are consciously finding and processing information [36]. Individuals "function as contributors to their own motivation, behavior, and development within a network of reciprocally interacting influences" [34]. While social cognitive theory encompasses many issues such as moral judgement and physiological anticipation, research has concentrated mainly on self-efficacy, or the assumptions about one's abilities to effectively achieve tasks or goals [37].

Social cognitive theory, according to [35], takes on an agent-like perspective for transition, development, and adaptation. Bandura defines an agent as someone who deliberately affects one's working and living circumstances; "In this view, people are self-organizing, proactive, self-regulating, and self-reflecting. They are contributors to their life circumstances not just products of them" [38].

Self-Efficacy Theory is component of Bandura's social cognitive theory, which she suggested that an individual's action, atmosphere and cognitive factors are all connected. Bandura proposed Social Cognitive Theory in reaction to his frustration with the clinical and psychoanalytical concepts.

The role of perception in motivation and the role of the condition was largely neglected in those two hypotheses [33]. "Unidirectional environmental determinism is carried to its extreme in the more radical forms of behaviorism" but humanists and existentialists, who emphasize human ability for moral thinking and deliberate intervention, contend that people decide what they are through their own free will.

Bandura [39] define the self-efficacy as: "People's judgments of their capabilities to organize and execute courses of action required to attain designated types of performance. It is not with the skills one has but with judgments of what one could do with whatever skills one possesses."

Self-efficacy is a person's prudence of their own ability to plan and take a course of action in order to achieve a goal. It is the natural belief of capability to doing a particular task to achieve a set goal [40].

Bandura [33] suggested the one's self-efficacy can be judgment based on four sources: mastery experiences, vicarious experiences, social persuasion, and physiological responses.

In this study we incorporate variables self-efficacy from Bandura's self-efficacy theory to help us understand the predict factors of behavioral intention for the school children to use mobile phones to promote their health outcome., then the study finally will examines the link between behavioral intention BI and actual using of technology.

Mastery experiences are is the first and most important sources of efficacy in Bandura theory. Which explained the source of information comes from the of past performance result interpretation.

Vicarious experiences: which is the second source of self-efficacy, this information comes and gained by observing others performance skills. The other meaning of this topic it's comes by observing the successes and failures of others.

Verbal Persuasion: This source of efficacy may be can effect in short time so it can't to be stay long time, so the Bandura stated in this point "the potency of the persuasion depends on the credibility, trustworthiness, and expertise of the persuader" [33].

Emotional and physiological states: this is States also from the sources of efficacy information. Anxiety, stress and other powerful emotional arousal can effectively change individuals' beliefs about their capabilities.

Bandura's four principles of learning process provide a suitable framework for this current study. The study implemented health education by mobile app that were attended by schoolchildren with chronic disease. They could learn from.

Thus, this framework may be utilized to (a) conceptualize the relationship among the individual's self-efficacy, consequences (response), and behavior, and (b) understand how an individual's self-efficacy may be influenced. Thus, acknowledging that there are several sources of influence on self-efficacy provided evidence regarding the complexity of the variables that can affect one's behavior and the need to control for potentially confounding variables within this study.

4.2   Technology Acceptance Model (TAM)

On 1989 Dr. Fred Davis has applied the Technology Acceptance Model
(TAM) which adapted from Theory of Reasoned Action (TRA). The TRA originated ten years prior to the TAM and concentrated primarily on the perspective of
the computer consumer postulating that one's values and behaviors influence the
probability that they will either adopt or oppose technology [41]. Theory of
Reasoned action (TRA) hypothesis, born in the field of social psychology, aims to
forecast the actions of a person by means of their behavioral expectations, interpreted as the statistical likelihood of an individual committing a given action, rather
than their attitudes, which reflect "the general feeling of favorability or
unfavourability of a person" [31].

The TAM advanced to the TRA, which was more comprehensive, and was
originally designed to measure the computer-related experience of the new user in
the workplace. These models are focused on psychological science, focusing on
beliefs, behaviors and expectations, and their relationship to adoptive behaviour.
Likewise, while the TAM is asking acceptance, it also contemplates denial [41].

TAM was initially developed with a focus on device architecture uses and
refuses to take into consideration other popular aspects on social media [42].
Initially, TAM assumed that information systems were used in organizational settings to improve the employee efficiency. TAM omitted the possibility that individual users may use the information system outside of the corporate environments,
and this usage could even include an "entertainment" aspect for these uses [43].

The TAM measured two psychological constructs: perceived usefulness
(PU) and perceived ease of use (PEU). Davis [44] defines perceived usefulness
(PU) as, "the degree to which a person believes that using a particular system would
enhance his or her job performance", Also he defines, perceived ease of use
(PEU) is understood in this model as "the degree to which a person believes that
using a particular system would be free from effort."

The key factor deciding the use and eventual implementation of technologies
was PU, or the degree to which a modern method claimed its work efficiency would
increase [45, 46]. A secondary determinant was the expected degree of mental and/
or physical commitment in ease of use (PEU) imposed upon one's job by the
proposed technique. The TAM posits a greater ease of use that specifically affects
expected usefulness. A total of 15 accepted Likert scaled questions have been
commonly updated and added to technology-specific acceptance research [47].

4.3   Research Model Development

In this study we will incorporate variables self-efficacy from Bandura's self-efficacy
theory and Technology Acceptance Model (TAM) to help understand the predict
factors of behavioral intention for the schoolchildren to use mobile phones to

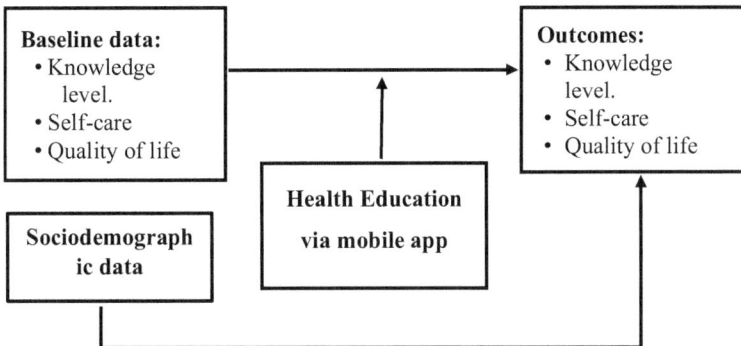

Fig. 1 Conceptual framework of the study

promote their health outcome; knowledge, self-care, and the quality of life level among schoolchildren. There are two types of variables in this study, dependent and independent variables. The dependent variables are patient knowledge, self-care and quality of life. The independent variables are demographic factors (age, gender and ethnic). The impact of mobile apps health education on the dependent variables is shown in Fig. 1.

5 Discussion and Conclusion

Smartphone applications are receiving rapid acceptance and wide distribution with the release of radically increasing number of smartphones and respective platforms [48]. This event makes mobile technology more attractive for new and further penetration into e-Health and Medical Informatics [49].

mHealth apps can help their users track themselves and inspire them to change their lifestyle in the short and long term. In fact, mHealth apps have the ability to solve adherence problems by communicating with the patient at a high level and by executing the action. Behavioral improvement treatments implemented by health apps reduce the need for face-to-face experiences and thereby improve cost efficiency via universal and lasting connectivity. An even higher return on investment from workplace health promotion services can also be expected if such technologies are successfully enforced [50].

The limit progress obtained by conventional health educational methods, along with insufficient clinic-time committed to implementing such initiatives, indicates the need for interesting ways to provide teenage health care. As indicated by teens and healthcare professionals, the use of mobile devices is one potential alternative that could satisfy the required elements of patient-centered treatment mentioned above. Mobile technology is on the rise, particularly among teenagers [8].

5.1 Practical Implications

Mobile applications for health education are valuable assets for patients and care-givers like each other, providing patients with immediate communication and those responsible for meeting their needs. There were 325,000 mobile health apps licensed to smartphone owners worldwide for early 2018, with a further 200 mobile new apps released daily [51].

The Web resources and mobile health (mHealth) software (apps) are regularly used by people to access information about their health. Past research showed that less than half of the mHealth applications available contained evidence-based guidelines, and none provided accurate knowledge and self-management supportive resources [52]. In comparison, the majority of health education-related applications currently on the market rely on patient self-monitoring and self-assessment, which could be less successful for caregivers [53].

5.2 Future Research

The World Health Organization (WHO) emphasized many challenges face the mobile health area such as the dominant form of mobile health today is isolated, small-scale pilot projects that determine the specific issues of information accessing and sharing [10].

Unluckily, there aren't a lot of information has been published regarding the integration of smartphone technology into patient care. Most of healthcare appli-cations consumers focus on how to monitoring their own condition rather than to improving knowledge or how to deal effectively with disease [54].

In fact, there are differences in mHealth applications that address teenage demographics or are specifically targeted to other user groups' tastes. Moreover, research evaluating the acceptability, efficacy and impact of mHealth among teenagers as a model of intervention for diseases are scarce [8].

References

1. Eysenbach, G.: What is e-health? J. Med. Internet Res. **3**(2), e20 (2001)
2. Free, C., et al.: The effectiveness of mobile-health technologies to improve health care service delivery processes: a systematic review and meta-analysis. PLoS Med. **10**(1), e1001363 (2013)
3. Huckvale, K., Car, M., Morrison, C., Car, J.: Apps for asthma self-management: a systematic assessment of content and tools. BMC Med. **10**(1), 144 (2012)
4. Ozdalga, E., Ozdalga, A., Ahuja, N.: The smartphone in medicine: a review of current and potential use among physicians and students (in Eng). J. Med. Internet Res. **14**(5), e128 (2012). https://doi.org/10.2196/jmir.1994

5. Ramsey, R.R., Plevinsky, J.M., Kollin, S.R., Gibler, R.C., Guilbert, T.W., Hommel, K.A.: Systematic review of digital interventions for pediatric asthma management. J. Allergy Clin. Immunol. Pract. **8**(4), 1284–1293 (2020). https://doi.org/10.1016/j.jaip.2019.12.013

6. Fedele, D.A., et al.: Applying interactive mobile health to asthma care in teens (AIM2ACT): development and design of a randomized controlled trial. Contemp. Clin. Trials **64**, 230–237 (2018). https://doi.org/10.1016/j.cct.2017.09.007

7. Ramsey, R.R., et al.: A systematic evaluation of asthma management apps examining behavior change techniques. J. Allergy Clin. Immunol. Pract. **7**(8), 2583–2591 (2019). https://doi.org/10.1016/j.jaip.2019.03.041

8. Schneider, T., Baum, L., Amy, A., Marisa, C.: I have most of my asthma under control and I know how my asthma acts: users' perceptions of asthma self-management mobile app tailored for adolescents. Health Inform. J. **26**(1), 342–353 (2020). https://doi.org/10.1177/1460458218824734

9. Iyengar, K., Upadhyaya, G.K., Vaishya, R., Jain, V.: COVID-19 and applications of smartphone technology in the current pandemic (in Eng). Diabetes Metab. Syndr. **14**(5), 733–737 (2020). https://doi.org/10.1016/j.dsx.2020.05.033

10. Iwaya, L.H., et al.: Mobile health in emerging countries: a survey of research initiatives in Brazil. Int. J. Med. Inform. **82**(5), 283–298 (2013). https://doi.org/10.1016/j.ijmedinf.2013.01.003

11. Choi, J.Y., Cho Chung, H.I.: Effect of an individualised education programme on asthma control, inhaler use skill, asthma knowledge and health-related quality of life among poorly compliant Korean adult patients with asthma (in Eng). J. Clin. Nurs. **20**(1–2), 119–126 (2011). https://doi.org/10.1111/j.1365-2702.2010.03420.x

12. Arpaci, I., Al-Emran, M., Al-Sharafi, M.A.: The impact of knowledge management practices on the acceptance of Massive Open Online Courses (MOOCs) by engineering students: a cross-cultural comparison. Telemat. Inform., 101468 (2020). https://doi.org/10.1016/j.tele.2020.101468

13. Rhee, H., Belyea, M.J., Sterling, M., Bocko, M.F.: Evaluating the validity of an automated device for asthma monitoring for adolescents: correlational design. J. Med. Internet Res. **17**(10), e234 (2015). https://doi.org/10.2196/jmir.4975

14. Carpenter, D.M., Geryk, L.L., Sage, A., Arrindell, C., Sleath, B.L.: Exploring the theoretical pathways through which asthma app features can promote adolescent self-management (in Eng). Transl. Behav. Med. **6**(4), 509–518 (2016). https://doi.org/10.1007/s13142-016-0402-z

15. Blaiss, M.S.: Asthma mobile applications: are they ready for prime time? Ann. Allergy Asthma Immunol. **120**(4), 347–348 (2018). https://doi.org/10.1016/j.anai.2018.02.002

16. Klasnja, P., Pratt, W.: Healthcare in the pocket: mapping the space of mobile-phone health interventions. J. Biomed. Inform. **45**(1), 184–198 (2012). http://dx.doi.org/10.1016/j.jbi.2011.08.017

17. Arnold, N., Paulus, T.: Using a social networking site for experiential learning: appropriating, lurking, modeling and community building. Internet High. Educ. **13**(4), 188–196 (2010). http://dx.doi.org/10.1016/j.iheduc.2010.04.002

18. Arpaci, I., Karataş, K., Baloğlu, M.: The development and initial tests for the psychometric properties of the COVID-19 Phobia Scale (C19P-S). Pers. Individ. Differ. **164**, 110108 (2020). https://doi.org/10.1016/j.paid.2020.110108

19. Dyer, O.: Covid-19: pandemic is having "severe" impact on non-communicable disease care, WHO survey finds. Ed: British Medical Journal Publishing Group (2020)

20. Pei, Y., et al.: COVID-19: children comparison with adults based on the latest data. Available at SSRN 3550063 (2020)

21. Oreskovic, N.M., Kinane, T.B., Aryee, E., Kuhlthau, K.A., Perrin, J.M.: The unexpected risks of COVID-19 on asthma control in children. J. Allergy Clin. Immunol. Pract. (2020). https://doi.org/10.1016/j.jaip.2020.05.027

22. Abrams, E.M., W't Jong, G., Yang, C.L.: RE: comment: asthma and COVID-19 (2020)

23. Kim, S.-H., et al.: Perceptions of severe asthma and asthma-COPD overlap syndrome among specialists: a questionnaire survey. Allergy Asthma Immunol. Res. **10**(3), 225–235. Available [online] http://synapse.koreamed.org/DOIx.php?id=10.4168%2Faair.2018.10.3.225 (2018)
24. Johnston, S.L.: Asthma and COVID-19: is asthma a risk factor for severe outcomes?. Allergy (2020)
25. mobiThinking: Global mobile statistics 2014 part A: mobile subscribers; handset market share; mobile operators. Available [online] http://mobiforge.com/research-analysis/global-mobile-statistics-2014-part-a-mobile-subscribers-handset-market-share-mobile-operators#subscribers (2014)
26. Varshney, U.: Mobile health: four emerging themes of research. Deci. Support Syst. **66**, 20–35 (2014). http://dx.doi.org/10.1016/j.dss.2014.06.001
27. Mickel, C.F., Shanovich, K.K., Evans, M.D., Jackson, D.J.: Evaluation of a school-based asthma education protocol: iggy and the inhalers. J. Sch. Nurs. (Sage Publications Inc.) **33**(3), 189–197 (2017). https://doi.org/10.1177/1059840516659912
28. Viner, R., et al.: School closure and management practices during coronavirus outbreaks including COVID-19: a rapid narrative systematic review. Available at SSRN 3556648 (2020)
29. Arpaci, I., et al.: Analysis of twitter data using evolutionary clustering during the COVID-19 pandemic. Comput. Mater. Contin. **65**(1), 193–204. Available [online] http://www.techscience.com/cmc/v65n1/39561 (2020)
30. Licari, A., et al.: Allergy and asthma in children and adolescents during the COVID outbreak: what we know and how we could prevent allergy and asthma flares. Allergy **n/a**(n/a) (2020). https://doi.org/10.1111/all.14369
31. Sánchez-Prieto, J.C., Olmos-Migueláñez, S., García-Peñalvo, F.J.: Informal tools in formal contexts: development of a model to assess the acceptance of mobile technologies among teachers. Comput. Hum. Behav. **55**, 519–528 (2016). https://doi.org/10.1016/j.chb.2015.07.002
32. Marcano Belisario, J.S., Huckvale, K., Greenfield, G., Car, J., Gunn, L.H.: Smartphone and tablet self management apps for asthma (in Eng). Cochrane Database Syst. Rev. **11**, Cd010013 (2013). https://doi.org/10.1002/14651858.cd010013.pub2
33. Bandura, A.: Self-efficacy: toward a unifying theory of behavioral change. Psychol. Rev. **84**(2), 191–215 (1977)
34. Bandura, A.: Social cognitive theory: an agentic perspective. Asian J. Soc. Psychol. **2**(1), 21–41 (1999)
35. Bandura, A.: Social cognitive theory: an agentic perspective. Annu. Rev. Psychol. **52**(1), 1–26 (2001)
36. Nevid, J.S., Rathus, S.A.: Psychology and the Challenges of Life. Wiley, New Jersey (2009)
37. Locke, E.A., Latham, G.P.: Building a practically useful theory of goal setting and task motivation: a 35-year odyssey. Am. Psychol. **57**(9), 705 (2002)
38. Bandura, A.: The evolution of social cognitive theory. Great Minds in Management, pp. 9–35 (2005)
39. Bandura, A.: Social Functions of Thought and Action: A Social Cognitive Theory. Prentice Hall Inc, New Jersey (1986)
40. Okhakhu, E., Emeka, P., Okhakhu-Okpodi, J.: Employee performance and self efficacy theory as tools for safe healthcare delivery in Nigerian hospitals: a literature based exploration. Insights to a Changing World Journal, Article vol. 2015, no. 2, p. 48. Available [online] http://ezproxy.um.edu.my:2048/login?url=http://search.ebscohost.com/login.aspx?direct=true&db=a9h&AN=109041280&site=eds-live (2015)
41. Couch, H.C.: Providers' acceptance of smartphone applications as a supportive strategy for adolescent asthma. PhD thesis, p. 1. Available [online] http://search.ebscohost.com/login.aspx?direct=true&db=ccm&AN=124649828&site=ehost-live (2017)
42. Al-Sharafi, M.A., Mufadhal, M.E., Arshah, R.A., Sahabudin, N.A.: Acceptance of online social networks as technology-based education tools among higher institution students: structural equation modeling approach. Sci. Iran. **26**(Special Issue on: Socio-Cognitive Engineering), 136–144 (2019)

43. Briz-Ponce, L., García-Peñalvo, F.J.: An empirical assessment of a technology acceptance model for apps in medical education. J. Med. Syst. **39**(11), 176 (2015)
44. Davis, F.D.: Perceived usefulness, perceived ease of use, and user acceptance of information technology. MIS Quart., 319–340 (1989)
45. Al-Emran, M., Al-Maroof, R., Al-Sharafi, M.A., Arpaci, I.: What impacts learning with wearables? An integrated theoretical model. Inter. Learn. Environ., 1–21 (2020)
46. Al-Sharafi, M.A., Arshah, R.A., Herzallah, F.A., Alajmi, Q.: The effect of perceived ease of use and usefulness on customers intention to use online banking services: the mediating role of perceived trust. Int. J. Innov. Comput. **7**(1) (2017)
47. Saadé, R., Bahli, B.: The impact of cognitive absorption on perceived usefulness and perceived ease of use in on-line learning: an extension of the technology acceptance model. Inf. Manag. **42**(2), 317–327 (2005)
48. Al-Qaysi, N., Mohamad-Nordin, N., Al-Emran, M., Al-Sharafi, M.A.: Understanding the differences in students' attitudes towards social media use: a case study from Oman. In: 2019 IEEE Student Conference on Research and Development (SCOReD), IEEE, pp. 176–179 (2019)
49. Paschou, M., Sakkopoulos, E., Tsakalidis, A.: easyHealthApps: e-Health apps dynamic generation for smartphones & tablets. J. Med. Syst. **37**(3), 1–12 (2013). https://doi.org/10.1007/s10916-013-9951-6
50. Melzner, J., Heinze, J., Fritsch, T.: Mobile health applications in workplace health promotion: an integrated conceptual adoption framework. Proc. Technol. **16**, 1374–1382 (2014). https://doi.org/10.1016/j.protcy.2014.10.155
51. Kagen, S., Garland, A.: Asthma and allergy mobile apps in 2018. **19**, ed (2019)
52. Farooqui, N., Phillips, G., Barrett, C., Stukus, D.: Acceptability of an interactive asthma management mobile health application for children and adolescents. Ann. Allergy Asthma Immunol. **114**(6), 527–529 (2015). https://doi.org/10.1016/j.anai.2015.03.006
53. Real, F.J., et al.: Dose matters: a smartphone application to improve asthma control among patients at an urban pediatric primary care clinic. Games Health J. **8**(5), 357–365 (2019). https://doi.org/10.1089/g4h.2019.0011
54. Haze, K.A., Lynaugh, J.: Building patient relationships: a smartphone application supporting communication between teenagers with asthma and the RN care coordinator (in Eng). Comput. Inform. Nurs. **31**(6), 266–271, quiz 272-3 (2013). https://doi.org/10.1097/nxn.0b013e318295e5ba

Predicting the Acceptance of Mobile Learning Applications During COVID-19 Using Machine Learning Prediction Algorithms

Mohammed Amin Almaiah, Omar Almomani, Ahmad Al-Khasawneh, and Ahmad Althunibat

Abstract The global spread of COVID-19 has motivated many universities to adopt online distance learning systems. Mobile learning applications could play a crucial role during this pandemic. Mobile learning applications are increasing popularity among learners due to their benefits and effectiveness. However, the acceptance of mobile learning system among university students is limited. Therefore, this study seeks to understand the main factors influencing the acceptance of mobile learning applications by proposing a hybrid model by combining the TAM with new constructs of TUT model. Machine learning algorithms were employed to analyze the hypothesized relationships among the constructs in the proposed model. The research findings found that RandomForest and IBK algorithms are the best two algorithms in predicting the main determinants of mobile learning acceptance as comparison with other machine learning algorithms with an accuracy of 81.3%. The results of machine learning predictive algorithms showed that constructs of perceived enjoyment, perceived ease of use, perceived usefulness, effectiveness, efficiency, behavioural intention to use and utilization could predict

M. A. Almaiah (✉)
Department of Computer Networks and Communications, College of Computer Sciences and Information Technology, King Faisal University, Al-Ahsa 31982, Hofuf, Saudi Arabia
e-mail: malmaiah@kfu.edu.sa

O. Almomani
Computer Network and Information Systems Department, The World Islamic Sciences and Education University, Amman, Jordan

A. Al-Khasawneh
CIS Department, President, Irbid National University, Irbid, Jordan

A. Al-Khasawneh
Professor, Hashemite University, Irbid, Jordan

A. Al-Khasawneh
CIS Department, Hashemite University, Zarqa, Jordan

A. Althunibat
Department of Software Engineering, Al-Zaytoonah University of Jordan, Amman 11947, Jordan

the acceptance of mobile learning within accuracy rate of 87%. The results of this paper will offer valuable directions for mobile learning designers and developers to better promote mobile learning application utilization in universities.

Keywords Mobile learning apps · Mobile learning acceptance · COVID-19 · Machine learning algorithms

1 Introduction

During the COVID-19 pandemic, several universities across the world have started to resume their lectures and exams through online learning tools such as e-learning and mobile learning applications [1–3]. In fact, mobile learning has been increasingly regarded as a promising tool to improve students' learning motivation. It provides learning environment in which students acquire information and knowledge from mobile devices [4–7]. Mobile learning not only offer students online learning space, but also enable them to quick access to learning activities and materials anytime, anywhere and anyhow and create pioneering opportunities for innovative learning [8, 9]. This kind of technology enables students to reach to the knowledge not only from teachers during the classroom, but also through their mobile device, this can develop their learning capability, and thereby achieve meaningful learning [10–13]. Hence, mobile learning has attracted many researchers' attention and have been introduced into many fields [14].

The use of mobile learning systems will help in reducing the transmission of COVID-19 between students with ensuring continuation the learning process during this lockdown. Mobile learning applications could play a crucial role during this pandemic because it has several features such as portability, where mobile learning applications can be taken in different locations at home, office and others by using mobile devices [15–17] instant connectivity, where mobile learning applications can be used to access a variety of information and learning activities anytime and any-where with instant connectivity facility between students and instructors [18, 19], context sensitivity, where mobile devices can be used to find and gather real or simulated data [2], interactivity and mobility [20].

Several studies have further indicated that mobile learning technology will improve students' learning effectiveness and performance [21–24]. Different from other educational technologies, mobile learning need to be designed carefully, so it is important to identify students' requirements and perceptions before implementing [25–27]; otherwise, they will fail. Accordingly, integrating students' perceptions into mobile learning development has become a crucial issue, for such integration can provide universities, designers and developers proper guidance to ensure the successful usage and acceptance of mobile learning in future [28–30]. Researchers have thus incorporated students' requirements and priorities into the development of mobile learning and examined their influences on acceptance of mobile learning.

However, the question as to whether the embedding of students' requirements' and perceptions will influence students' acceptance of mobile learning at different stages of mobile learning usage that has not been specifically addressed. Several studies have clearly indicated that a successful mobile learning technology should be accepted by students wholeheartedly, otherwise it will fail [31–33]. Accordingly, investigation into students' acceptance of mobile learning technology has been considered as a critical step for ensuring the success of mobile learning technology in educational environment [34, 35]. More interestingly, this kind of investigation will help designers and developers to optimise the mobile learning system in a more effective manner as well as enables students to the full potential of the mobile learning technology.

Although prior studies have highlighted the importance of mobile learning applications in university settings [10–12], there is a limited of knowledge regarding the understanding of the factors affecting the use of mobile learning applications in educational and learning activities among university students. To verify the actual usage of mobile learning applications by university students, this research extends the technology acceptance model (TAM) to examine the students' acceptance of mobile learning applications. This study selected the TAM model for several reasons are: (1) prior studies have confirmed that TAM has been successfully adopted to study the usage and acceptance of many types of educational technologies, (2) the integration of TAM model with TUT model have not been used to study the actual use of mobile learning applications among university students. In accordance with these reasons, this study aims to fill this research gap by integrating the TAM with TUT constructs to determine the main factors affecting the students' acceptance of mobile learning applications at the university level.

2 Theoretical Framework and Background

2.1 Technology Acceptance Model (TAM)

TAM model developed by Davis [36], which it has been widely used to study students' acceptance of educational technologies [37–40]. In fact, investigating the critical factors behind users' choices of mobile educational technologies has been proven helpful in providing users with more acceptable mobile learning applications, and therefore has been widely regarded as a vital issue [10–12]. Davis [36] developed the TAM based on five main constructs namely perceived usefulness (PU), perceived ease of use (PEU), attitude toward to use (ATU), behavioural intention (BI) and actual use (AU). PU means that "the degree to which a person believes that using a particular system would enhance his or her job performance", and PEU signifies "the degree to which a person believes that using a particular system would be free from effort," [36]. ATU defined as the degree to which a user holds positive or negative feelings about using a particular technology, and BI defined as the degree to which a user is willing to use a particular technology [36].

Due to the success TAM model in exploring the user acceptance of technology, many researchers have used the TAM model extensively to clarify the factors that affect students' acceptance of mobile learning. For example, Almaiah et al. [11] used the TAM to identify the factors that influence intention to use mobile learning system among students. Their results showed that the users' acceptance of mobile learning was positively influenced by perceived ease of use, perceived usefulness and seven quality factors. They also revealed that TAM model is a useful instrument for exploring the attitude of learners to accepting mobile learning system. Al-Emran et al. [8] also employed the TAM, aiming to examine students' acceptance of mobile learning. Their results suggested that the students' preference for using mobile learning was most significantly affected by their perception of usefulness and ease of use. Overall, these studies not only signified the extensive use of TAM in examining students' acceptance of mobile learning, but also indicated the lack of scholarly attention to the critical role of important other factors such as perceived enjoyment, predictive effectiveness and predictive efficiency behind students' acceptance of mobile learning applications.

2.2 Technology Utilization Theory (TUT)

TUT is a new model developed by Ghapanchi and Talaei-Khoei [41] to study the technology acceptance among users by using three main constructs (effectiveness, efficiency and utilization). In this model, utilization of new technology is measured by two main constructs are predictive effectiveness and predictive efficiency. Effectiveness defined as getting the right things done [41]. Predictive effectiveness means the expected effect or impact of the specific technology. On the other hand, efficiency means doing things in the most economical way. Predictive efficiency means the expected output created out of particular amount of input (e.g. cost, time) for the specific technology.

3 Research Model

To understand the acceptance of mobile learning applications, it is important to determine whether students are willing to adopt mobile learning apps and which factors influence their decision to use the mobile learning apps. This research proposes a comprehensive model, which incorporates TAM model with adding new constructs related to enjoyment and construct related to TUT model, which have not been examined previously. According to Fig. 1, the research model proposes the following hypotheses:

H1: Perceived enjoyment is positively influenced by perceived ease of use.
H2: Perceived usefulness is positively influenced by perceived ease of use.

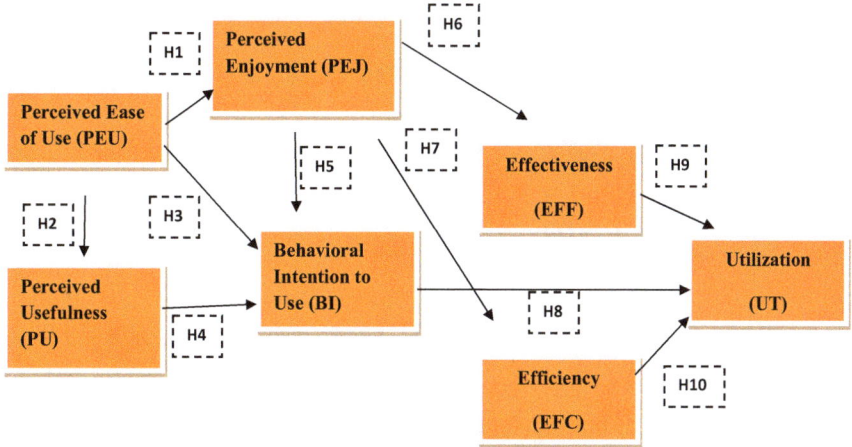

Fig. 1 The proposed research model

H3: Behavioral intention to use is positively influenced by perceived ease of use.
H4: Behavioral intention to use is positively influenced by perceived usefulness.
H5: Behavioral intention to use is positively influenced by perceived enjoyment.
H6: Effectiveness is positively influenced by perceived enjoyment.
H7: Efficiency is positively influenced by perceived enjoyment.
H8: Utilization is positively influenced by perceived behavioral intention to use.
H9: Utilization is positively influenced by effectiveness.
H10: Utilization is positively influenced by efficiency.

4 Research Methodology

4.1 Data Collection

To clarify the main determinants of mobile learning acceptance among students, online questionnaires were distributed for both undergraduate and postgraduate students, who they play a key role in actual use of mobile learning systems at five universities in Jordan. The employ of online questionnaire in this study, specifically in Corona virus time is considered the best method to collect the data. In addition, previous studies pointed out that it is an effective method to measure the hypotheses in the proposed model [4]. These universities have already developed mobile learning systems in their settings. Using online survey questionnaire, students were invited to participate in this study through online classes, during the second semester 2020. In total, 487 online questionnaires were distributed, with 397 questionnaires being returned, indicating an 81.52% response rate. Most of responses had incomplete or invalid answers and therefore were excluded. Hence,

397 responses were considered valid for further analysis. Among 397 valid responses, 60.7% of respondents were female, while 39.3% were male. Moreover, 52.6% of respondents who responded were undergraduate; 47.4% were postgraduate students.

4.2   Research Measurements

The items and scales for testing the constructs in the developed model were adopted from current research in the literature. A 5 point scale similar to Likert model was utilized for testing every item, ranging from "strongly disagree = 1" to "strongly agree = 5". We invited six university lectures, each of them holding significant expertise in the mobile learning felid, to examine the appropriateness and clarify of the questionnaire. After that, pre-tested was carried out with 25 post-graduate students from University of Jordan, with the results indicating that the instructions and questions were completely understood. The survey questionnaire consists of seven constructs (perceived ease of use, perceived usefulness, perceived enjoyment, behavioural intention to use, effectiveness, efficiency and utilization) and includes demographic information (e.g., gender and age). The items for measuring perceived ease of use, perceived usefulness, behavioural intention to use, were developed from the measurements used by Almaiah and Al-Khasawneh [35]. The measurement items for perceived enjoyment was drawn from Al-Shihi et al. [42]. Effectiveness, efficiency and utilization were adapted from Ghapanchi and Talaei-Khoei [41].

4.3   Machine Learning Prediction Algorithms

Machine learning prediction is an analysis technique that is used to predict future events based on current and historical data. For the context of this study, machine learning prediction algorithms could predict the main determinants of mobile learning applications acceptance. Machine learning techniques have become popular among researchers and analysts as it helps them to understand more about any system and it is easy to develop the prediction model [43]. Among the most popular machine learning techniques used in prediction are rule-learner (PART), meta-classifier (Bagging), decision-tree (RandomForest), lazy-classifier (IBk), logistic regression classifier (SMO) and bayesian classifier (NaiveBayes) [44, 45].

In this research, machine learning classification approaches were used to analyze the proposed model based on the distribution of the class scales with regard to predictor features. Thus, five machine-learning classifiers were employed to predict the acceptance of mobile learning applications based on seven constructs of perceived ease of use, perceived usefulness, perceived enjoyment, effectiveness, efficiency, behavioural intention to use and utilization. The methodology used in this research was adopted from a study conducted by Al-Maroof et al. [44] for studying the acceptance of WhatsApp stickers among students.

5 Data Analysis and Findings

In this paper, we have used two primary methods to analyze the data and evaluate the developed model of this research. The first method is the confirmatory factor analysis (CFA) in order to evaluate the measurement model in terms of reliability, convergent validity, and discriminant validity. In the second method, machine learning classification techniques were applied to test the hypotheses in the proposed model. In this research, the methodology used to analyze the data was adopted from a study conducted by Al-Maroof et al. [44] for analysing the acceptance of WhatsApp stickers among students.

5.1 Results of Confirmatory Factor Analysis

(1) **Reliability Analysis**

The Cronbach's alpha coefficient was applied to determine the reliability of measures for each construct in the proposed research model. As presented in Table 1, the value of this coefficient ranged between 0.795 and 0.934, exceeding the critical value of 0.7 as suggested by Christmann and Van Aelst [46], and indicating satisfactory reliability for all constructs in the proposed research model.

(2) **Validity Analysis**

For the current study, each construct was assessed in terms of its convergent and discriminant validity. For convergent validity analysis, Table 1 shows that the average variance extracted (AVE) was above (0.5). According to Hair et al. [47], specify that a variance greater than 0.5 is acceptable. Therefore, the convergent

Table 1 Reliability and convergent validity analysis

Constructs	Cronbach's alpha	Average variance extracted (AVE > 0.5)
PEU	0.894	0.773
PU	0.795	0.731
PEJ	0.887	0.796
BI	0.865	0.801
EFF	0.934	0.704
EFC	0.897	0.889
UT	0.832	0.841

Table 2 Discriminant validity analysis

	PEU	PU	PEJ	BI	EFF	EFC	UT
PEU	**0.936**						
PU	0.797	**0.958**					
PEJ	0.630	0.758	**0.964**				
BI	0.646	0.684	0.545	**0.978**			
EFF	0.759	0.769	0.563	0.689	**0.963**		
EFC	0.769	0.792	0.643	0.707	0.790	**0.943**	
UT	0.530	0.623	0.506	0.643	0.527	0.614	**0.988**

validity values for the research constructs are acceptable. Concerning the discriminant validity analysis, the square root of AVE was obtained to correlate the latent constructs. Table 2 highlights that the square root of the AVE for each construct is greater than the pairwise correlations. This result means that the psychometric characteristics of the instrument are also deemed acceptable in terms of their discriminant validity Hair et al. [47].

5.2 Results of Research Model Analysis Using Machine Learning Classifiers

This research employed machine learning classification techniques to test the relationships among the constructs in the proposed research model. This study used Weka (version 3.8.3) to analyze the collected data by applying the percentage split (66%) test mode based on using five machine learning techniques including, a rule-learner (PART), a meta-classifier (Bagging), a decision-tree (RandomForest), a lazy-classifier (IBk), a logistic regression classifier (SMO) and a bayesian classifier (NaiveBayes) [44, 45].

Based on Table 3, the findings showed that IBK algorithm have a highest score in predicting the perceived enjoyment (PEJ) by the construct of perceived ease of use (PEU) in accuracy rate of 83.75% as comparison with other machine learning algorithms. The IBK algorithm also have a high performance score in terms of True Positive (TP = 0.836), precision (0.825) and ROC area (0.942). Thus, this result support H1.

According to the results in Table 4, which indicated that both IBK and RamdomForest algorithms had the best results in predicting behavioral intention to use (BI) by the constructs of perceived ease of use (PEU), perceived usefulness (PU) and perceived enjoyment (PEJ). This means that both classifiers predict the perceived enjoyment at the highest accuracy scores as comparison with other classifiers (IBK = 86.72% and RamdomForest = 85.17%). In addition, Further, both algorithms have a better performance in precision (IBK = 0.827 and RamdomForest = 0.809) and TP rate (IBK = 0.867 and RamdomForest = 0.851). These results imply that hypotheses H3, H4 and H5 were supported.

Table 3 Predicting the perceived enjoyment by perceived ease of use

Algorithms	CCI[a] (%)	TP[b] rate	FP[c] rate	Precision	Recall	F-measure	ROC area
Bagging	75.40	0.754	0.137	0.789	0.754	0.757	0.889
PART	56.35	0.563	0.289	0.634	0.563	0.559	0.648
IBk	**83.75**	**0.838**	**0.287**	**0.825**	**0.833**	**0.829**	**0.946**
RandomForest	72.22	0.722	0.156	0.708	0.722	0.710	0.885
SMO	74.22	0.742	0.164	0.794	0.722	0.718	0.873
NaiveBayes	84.92	0.849	0.082	0.846	0.849	0.844	0.949

[a]*CCI* correctly classified instances, [b]*TP* true positive, [c]*FP* false positive, [d]*ROC* receiver operating characteristic

Table 4 Predicting the behavioral intention to use by perceived ease of use, perceived usefulness and perceived enjoyment

Algorithms	CCI (%)	TP rate	FP rate	Precision	Recall	F-measure	ROC area
Bagging	78.57	0.786	0.113	0.829	0.786	0.788	0.923
PART	76.98	0.770	0.136	0.784	0.770	0.772	0.918
IBk	**86.72**	**0.867**	**0.487**	**0.827**	**0.848**	**0.861**	**0.968**
RandomForest	**85.17**	**0.851**	**0.393**	**0.851**	**0.784**	**0.801**	**0.938**
SMO	72.22	0.722	0.164	0.794	0.722	0.718	0.873
NaiveBayes	80.95	0.810	0.124	0.814	0.810	0.810	0.909

Table 5 Predicting the perceived usefulness by perceived ease of use

Algorithms	CCI (%)	TP rate	FP rate	Precision	Recall	F-measure	ROC area
Bagging	75.40	0.754	0.137	0.789	0.754	0.757	0.889
PART	56.35	0.563	0.289	0.634	0.563	0.559	0.648
IBk	80.75	0.807	0.287	0.825	0.833	0.829	0.946
RandomForest	**84.51**	**0.845**	**0.295**	**0.831**	**0.821**	**0.833**	**0.954**
SMO	74.22	0.742	0.164	0.794	0.722	0.718	0.873
NaiveBayes	79.92	0.780	0.182	0.792	0.781	0.805	0.901

The results in Table 5 revealed that RandomForest algorithm had the best performance in predicting the perceived usefulness by the construct of perceived ease of use than other algorithms at 84.51% of accuracy. In addition, the results showed that RandomForest algorithm had a better performance in terms of precision (0.831) and TP rate (0.845). Thus, hypothesis H2 was supported.

The findings in Table 6 presented that NaiveBayes algorithm had the highest score of accuracy (85.72%) in predicting the effectiveness by the attribute of perceived enjoyment as comparison with other algorithms. In addition, it had a better

Table 6 Predicting the effectiveness by perceived enjoyment

Algorithms	CCI (%)	TP rate	FP rate	Precision	Recall	F-measure	ROC area
Bagging	56.35	0.563	0.289	0.634	0.563	0.559	0.648
PART	72.22	0.722	0.164	0.794	0.722	0.718	0.873
IBk	80.75	0.807	0.287	0.825	0.833	0.829	0.946
RandomForest	80.95	0.810	0.124	0.814	0.810	0.810	0.909
SMO	74.22	0.742	0.164	0.794	0.722	0.718	0.873
NaiveBayes	**85.72**	**0.857**	**0.321**	**0.874**	**0.851**	**0.864**	**0.984**

Table 7 Predicting the efficiency by perceived enjoyment

Algorithms	CCI (%)	TP rate	FP rate	Precision	Recall	F-measure	ROC area
Bagging	74.22	0.742	0.164	0.794	0.722	0.718	0.873
PART	72.22	0.722	0.164	0.794	0.722	0.718	0.873
IBk	76.98	0.770	0.136	0.784	0.770	0.772	0.918
RandomForest	80.95	0.810	0.124	0.814	0.810	0.810	0.909
SMO	56.35	0.563	0.289	0.634	0.563	0.559	0.648
NaiveBayes	**85.66**	**0.856**	**0.315**	**0.864**	**0.844**	**0.859**	**0.976**

performance in precision (0.874) and TP rate (0.875). Thus, hypothesis H6 was supported. Furthermore, the results indicated that NaiveBayes algorithm also had a better performance in predicting the efficiency by the attribute of perceived enjoyment with an accuracy rate (85.66%) as shown in Table 7.

Finally, the findings in Table 8 showed that both algorithms of IBK and RamdomForest had the best performance in predicting the utilization by the constructs of behavioural intention to use, effectiveness and efficiency with an accuracy rate of 87.06% as comparison with other machine learning algorithms. The IBK and RamdomForest algorithms also had a high performance score in terms of True Positive (TP = 0.870), precision (0.863) and ROC area (0.964). Thus, these results support H8, H9 and H10.

Table 8 Predicting the utilization by behavioural intention to use, effectiveness and efficiency

Algorithms	CCI (%)	TP rate	FP rate	Precision	Recall	F-measure	ROC area
Bagging	64.22	0.642	0.164	0.794	0.722	0.718	0.795
PART	72.22	0.722	0.164	0.794	0.722	0.718	0.873
IBk	**87.06**	**0.870**	**0.295**	**0.835**	**0.815**	**0.827**	**0.985**
RandomForest	**87.06**	**0.870**	**0.295**	**0.835**	**0.815**	**0.827**	**0.985**
SMO	84.92	0.849	0.082	0.846	0.849	0.844	0.949
NaiveBayes	76.98	0.770	0.136	0.784	0.770	0.772	0.918

6 Discussions and Conclusions

In fact, COVID-19 pandemic has affected many universities over the world. Specifically, this pandemic has changed the form of education process form face to face to online distance learning. Online distance learning is aimed to minimize the community transmission of COVID-19, which can rapidly spread in densely populated places such as universities and schools. This transition has motivated many universities to adopt many types of online learning systems in order to ensure continuation the learning process during COVID-19 pandemic. One of these tools, which known as mobile learning applications. Mobile learning applications can play a significant role during this pandemic, it aims to help instructors, and universities facilitate student learning during periods of universities closure [48]. Besides, most of these applications in mobile devices are free which can help ensure continuous learning during Coronavirus pandemic [1]. Therefore, this study was empirically applied with the purpose of identifying and understanding the main factors influencing the students' acceptance of mobile learning applications. To achieve this objective, this study proposes a predictive model by integrating the TAM model with the TUT constructs to understand the main determinants influencing the students' decisions to accept mobile learning applications.

In order to test the proposed research model, machine learning algorithms were applied using five classifiers are rule-learner (PART), a meta-classifier (Bagging), a decision-tree (RandomForest), a lazy-classifier (IBk), a logistic regression classifier (SMO) and a bayesian classifier (NaiveBayes) by using Weka version 3.8.3. The results of machine learning predictive algorithms showed that constructs of perceived ease of use, perceived usefulness, perceived enjoyment, effectiveness, efficiency, behavioural intention to use and utilization could predict the acceptance of mobile learning applications within accuracy rate of 87%. The results also found that RandomForest and IBK algorithms are the best two algorithms in predicting the main determinants of mobile learning acceptance as comparison with other machine learning algorithms with an accuracy of 81.3%.

7 Research Implications

The findings of this research offer both theoretical and practical implications. First, this study is among of the few studies that seek to understand the acceptance of mobile learning applications among university students in one of the developing countries like Jordan during COVID-19 pandemic. Second, this study proposes a predictive model by combining TAM with constructs of TUT model for predicting the essential determinants of mobile learning acceptance among students. Third, this research employs a powerful method for testing the proposed research model and hypotheses by using machine learning algorithms. This novel method was rarely employed in mobile learning acceptance literature, and thereby, it is confirmed that this technique will add an important contribution to the literature of mobile learning. Finally, the findings of this research can offer for both designers and developers of mobile learning applications important recommendations to better promote mobile learning application utilization in universities during COVID-19 pandemic.

References

1. Almaiah, M.A., Al-Khasawneh, A., Althunibat, A.: Exploring the critical challenges and factors influencing the E-learning system usage during COVID-19 pandemic. Educ. Inf. Technol. **1** (2020)
2. Al-Emran, M., Salloum, S.A.: Students' attitudes towards the use of mobile technologies in e-Evaluation. Int. J. Interact Mob. Technol. (IJIM) **11**(5), 195–202 (2017)
3. Almaiah, M.A., Al Mulhem, A.: Analysis of the essential factors affecting of intention to use of mobile learning applications: a comparison between universities adopters and non-adopters. Educ. Inf. Technol. **24**(2), 1433–1468 (2019)
4. Al-Emran, M., Mezhuyev, V., Kamaludin, A., ALSinani, M.: Development of M-learning application based on knowledge management processes. In: Proceedings of the 2018 7th international conference on software and computer applications, pp. 248–253 (2018)
5. Almaiah, M.A.: Acceptance and usage of a mobile information system services in University of Jordan. Educ. Inf. Technol. **23**(5), 1873–1895 (2018)
6. Arpaci, I., Alshehabi, S., Al-Emran, M., Khasawneh, M., Mahariq, I., Abdeljawad T., Hassanien, A.E.: Analysis of Twitter data using evolutionary clustering during the COVID-19 pandemic. Comput. Mater. Contin. **65**(1), 193–204 (2020). https://doi.org/10.32604/cmc.2020.011489
7. Arpaci, I., Karataş, K., Baloğlu, M.: The development and initial tests for the psychometric properties of the COVID-19 Phobia Scale (C19P-S). Pers. Indiv. Differ. **164**, 110108 (2020). https://doi.org/10.1016/j.paid.2020.110108
8. Al-Emran, M., Arpaci, I., Salloum, S.A.: An empirical examination of continuous intention to use m-learning: an integrated model. Educ. Inf. Technol. 1–20 (2020)
9. Almaiah, M.A., Jalil, M.A.: Investigating students' perceptions on mobile learning services. Int. J. Interact. Mob. Technol. (IJIM) **8**(4), 31–36 (2014)
10. Al-Emran, M., Elsherif, H.M., Shaalan, K.: Investigating attitudes towards the use of mobile learning in higher education. Comput. Hum. Behav. **56**, 93–102 (2016)
11. Almaiah, M.A., Jalil, M.A., Man, M.: Extending the TAM to examine the effects of quality features on mobile learning acceptance. J. Comput. Educ. **3**(4), 453–485 (2016)
12. Althunibat, A.: Determining the factors influencing students' intention to use m-learning in Jordan higher education. Comput. Hum. Behav. **52**, 65–71 (2015)
13. Almaiah, M.A., Man, M.: Empirical investigation to explore factors that achieve high quality of mobile learning system based on students' perspectives. Eng. Sci. Technol. Int. J. **19**(3), 1314–1320 (2016)
14. Almaiah, M.A., Alismaiel, O.: A examination of factors influencing the use of mobile learning system: an empirical study. Educ. Inf. Technol. **24**(1), 885–909 (2019)
15. Uğur, N.G., Koç, T., Koç, M.: An analysis of mobile learning acceptance by college students. J. Educ. Instr. Stud. World **6**(2) (2016)
16. Almaiah, M.A., Jalil, M.A., Man, M.: Preliminary study for exploring the major problems and activities of mobile learning system: a case study of Jordan (2016)
17. Al-Emran, M., Shaalan, K.: Academics' awareness towards mobile learning in Oman. Int. J. Comput. Dig. Syst. **6**(01), 45–50 (2017)

18. Almaiah, M.A., Alamri, M.M., Al-Rahmi, W.: Applying the UTAUT model to explain the students' acceptance of Mobile learning system in higher education. IEEE Access **7**, 174673–174686 (2019)
19. Al-Emran, M., Mezhuyev, V.: Examining the effect of knowledge management factors on mobile learning adoption through the use of Importance-Performance Map Analysis (IPMA). In: International conference on advanced intelligent systems and informatics, pp. 449–458. Springer, Cham (2019)
20. Almaiah, M.A., Almulhem, A.: A conceptual framework for determining the success factors of e-learning system implementation using Delphi technique. J. Theor. Appl. Inf. Technol. **96** (17) (2018)
21. Heflin, H., Shewmaker, J., Nguyen, J.: Impact of mobile technology on student attitudes, engagement, and learning. Comput. Educ. **107**, 91–99 (2017)
22. Jeno, L.M., Vandvik, V., Eliassen, S., Grytnes, J.-A.: Testing the novelty effect of an m-learning tool on internalization and achievement: a self-determination theory approach. Comput. Educ. **128**, 398–413 (2019)
23. Al-Emran, M., Mezhuyev, V., Kamaludin, A.: Technology acceptance model in m-learning context: a systematic review. Comput. Educ. **125**, 389–412 (2018)
24. Almaiah, M.A., Alamri, M.M., Al-Rahmi, W.M.: Analysis the effect of different factors on the development of Mobile learning applications at different stages of usage. IEEE Access **8**, 16139–16154 (2019)
25. Hamidi, H., Jahanshaheefard, M.: Essential factors for the application of education information system using mobile learning: a case study of students of the university of technology. Telemat. Inform. **38**, 207–224 (2019)
26. Almaiah, M.A., Alamri, M.M.: Proposing a new technical quality requirements for mobile learning applications. J. Theor. Appl. Inf. Technol. **96**, 19 (2018)
27. Shawai, Y.G., Almaiah, M.A.: Malay language mobile learning system (MLMLS) using NFC technology. Int. J. Educ. Manage. Eng. **8**(2), 1 (2018)
28. Criollo-C, S., Luján-Mora, S., Jaramillo-Alcázar, A.: Advantages and disadvantages of M-learning in current education. In: 2018 IEEE world engineering education conference (EDUNINE), pp. 1–6. IEEE (2018)
29. Al-Emran, M., Mezhuyev, V., Kamaludin, A.: Towards a conceptual model for examining the impact of knowledge management factors on mobile learning acceptance. Technol. Soc., 101247 (2020)
30. Almaiah, M., Al-Khasawneh, A., Althunibat, A., Khawatreh, S.: Mobile government adoption model based on combining GAM and UTAUT to explain factors according to adoption of mobile government services (2020)
31. Chung, C.-J., Hwang, G.-J., Lai, C.-L.: A review of experimental mobile learning research in 2010–2016 based on the activity theory framework. Comput. Educ. **129**, 1–13 (2019)
32. Almaiah, M.A., Alyoussef, I.Y.: Analysis of the effect of course design, course content support, course assessment and instructor characteristics on the actual use of E-learning system. IEEE Access **7**, 171907–171922 (2019)
33. Alamri, M.M., Almaiah, M.A., Al-Rahmi, W.M.: Social media applications affecting students' academic performance: a model developed for sustainability in higher education. Sustainability **12**(16), 6471 (2020)
34. Moorthy, K., Yee, T.T., T'ing, L.C., Kumaran, V.V.: Habit and hedonic motivation are the strongest influences in mobile learning behaviours among higher education students in Malaysia. Australas. J. Educ. Technol. **35**(4) (2019)
35. Almaiah, M.A., Al-Khasawneh, A.: Investigating the main determinants of mobile cloud computing adoption in university campus. Educ. Inf. Technol. 1–21 (2020)
36. Davis, F.D.: Perceived usefulness, perceived ease of use, and user acceptance of information technology. MIS Q. 319–340 (1989)
37. Liu, C.-H., Huang, Y.-M.: An empirical investigation of computer simulation technology acceptance to explore the factors that affect user intention. Univ. Access Inf. Soc. **14**(3), 449–457 (2015)

38. Salloum, S.A., Al-Emran, M., Shaalan, K., Tarhini, A.: Factors affecting the E-learning acceptance: a case study from UAE. Educ. Inf. Technol. **24**(1), 509–530 (2019)
39. Alksasbeh, M., Abuhelaleh, M., Almaiah, M.: Towards a model of quality features for Mobile social networks apps in learning environments: an extended information system success model (2019)
40. Almaiah, M.A., Nasereddin, Y.: Factors influencing the adoption of e-government services among Jordanian citizens. Electr. Gov. Int. J. **16**(3), 236–259 (2020)
41. Ghapanchi, A.H., Talaei-Khoei, A.: Rethinking technology acceptance: towards a theory of technology utilization (2018)
42. Al-Shihi, H., Sharma, S.K., Sarrab, M.: Neural network approach to predict mobile learning acceptance. Educ. Inf. Technol. **23**(5), 1805–1824 (2018)
43. Arpaci, I.: A hybrid modeling approach for predicting the educational use of mobile cloud computing services in higher education. Comput. Hum. Behav. **90**, 181–187 (2019). https://doi.org/10.1016/j.chb.2018.09.005
44. Al-Maroof, R.A., Arpaci, I., Al-Emran, M., Salloum, S.A., Shaalan, K.: Examining the acceptance of whatsapp stickers through machine learning algorithms. In: Recent advances in intelligent systems and smart applications, pp. 209–221. Springer, Cham
45. Arpaci, I., Al-Emran, M., Al-Sharafi, M. A., Shaalan, K.: A novel approach for predicting the adoption of smartwatches using machine learning algorithms. In: Recent advances in intelligent systems and smart applications, pp. 185–195. Springer, Cham (2021)
46. Christmann, A., Van Aelst, S.: Robust estimation of Cronbach's alpha. J. Multivar. Anal. **97**(7), 1660–1674 (2006)
47. Hair, J.F., Black, J.W., Babin, B.J., Anderson, R.E.: Multivariate data analysis, 7th edn. Prentice-Hall, Englewood Cliffs, NJ, USA (2010)
48. Almaiah, M.A., Al-Khasawneh, A., Althunibat, A., Khawatreh, S.: Mobile government adoption model based on combining GAM and UTAUT to explain factors according to adoption of mobile government services, pp. 199–225 (2020)

How COVID-19 Pandemic Is Accelerating the Transformation of Higher Education Institutes: A Health Belief Model View

Ali Nasser Al-Tahitah, Mohammed A. Al-Sharafi, and Mohammed Abdulrab

Abstract At the beginning of the year 2020, massive fast spread Coronavirus or (COVID-19) pandemic caused a serious impact on the education system worldwide. This chapter aims to explore the students' attitude to use social media as a learning tool during COVID-19 pandemic based on the view of the Health Belief Model (HBM). A total of 504 students in Malaysian universities were involved in this study. The partial least squares structural equation modelling (PLS-PM) has been employed to analyse the data collected in this study. The results indicated that perceived susceptibility, perceived severity, perceived barriers, perceived (health) motivation, perceived benefits and self-efficacy were significant in predicting the students' attitude to use social media as a learning tool during COVID-19 pandemic. The results of this study has been contributed to the existing literature by validating HBM in the Malaysian context and provide theoretical contributions and practical implications to the theory, and practice.

Keywords Social media · Health belief model · COVID-19 pandemic · Higher education institutes

A. N. Al-Tahitah
Faculty of Economics and Muamalat, Universiti Sains Islam Malaysia, Nilai, Malaysia
e-mail: alinasser@usim.edu.my

M. A. Al-Sharafi (✉)
Institute for Artificial Intelligence and Big Data, Universiti Malaysia Kelantan, City Campus, Pengkalan Chepa, 16100 Kota Bharu, Kelantan, Malaysia
e-mail: alsharafi@ieee.org

M. Abdulrab
Department of Management and Information System, College of Business Administration, University of Hail, Hail, Saudi Arabia
e-mail: abdulrabd@gmail.com

© The Author(s), under exclusive license to Springer Nature Switzerland AG 2021
I. Arpaci et al. (eds.), *Emerging Technologies During the Era of COVID-19 Pandemic*, Studies in Systems, Decision and Control 348,
https://doi.org/10.1007/978-3-030-67716-9_21

1 Introduction

At the beginning of year 2020, massive fast spread Coronavirus or (COVID-19) pandemic caused a serious impact on the education system worldwide. From that moment, the traditional methods of teaching and learning in all schools and universities of the world have almost changed. To the extent that this pandemic has inflicted on harming all countries in many fields, especially the economic and educational domains, however, it has contributed to raising a new philosophy and concepts where the traditional methods of education could be changed to new orientations that may become essential styles in the nearest future. Recently, according to the United Nations Education, Scientific and Cultural Organization (UNESCO), over 91 % of the world's students effected and most of schools and universities have suspended the teaching and the learning process. Alongside, they claimed that the (UNESCO) started supporting all countries, especially those in need of assistance through strive to find other alternative ways to maintain the educational process through many solutions such as remote learning.

Researchers, in the current chapter, would slightly criticize the claim that said the change in any field will not succeed unless the organizations or individuals are fully ready for change [1–5]. The main evidence for this claim is what we are witnessing in the current time with COVID-19 pandemic which forces educational institutions to accept the new method of teaching and learning. It can be said that most of those organizations, students, lecturers, and even staff were not ready for change, rather than that, they accepted it and it becomes a reality where became an alternative solution to proceed educational process progress using several social media platforms which become essential tools used by people across the globe [6, 7]. In spite, some previous studies and reports such as LCIBS [8] expected that the e-learning tools such as using social media platforms would be as the mechanism and effective tools in teaching/learning process in terms of sharing and delivering the educational materials through these tools to students.

According to Huang et al. [9], there is still a lack of further empirical studies to examine the impact of health belief model on attitude. Thus, this study aims to explore the students' attitude to use social media as a learning tool during COVID-19 pandemic based on the view of the Health Belief Model (HBM). According to literature, social media could be used as a novel educational technic in learning, where that study was clarified on how the possibility of social media employing for that purpose internationally [10, 11]. On the other hand, the students' engaging a with social media long hours a day are less motivated to achieve their academic performance than those who use it for educational purposes [12, 13]. That claim also supported by Giunchiglia et al. [14], who demonstrated that the constant using social media through their computers, or smartphones will affect their study outcome, however, they encourage students to stay away of social media during studying to avoid the unwilling consequences.

Regarding the possibility of students interacting in the educational process through available social media such as Facebook, Twitter, Instagram, various

Google applications, etc., a study in Australia and New Zealand (ANZ), conducted by some researchers at the beginning of the year 2020 confirms that students of the surgical department have a positive attitude with posts with educational content [6]. The study recommended conducting more studies in this regard and in different societies to find out the extent of interaction with non-traditional educational methods in the teaching and learning process. We conclude from such studies the extent to which these platforms are used in raising awareness to reduce the spread of diseases and risks to people as illustrated by the Health Behaviour Model and also to realize the possibility of employing these methods in the educational process in addition to the basic traditional educational process (face to face) that is indispensable, whatever the reasons. In line with that, the e-learning could be work properly through certain social media platforms as an effective tool, however, it could not completely replace the tradition teaching/ learning styles [15].

2   Research Model and Hypotheses Development

A Health Belief Model (HBM) which was introduced in 1950 by social psychologists in the United States, the public health sector. HBM used as a systematic way to identify, explain the preventive health behaviours and to predict factors involved in the failure of screening programs for early detection of tuberculosis [16]. It has been employed to explain and to enhance the effectiveness of health education programs [17]. That style was the health belief model the one that was the first social cognition models which able to explain health behaviour change [18]. Moreover, the HBM model was based on three assumptions that must be taken into consideration, Rosenstock et al. [19], (i) the presence of anxiety and health concerns that individuals feel due to some logical motivations about the probability of an occurring some events, (ii) establishing the belief in people that they may suddenly be exposed to potential risks, so they feel in constant danger. (iii) HBM model prompts the perpetual belief that a healthy culture and taking precautions that make society alert is of great benefit. From this standpoint, this study highlights the relationship between the components of the HBM theory and students' attitudes about using the social media platforms as an educational method in exceptional circumstances to alleviate panic that caused by COVID-19 pandemic and employing it in raising awareness instead. According to Huang et al. [9], there is still a lack of further imperial studies to examine the impact of health belief model on attitude. Thus, the current study highlighted the effect of perceived susceptibility, perceived severity, perceived barriers, perceived (health) motivation, perceived benefits and self-efficacy, with students' attitude toward COVID-19 pandemic. HBM's constructs have found appropriate for diverse contexts and importantly explain the health-related behaviours [9, 20].

Furthermore, HBM through its main dimensions' focuses on the behaviour of health beliefs of the individuals to identify the impact on student s' attitudes during COVID-19 pandemic. This study aims to explore the students' attitude to use social media as a learning tool during COVID-19 pandemic based on the view of the HBM.

Perceived Susceptibility is as a personal perception of the risk of illness [21]. And according to Glanz et al. [22] it refers to "beliefs about the likelihood of getting a disease or condition". Since the world is currently and since the beginning of 2020 is facing the fiercest contagious viral pandemic called COVID-19 which has caused fear among the world's population. However, the rapid transmission of this epidemic from one person to another led to increased perceived susceptibility to infection and raising the level of caution and taking precautions to save themselves away from infection [23].

In Malaysia, Dengue fever is endemic which consider as a public health concern. Thus, Othman et al. [24], found that most of the respondents answered that they were highly convinced that they might be affected by the epidemic at any moment. This is what proves that the perceived susceptibility plays a prominent role in an individual's attitude especially among the adults towards dangers. In line with that, they have a belief that mosquito bites the main reason behind the spread disease and then they are taking care of themselves by avoiding that dangerous insect, especially in the affected areas. We can realize here that the perceived susceptibility towards something has a significant effect on people attitude. Moreover, Huang et al. [9] demonstrated that the perceived susceptibility has an essential role in predicting individuals' attitudes that it boost taking preventive measures towards a serious situation that could be happened any time. Thus, the researchers intend to find out the impact of perceived susceptibility to e-learning tools on students' attitude during COVID-19.

Regarding the perceived severity, according to Glanz et al. [22] "the feelings about the seriousness of contracting an illness or of leaving it untreated include evaluations of both medical and clinical consequences (e.g., death, disability, and pain) and possible social consequences (e.g., the effect of the condition on work, family life, and social relations)".

As mentioned earlier, a study of Othman et al. [24] conducted on Malaysian context, which adopted the health belief model (HBM) to highlight and clarify the level of target participants' attitude in terms of dengue disease issue. the study concluded that the perceived severity has a positive significant impact on the individuals' attitude towards dengue fever since most respondents were strongly agreed and agreed that dengue fever is a serious illness and few of them believe it probably causes death. Moreover, people during COVID-19 pandemic are quite afraid of being infected with any type of diseases, thus the probability of getting infected during the COVID-19 outbreak time becomes as a source of concern for many people, that proves the perceiving severity could change and affect the individuals' attitudes [25, 26]. Hence, all previous researches indicate and focus on that the social media has also significant impact in increasing the awareness and sometimes the anxious among people, however, we can assume that there is social media's impact as a learning tool on students' attitude and behavioural intention during COVID-19.

Perceived barriers are the second construct of the health belief model which refer to the seriousness and negative outcomes of personal behaviour. According to Mohamed et al. [21], perceived barriers as one of the negative aspects of individuals' behaviour that prevents necessary health measures. A study conducted in Pakistan during COVID-19 period by Saqlain et al. [27], on the healthcare workers, found the overcrowding in emergency rooms was one of the barriers that prevent workers to controlling the infection spreading based on the majority of study respondents. However, on the other hand, some of the healthcare workers in that context thought that not performing handwashing and not wearing a mask was not a barrier to infection control. Thus, these consequences indicate that the perceived barriers could change and affect some people attitude whether negatively or positively. Another empirical study in Iran by Hatefnia et al. [28] confirmed that the education program on promoting mammography behaviours perceived barriers have a significant impact in decreased in case group but not in the control group. Consequently, perceived barriers are the actual cost associated with the health behaviour change and the negative outcomes that could interrupt individuals' desire to change

Health motivation has found it to be associated with most health behaviours, where helps to improve the effects of education on health behaviours and health knowledge [29]. In that study, researchers found that health motivation related to most health behaviours. Besides, according to several previous studies that health motivation refers to individuals' interest in their health in general and constantly [30, 31]. Recently, Kocoglu-Tanyer et al. [30] claimed that most of health belief model dimensions are associated with the individuals' attitudes in terms of health awareness. Spite of the people has sufficient knowledge about coronavirus disease and they have become more educated, however, we can realize that health motivation plays a vital role in making the acquisition the health information and will significantly affect their attitude toward adopting social media in raise the awareness among students and also as an alternative tool can be used in the critical circumstances such as the COVID-19 pandemic.

Regarding perceived benefits in health belief model, it is one of the main objectives in this chapter to examine the effect of the students' attitudes towards the e-learning process during COVID-19. To highlight this matter, there are a few previous studies have in-depth examined that relationship between these variables in different contexts and on different topics. Based on the (HBM), perceived benefits are one of the positive components that can occur from behaviour change. Initially, perceived benefits refer to the belief of the individuals that a particular behaviour will occur positively based on some positive believes of the individuals [32]. Furthermore, in terms of the relationship between perceived benefits and individual attitude, Liu et al. [33], claimed that the perceived benefits can be an element factor that influences people attitude. According to Han et al. [34] an empirical study examined the impact of playing screen golf on player attitudes and the hypothesis was supported where perceived benefits of golf have a positive and significant effect on customer's attitudes about screen golf. This study strives to highlight to which perceived benefits of the social media that could be used in teaching and learning properly in the exceptional circumstances especially in natural disasters or pandemic crises. Thus, at the end of this part, the main hypotheses and the results of those assumptions have been shown.

The last construct of health belief model shown in Fig. 1 is Self-efficacy. According to Schneider [35], this theory has been defined as the level of confidence for the patient in taking the right action in their healthcare journey. Furthermore, Self-efficacy is defined as the actions taken by individuals toward particular behaviour successfully and carefully to meet the proposed goals [36]. Also, it is defined as "beliefs in one's capabilities to organize and execute the courses of action required to produce given levels of attainments" [37]. In terms of the relationship between self-efficacy attitude, Özokcu [38], reported in his study that there is a positive significant relationship between attitudes of the targeted teachers and self-efficacy for inclusive practices where teachers' self-efficacy has been found as a statistically significant predictor of their attitudes towards inclusive education. Moreover, another study has been done by Uyanik [39], in one of the Turkish universities, investigated the relationship between teachers' self-efficacy beliefs in teaching science and attitudes towards teaching profession at the teacher candidates. The result revealed that there is a positive relationship between self-efficacy of teachers' beliefs and attitudes to the teaching profession.

Lastly, in this part, researchers reviewed some of the previous studies that discussed the relationship and the effect of health belief model dimensions "Perceived severity, Perceived susceptibility, Perceived barriers, Perceived (health) motivation, Self-efficacy, Perceived benefits" on the individual's attitudes and behaviour, in order to enable us in formulating the assumptions of the current research that aimed to investigate the effect of the constructs of health belief model to using social media platforms as an effective tool in teaching and learning process on the attitude of the students in the Malaysian universities during COVID-19 period. Based on all the above arguments, the researchers hypothesize as a following:

H1: Perceived susceptibility has a significant positive effect on students' attitudes to use social media as a learning tool during the COVID-19 outbreak.

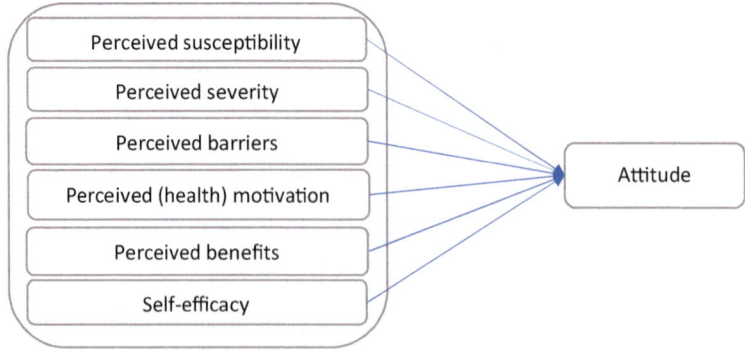

Fig. 1 Research model

H2: Perceived severity has a significant positive effect on students' attitudes to use social media as a learning tool during the COVID-19 outbreak.

H3: Perceived barriers has a significant positive effect on students' attitudes to use social media as a learning tool during the COVID-19 outbreak.

H4: Health motivation has a significant positive effect on students' attitudes to use social media as a learning tool during the COVID-19 outbreak.

H5: Perceived benefits has a significant positive effect on students' attitudes to use social media as a learning tool during the COVID-19 outbreak.

H6: Self-efficacy has a significant positive effect on students' attitudes to use social media as a learning tool during the COVID-19 outbreak.

3 Research Methodology

Due to COVID-19 outbreak, all universities, colleges, and schools in Malaysia have been suspended the study for various period of times during 2020. Almost, all those institutions replaced face to face teaching and learning process with virtual methods as a temporary to keep the education process ongoing using different social media platforms. Hence, to meet study objectives which were based on the situation, the targeted students in Malaysian universities have been asked about the role of e-learning during COVID-19. The time for collecting data was between February 2020—May 2020 using an online questionnaire where the questionnaires distributed online universities and lecturers WhatsApp groups. a total of 504 participants were considered out of 549 participants responded to the online survey.

The online survey questionnaire consisted of two sections. The first section includes 25 items measured university students' shifting to using social media as a learning tool during the COVID-19 pandemic while the second represents the demographic data which consisted of 10 questions. To measure the extent of agreeing or disagree with the items given in the questionnaire, the researchers employed a 5-point Likert scale that ranged from 'strongly disagree' (1) to 'strongly agree' (5). The questionnaire used in this study was developed based on the Health Belief Model, which derived from several previous studies as shown in Table 1.

Table 1 Questionnaire sources

SI. No.	Factor	Items	Source
1	Perceived Susceptibility	4	[40, 41] [42]
2	Perceived Severity	3	[41–43]
3	perceived barriers	3	[40–43]
4	Health Motivation	4	[40, 42]
5	Perceived benefits	4	[40–43]
6	Self-Efficacy	3	[41, 43–47],
7	Attitude	4	[48–50]

To make sure of the quality and content validity the questionnaire as it is an essential requirement during designing and developing the study's questionnaire [51], every single item of the questionnaire was assessed by several experts in the related area of research. Then the last version of the questionnaires has been adopted based on draft given back by experts with the proper suggestions. Finally, in terms of data analysis, the current research utilised both software programs "SPSS version 25.0 and structural equation modelling (PLS-SEM) through SmartPLS V.3.2.8". According to Hair et al. [52], Structural Equation Modelling (SEM) are one of the most powerful statistical tools in the area of social science due to its ability to test several relationships simultaneously. Therefore, PLS-SEM has been chosen to be the statistical program software due to the current research is an exploratory and its conceptual model is a quite complex with 7 constructs and 25 indicators.

4 Results

4.1 Descriptive Analysis

The percentage of females 290 (57.5 %) out of 504 respondents was over 214 (42.5 %) males. Their total was 290. The majority of respondents are relatively young, of them (67.7 %) aged 18–25 and 17.5 % of total respondents are aged 26–35, of which 9.1 % are aged between 36 and 45 and 5.8 % are aged 46 or higher. With regards to the education level, the largest percentage of bachelors graduates with a degree (66.3 %) leads by a diploma (18.8 %), master's degree (8.3 %) and Ph.D. (6.5 %). The largest group in terms of the university in which the participants are connected is from UMP (124), and the smallest group is UTP (89). USIM, UiTM and INTI members have 103, 98 and 90 respectively. About social media usage, data shows that (93.1 %) of students use social media every day, (3.2 %) three times a week, (2.0 %) weekly and (1.8 %) two days a week. Once asked on the social media service they mainly use, the respondents indicated that they used Whatsapp (59.7 %), Youtube (14.9 %), Facebook (10.3 %), Twitter (9.1 %) and Instagram (6.0 %) (9.1 %).

4.2 Measurement Model Assessment

Models for measuring reliability and validity are developed [53]. The internal consistency of the measurement item is evaluated using Cronbach's alpha (CA) and composite reliability (CR) [54]. Alpha value 0.70 for quantitative research is deemed adequate, and CR of 0.70 in exploratory research considered acceptable [55]. As in Table 2, the alpha value of Cronbach ranged from 0.8219 to 0.9388,

Table 2 Loading, Cronbach's alpha, CR and AVE

Constructs	Indicators	Loading	Cronbach's alpha	Composite reliability	AVE
Perceived susceptibility	PSUS1	0.9210	0.9388	0.9561	0.8449
	PSUS2	0.9311			
	PSUS3	0.9237			
	PSUS4	0.9006			
Perceived severity	PS1	0.8795	0.8545	0.9115	0.7744
	PS2	0.8815			
	PS3	0.8790			
Perceived barriers	PBA1	0.8894	0.8340	0.8995	0.7494
	PBA2	0.8912			
	PBA3	0.8141			
Health motivation	HM1	0.7912	0.8282	0.8862	0.6619
	HM2	0.7237			
	HM3	0.8606			
	HM4	0.8701			
Perceived benefits	PBE1	0.8908	0.8946	0.9269	0.7605
	PBE2	0.8966			
	PBE3	0.8184			
	PBE4	0.8802			
Self-efficacy	SE1	0.9097	0.9090	0.9425	0.8454
	SE2	0.9268			
	SE3	0.9218			
Attitude	ATT1	0.7373	0.8219	0.8811	0.6502
	ATT2	0.8460			
	ATT3	0.8034			
	ATT4	0.8342			

PSUS perceived susceptibility, *PS* perceived severity, *PBA* perceived barriers, *HM* health motivation, *PBE* perceived benefits, *SE* self-efficacy, *ATT* attitude, *AVE* average variance extracted

which was higher than the standard of 0.7. Besides, the CR values ranged between 0.784 and 0.933. This then set in place adequate steps of internal reliability.

Convergent validity and discriminant validity should be examined as far as validity is concerned [55]. The validity of a model convergent validity was assured by the measurement of factor loadings and the average variance extracted. This was done by looking at the loading of the items and all items had a loading of more than 0.7, which is sufficient in the multivariate analysis [54]. The AVE values for the constructs also ranged from 0.6502 to 0.8454, above the estimated value of 0.5 [56]. HTMT (Heterotrait-monothrait ratio), as suggested by Henseler et al. [57] has been used to test discriminant validity for the measures in this study. The mean value of the indicator correlations between constructs was defined, as indicated by Ringle et al. [58] HTMT criterion, relative to the (geometric) mean of average correlations between indicators that measure the same construct. Henseler et al. [57] stated that if values greater than 0.85 HTMT values may be an issue. Table 3 showed that all values were below the 0.85 standards, indicating adequate

Table 3 HTMT results

		1	2	3	4	5	6	7
1	Attitude							
2	Health motivation	**0.4773**						
3	Perceived barriers	0.3795	**0.2029**					
4	Perceived benefits	0.4948	0.5992	**0.2192**				
5	Perceived severity	0.4625	0.6409	0.1425	**0.6674**			
6	Perceived susceptibility	0.4359	0.3385	0.8547	0.3789	**0.3476**		
7	Self efficacy	0.4903	0.5145	0.5257	0.5832	0.5023	**0.5351**	

discriminatory validity. The standardized root mean square residual (SRMR) is the model fit criterion for PLS path modelling that was used to determine the model fit [59]. SRMR is defined as the average root square difference between the correlations observed and the model-involved correlations [60]. SRMR is an index of malfunction (greater signal values are worse fit), ranging from 0.0 to 1.0. When model predictions fit perfectly with data, SRMR is zero. In the clean (high-factor load) model [61], the adequate cut off value for PLS path models is enhanced (0.08) Henseler et al. [57], SRMR is increased (lowered).

4.3 Structural Model Assessment

The structural model used to analytically test whether data support the hypotheses presented by the structural model. Only after a successful evaluation of the measurement model can the structural model be analysed. A structural model can be evaluated using the determination coefficient ($R2$) and path coefficients in PLS. The structural model evaluation shown in Fig. 2 proves a hypothesis test indication with six direct hypotheses. The results indicated that all the study hypotheses are supported. Perceived Susceptibility was found to have significant influence on attitude ($\beta = 0.1195$, $t = 2.3608$, $p < 0.05$). Hence, H1 is accepted. Further, the result found that perceived severity ($\beta = 0.0928$, $t = 2.1176$, $p < 0.05$) has a significant effect on attitude. Therefore, H2 is accepted. Similarly, the result indicated that perceived barriers have significant effect on attitude ($\beta = 0.1029$, $t = 1.9761$, $p < 0.05$). Hence, H3 is accepted. Likewise, health motivation found to have a significant effect on attitude ($\beta = 0.1015$, $t = 2.4245$, $p < 0.001$). Hence, H4 is accepted. Furthermore, the result found that perceived benefits have a significant influence on attitude ($\beta = 0.1317$, $t = 2.6167$, $p < 0.05$). Therefore, H5 is accepted.

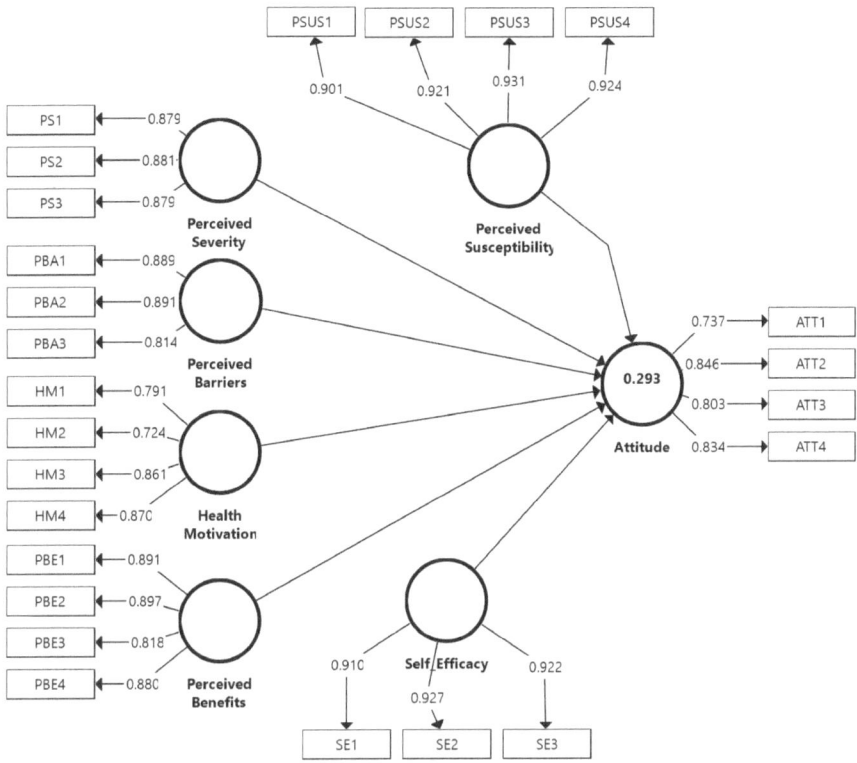

Fig. 2 Research model results

Additionally, self-efficacy significantly predicts attitude ($\beta = 0.1796$, t = 4.1969, p < 0.001). Therefore, H6 is accepted. However, the thumb rule for an appropriate R2 differs but the R2 value of 0.26 and higher is considered significant in accordance with Cohen [62]. As shown in Fig. 2, 29.3 % of the variation in attitude is explained by all independent variables (Table 4).

Table 4 Structural assessment results

Hypothesis	Relationship	Std. Beta	t-value	p-value	Decision
H1	PSUS → ATT	0.1195	2.3608	0.0186	Supported
H2	PS → ATT	0.0928	2.1176	0.0347	Supported
H3	PBA → ATT	0.1029	1.9761	0.0487	Supported
H4	HM → ATT	0.1015	2.4245	0.0157	Supported
H5	PBE → ATT	0.1317	2.6167	0.0091	Supported
H6	SE → ATT	0.1796	4.1969	0.0000	Supported

PSUS perceived susceptibility, *PS* perceived severity, *PBA* perceived barriers, *HM* health motivation, *PBE* perceived benefits, *SE* self-efficacy, *ATT* attitude

5 Discussion and Conclusion

Many instructors and students in higher education institutions and universities, in particular, have being been affected by the COVID19 pandemic. Social media has proven to be a novel educational technic in learning during the COVID19 pandemic. Although in the dark side of social media which spread panic and affected the mental health of social media users and it could be described as a double-edged sword [63, 64], in the other hand, it can play an important role in terms of growing the awareness and educating people about the epidemic's dangers through the flow of the information and constantly update the information. This study aimed to explore the students' attitude to use social media as a learning tool during COVID-19 pandemic based on the view of the Health Belief Model (HBM). By collecting and analysing data about different level of students from Malaysian Universities, it was found that perceived susceptibility, perceived severity, perceived barriers, health motivation, perceived benefits, and self-efficacy constructs have a positive effect on the students' attitude to use social media as a learning tool during COVID-19 pandemic. More interesting, the developed model explains a substantial variance (29.3 %) in the students' attitude to use social media as a learning tool during COVID-19 pandemic, which clearly shows that the proposed research model is sound and valid. Based on the findings of the present study, it is recommended that social media as a learning tools should have more focus especially in the pandemics era.

References

1. Al Tahitah, A.N.A.: The effect of transformational and transactional leadership on readiness for change in the educational ministries in Yemen: learning organizational culture as a mediator. Universiti Sains Islam Malaysia (2019)
2. Al-Tahitah, A., Abdulrab, M., Alwaheeb, M.A., Al-Mamary, Y.H.S., Ibrahim, I.: The effect of learning organizational culture on readiness for change and commitment to change in educational sector in Yemen. J. Crit. Rev. 7(9), 1019–1026 (2020)
3. Bateh, J., Castaneda, M.E., Farah, J.E.: Employee resistance to organizational change. Int. J. Manage. Inf. Syst. 17(2), 113 (2013)
4. Kotter, J.P.: Leading change, p. 208. Harvard business press, USA (2012)
5. Miller, D., Madsen, S.R., John, C.R.: Readiness for change: implications on employees' relationship with management, job knowledge and skills, and job demands. J. Appl. Manage. Entrep. 11(1), 3 (2006)
6. Larkins K., Murphy V., Loveday B.P.: Use of social media for surgical education in Australia and New Zealand. ANZ J. Surg. (2020)
7. Xu, X., Wang, J., Peng, H., Wu, R.: Prediction of academic performance associated with internet usage behaviors using machine learning algorithms. Comput. Hum. Behav. 98, 166–173 (2019)
8. LCIBS: The role of social media in education. London College of International Business Studies, UK. https://www.lcibs.co.uk/the-role-of-social-media-in-education/

9. Huang, X., Dai, S., Xu, H.: Predicting tourists' health risk preventative behaviour and travelling satisfaction in Tibet: combining the theory of planned behaviour and health belief model. Tour. Manage. Perspect. **33**, 100589 (2020)
10. Al-Sharafi, M.A., Mufadhal, M.E., Arshah, R.A., Sahabudin, N.A.: Acceptance of online social networks as technology-based education tools among higher institution students: structural equation modeling approach. Sci. Iran. **26**(Special Issue on: Socio-Cognitive Engineering), 136–144 (2019)
11. Mufadhal, M.E., Sahabudin, N.A., Al-Sharafi, M.A.: Conceptualizing a model for adoption of online social networks as a learning tool. Presented at the 5th international Conference on Software Engineering and Computer Systems (ICSECS), Langkaw (2017)
12. Swansea, U.: Internet use reduces study skills in university students. In ScienceDaily (ed.), USA (2020)
13. Reames, B.N., Sheetz, K.H., Englesbe, M.J., Waits, S.A.: Evaluating the use of twitter to enhance the educational experience of a medical school surgery clerkship. J. Surg. Educ. **73**(1), 73–78 (2016)
14. Giunchiglia, F., Zeni, M., Gobbi, E., Bignotti, E., Bison, I.: Mobile social media usage and academic performance. Comput. Hum. Behav. **82**, 177–185 (2018)
15. Maertens, H., Madani, A., Landry, T., Vermassen, F., Van Herzeele, I., Aggarwal, R.: Systematic review of e-learning for surgical training. Br. J. Surg. **103**(11), 1428–1437 (2016)
16. Khiyali, Z., Aliyan, F., Kashfi, S.H., Mansourian, M., Jeihooni, A.K.: Educational intervention on breast self-examination behavior in women referred to health centers: application of health belief model. Asian Pac. J. Cancer Prev. APJCP **18**(10), 2833 (2017)
17. Steckler, A., McLeroy, K.R., Holtzman, D.: Godfrey H. Hochbaum (1916–1999): from social psychology to health behavior and health education. Am. J. Public Health **100**(10), 1864 (2010)
18. Kim, J., Park, H.-A.: Development of a health information technology acceptance model using consumers' health behavior intention. J. Med. Internet Res. **14**(5), e133 (2012)
19. Rosenstock, I.M., Strecher, V.J., Becker, M.H.: Social learning theory and the health belief model. Health Educ. Q. **15**(2), 175–183 (1988)
20. Jones, C.L., Jensen, J.D., Scherr, C.L., Brown, N.R., Christy, K., Weaver, J.: The health belief model as an explanatory framework in communication research: exploring parallel, serial, and moderated mediation. Health Commun. **30**(6), 566–576 (2015)
21. Mohamed, H.A.E.-A., Ibrahim, Y.M., Lamadah, S.M., Hassan, M., El-Magd, A.: Application of the health belief model for breast cancer screening and implementation of breast self-examination educational program for female students of selected medical and non-medical faculties at Umm al Qura University. Life Sci. J. **13**(5), 21–33 (2016)
22. Glanz, K., Rimer, B.K., Viswanath, K.: Health behavior and health education: theory, research, and practice. Wiley (2008)
23. Salzberg, S.: Coronavirus: there are better things to do than panic. https://www.forbes.com/sites/stevensalzberg/2020/02/29/coronavirus-time-to-panic-yet/#64ebc2867fa6
24. Othman, H., et al.: Applying health belief model for the assessment of community knowledge, attitude and prevention practices following a dengue epidemic in a township in Selangor, Malaysia. Int. J. Commun. Med. Pub. Health **6**(3), 958 (2019)
25. Tweneboah-Koduah, E.Y.: Social marketing: using the health belief model to understand breast cancer protective behaviours among women. Int. J. Nonprofit Volunt. Sect. Mark. **23**(2), e1613 (2018)
26. Ahadzadeh, A.S., Sharif, S.P., Ong, F.S., Khong, K.W.: Integrating health belief model and technology acceptance model: an investigation of health-related internet use. J. Med. Internet Res. **17**(2), e45 (2015)
27. Saqlain, M., et al.: Knowledge, attitude, practice and perceived barriers among healthcare professionals regarding COVID-19: a cross-sectional survey from Pakistan. J. Hosp. Infect. (2020)

28. Hatefnia, E., Niknami, S., Mahmoudi, M., Ghofranipour, F., Lamyian, M.: The effects of health belief model education on knowledge, attitude and behavior of Tehran pharmaceutical industry employees regarding breast cancer and mammography (in Persian). Behbood J. **14**(1), Pe42–Pe53, En6 (2010)
29. Moorman, C., Matulich, E.: A model of consumers' preventive health behaviors: the role of health motivation and health ability. J. Consum. Res. **20**(2), 208–228 (1993)
30. Kocoglu-Tanyer, D., Dengiz, K.S., Sacikara, Z.: Development and psychometric properties of the public attitude toward vaccination scale-health belief model. J. Adv. Nurs. (2020)
31. Champion, V.L., Skinner, C.S.: The health belief model. Health Behav. Health Educ. Theory Res. Pract. **4**, 45–65 (2008)
32. Farah, M.F.: Application of the theory of planned behavior to customer switching intentions in the context of bank consolidations. Int. J. Bank Market. (2017)
33. Liu, M.T., Chu, R., Wong, I.A., Zúñiga, M.A., Meng, Y., Pang, C.: Exploring the relationship among affective loyalty, perceived benefits, attitude, and intention to use co-branded products. Asia Pac. J. Market. Logist. (2012)
34. Han, H., Baek, H., Lee, K., Huh, B.: Perceived benefits, attitude, image, desire, and intention in virtual golf leisure. J. Hosp. Market. Manage. **23**(5), 465–486 (2014)
35. Schneider, M.-J.: Introduction to public health. Jones & Bartlett Publishers (2016)
36. Bandura, A.: Self-efficacy: the exercise of control. WF Freeman, USA (1997)
37. Bandura, A.: Health promotion from the perspective of social cognitive theory. Psychol. Health **13**(4), 623–649 (1998)
38. Özokcu, O.: The relationship between teacher attitude and self-efficacy for inclusive practices in Turkey. J. Educ. Train. Stud. **6**(3), 6–12 (2018)
39. Uyanik, G.: Investigation of the self-efficacy beliefs in teaching science and attitudes towards teaching profession of the candidate teachers. Univ. J. Educ. Res. **4**(9), 2119–2125 (2016)
40. Champion, V.L.: Instrument development for health belief model constructs. Adv. Nurs. Sci. **6**(3), 73–85 (1984)
41. Ng, B.-Y., Kankanhalli, A., Xu, Y.C.: Studying users' computer security behavior: a health belief perspective. Decis. Support Syst. **46**(4), 815–825 (2009)
42. Kocoglu-Tanyer, D., Dengiz, K., Sacikara, Z.: Development and psychometric properties of the public attitude towards vaccination scale-Health belief model. J. Adv. Nurs. (2020)
43. Ng, B.-Y., Xu, Y.: Studying users' computer security behavior using the health belief model. PACIS 2007 Proceedings, p. 45 (2007)
44. Woon, I., Tan, G.-W., Low, R.: A protection motivation theory approach to home wireless security. ICIS 2005 proceedings, p. 31 (2005)
45. Ifinedo, P.: Understanding information systems security policy compliance: an integration of the theory of planned behavior and the protection motivation theory. Comput. Secur. **31**(1), 83–95 (2012)
46. Gao, Y., Li, H., Luo, Y.: An empirical study of wearable technology acceptance in healthcare. Ind. Manage. Data Syst. **115**(9), 1704–1723 (2015). https://doi.org/10.1108/IMDS-03-2015-0087
47. Thompson, N., McGill, T.J., Wang, X.: "Security begins at home": determinants of home computer and mobile device security behavior. Comput. Secur. **70**, 376–391 (2017)
48. Davis, F.D.: Perceived usefulness, perceived ease of use, and user acceptance of information technology. MIS Q. 319–340 (1989)
49. Ajzen, I.: The theory of planned behavior. Organ. Behav. Hum. Decis. Process. **50**(2), 179–211 (1991)
50. Venkatesh, V., Bala, H.: Technology acceptance model 3 and a research agenda on interventions (in English). Decis. Sci. **39**(2), 273–315 (2008). https://doi.org/10.1111/j.1540-5915.2008.00192.x
51. Almanasreh, E., Moles, R., Chen, T.F.: Evaluation of methods used for estimating content validity. Res. Soc. Adm. Pharm. **15**(2), 214–221 (2019)
52. Hair, J.F., Anderson, R.E., Babin, B.J., Black, W.C.: Multivariate data analysis: a global perspective. Pearson Upper Saddle River, NJ (2010)

53. Sarstedt, M., Ringle, C.M., Hair, J.F.: Partial least squares structural equation modeling. Handb. Market Res. **26**, 1–40 (2017)
54. Hair Jr, J.F., Sarstedt, M., Ringle, C.M., Gudergan, S.P.: Advanced issues in partial least squares structural equation modeling. saGe publication (2017)
55. Hair Jr, J.F., Sarstedt, M., Hopkins, L., Kuppelwieser, V.G.: Partial least squares structural equation modeling (PLS-SEM). Eur. Bus. Rev. (2014)
56. Fornell, C., Larcker, D.F.: Evaluating structural equation models with unobservable variables and measurement error. J. Market. Res. 39–50 (1981)
57. Henseler, J., Ringle, C.M., Sarstedt, M.: A new criterion for assessing discriminant validity in variance-based structural equation modeling. J. Acad. Market. Sci. **43**(1), 115–135 (2015)
58. Ringle, C.M., Sarstedt, M., Mitchell, R., Gudergan, S.P.: Partial least squares structural equation modeling in HRM research. Int. J. Hum. Resour. Manage. 1–27 (2018)
59. Henseler, J., Hubona, G., Ray, P.A.: Using PLS path modeling in new technology research: updated guidelines. Ind. Manage. Data Syst. **116**(1), 2–20 (2016)
60. Henseler, J., et al.: Common beliefs and reality about PLS: comments on Rönkkö and Evermann (2013). Organ. Res. Methods **17**(2), 182–209 (2014)
61. Anderson, J.C., Gerbing, D.W.: The effect of sampling error on convergence, improper solutions, and goodness-of-fit indices for maximum likelihood confirmatory factor analysis. Psychometrika **49**(2), 155–173 (1984)
62. Cohen, J.: Statistical power analysis for the behavioral sciences, p. 23. Hilsdale (NJ: Lawrence Earlbaum Associates, no. 1). Lawrence Erlbaum Associates, Publishers, New York (1988)
63. Baccarella, C.V., Wagner, T.F., Kietzmann, J.H., McCarthy, I.P.: Social media? It's serious! Understanding the dark side of social media. Eur. Manage. J. **36**(4), 431–438 (2018)
64. Sands, S., Campbell, C., Ferraro, C., Mavrommatis, A.: Seeing light in the dark: investigating the dark side of social media and user response strategies. Eur. Manage. J. **38**(1), 45–53 (2020)

Online Testing in Higher Education Institutions During the Outbreak of COVID-19: Challenges and Opportunities

Mohammed Adulkareem A. Alkamel, Santosh S. Chouthaiwale, Amr Abdullatif Yassin, Qasim AlAjmi, and Hanan Yahia Albaadany

Abstract This study aimed to investigate the challenges and opportunities of online testing during the outbreak of COVID-19. The main focus was on the under-graduate and post-graduate international students in various universities of India. This study employed a qualitative design as the data were collected through semi-structured interviews. The findings showed that the attitude of students towards online testing was positive even though they experienced worry and anxiety. Moreover, students used self-strategies to overcome such challenges such as improving their self-confident and motivating themselves during online tests. In terms of ICT skills, the findings showed that students did not face any challenges during the online testing, which supported the idea that is online testing does not require professional skills, and students could improve their ICT skills as a main requirement for 21st century education. Additionally, the findings of the study showed that there were other online testing challenges faced by students, including the transmission from traditional tests into online tests, and the poor connection of the internet. However, online testing advantages outweighed the challenges, especially when online testing is considered to be a good alternative method in such

M. A. A. Alkamel · S. S. Chouthaiwale
Department of English, Indraraj Arts, Commerce and Science College,
Dr. BAM University, Aurangabad, India
e-mail: alkamel2030@gmail.com

S. S. Chouthaiwale
e-mail: dr.sschothaiwale@gmail.com

A. A. Yassin (✉)
Center of Languages and Translation, Ibb University, Ibb, Yemen
e-mail: amryassin84@gmail.com

Q. AlAjmi
Department of Education, College of Arts and Humanities,
A' Sharqiyah University, Ibra, Oman
e-mail: qasim.alajmi@asu.edu.om

H. Y. Albaadany
Department of Language and Humanities Education, Faculty of Educational Studies,
Universiti Putra Malaysia, 43400 Serdang, Selangor, Malaysia
e-mail: h.albaadany@gmail.com

critical stations like the outbreak of COVID-19, and it would be suitable for future online learning. Accordingly, online testing should receive more focus by Higher Education Institutions (HEIs) to get along with the educational requirements of the 21st century.

Keywords Online testing · E-learning · HEIs · Challenges · Opportunities · COVID-19

1 Introduction

COVID-19 is an infectious disease caused by a newly discovered coronavirus, appeared in 2019 as a worldwide pandemic. The novel coronavirus disease has infected over nineteen million people globally and is taking a severe toll on individuals, families, and all aspects of life, including education [1–5] which is considered to be one of the most important aspects in our life, and it has been greatly affected by the outbreak of COVID-19 [6]. Many schools, universities and educational institutions all over the world have stopped their teaching and learning process for many months which highly effected students, educators/teachers as well as the educational institutions themselves and the whole educational process. For other educational institutions, they chose to suspend the final exams until this pandemic crisis subsides as final testing is very important aspect to evaluate students' performance. Otherwise, so many Higher Education Institutions tried to find alternative solutions to adapt to the status quo and reduce the passive effects on the education sector such as using online learning and testing, which became necessary to make students linked with their institutions [7–9]. With all of those decisions made due to the outbreak of COVID-19 pandemic, the authors have been motivated to investigate more about the online testing, its challenges and opportunities from the perceptions of the HEIs' students which is the main focus of the study. Besides, the area of investigation did not attract scholars during this pandemic period which motivated the authors of this study to fill the gap and investigate in the area of online testing.

Online learning can be defined as "learning that takes place partially or entirely over the Internet" [10], and it enables those who live in a distance from campus and in rural areas to join university courses and programs. Online learning systems are web-based software for distributing, tracking, and managing courses over the Internet [11]. They involve the implementation of advancements in technology to direct, design and deliver the learning content, and to facilitate two-way communication among students and their faculties [12, 13]. It is often referred to as "e-learning" among other terms. However, online learning is only one type of "distance learning" which is the umbrella term of any learning taking place over distance and not in a traditional classroom.

With the advancement of computer, multimedia and network technologies, alternatives to traditional classroom learning have been developed. Online testing is one of such alternatives where students can access exams via online systems, especially from remoted places and due to critical situations. The outbreak of COVID-19 shows that online testing might become a must in critical situations, which requires students to carry out their tests online to accomplish their studies [14]. However, online testing might be a real challenge for all including students, teachers and educational institutions in terms of implementing the exams, making tests more reliable and protected to avoid any misconduct or dishonesty from the students' side [15].

Even though COVID-19 has attracted many researchers from different disciplines, online testing, as an essential part of e-learning and education, has not received enough attention from scholars during the outbreak of COVID-19 like other issues, particularly the psychological well-being of students associated with the online testing. Hence, the present research study attempts to investigate the challenges and opportunities of online learning during the outbreak of COVID-19 pandemic. The main objective is to study the attitudes of students towards online testing; therefore, the study aims to answer the following questions:

1. What advantages do the learners realize from online exams?
2. What challenges do the learners face while taking online exams?
3. What are the possible solutions for online testing challenges from the perspective of students?

The focus of the study is centred on the students' perceptions about the challenges and advantages of online testing. In this study, the authors begin with describing the contextual factors that motivated the study. Then, the study objectives and questions are presented, and briefly overview the study's approach to investigate the research questions, with relevant research literature as well as the research methodology. Then, the discussion section is presented, followed by the findings and conclusion. Finally, the authors end with the implications and directions for future studies.

2 Literature Review

2.1 Online Learning

Online learning has become an essential learning and teaching method. The present situation caused by COVID-19 pandemic makes various higher education institutions all over the globe adopt online learning and testing. This makes students more positive towards online learning. A study was conducted to investigate the attitudes of students towards online learning in Australia [16]. The study included 120 university students as participants. Two phases were used in the study, including pre- and post-test, course participation in phase one and interviews in phase two. At the end of the course, the results showed that the attitudes of students towards online learning were positive. Thus, online learning is very helpful for learners of various specializations, and it should be adopted in education. Joshi and Samir Thakkar [17] supported the same idea of students being positive during the online learning. In their study, they stated that E-learning would not be meaningful if the students do not adopt it in their learning process. The data were collected from fifty-four diploma students of information technology under the engineering program. So, it was found that the attitude of students toward e-learning was positive and not affected by differences in gender, locality or social category of students. The same students' positive attitude towards online learning was found in different previous studies [16, 18, 19].

Moreover, some scholars discussed the awareness and advantages of online education. Kumar [20] examined and discussed the awareness, benefits and challenges of e-learning in Kurukshetra University, India. The findings showed that the awareness about e-learning among university students was good while their knowledge on the e-learning was poor as the main source of their knowledge is the internet. Students thought that e-learning enable them to acquire new ideas and provide them an alternative learning environment. Results showed that lack of support from e-learning, students' non-familiarity with English Language, infrastructure problem, students' low knowledge of computer skills, and lack of funding for research and encouragement are part of barrier elements of E-learning in KUK. This argument was supported by K. Mukhtar and others [21], who explored the perception of teachers and students regarding advantages, limitations and recommendations of online learning during COVID-19 in Pakistan. According to the teachers and students, online learning is flexible and effective source of teaching and learning. The results of the study showed that the advantages of online learning included remote learning, comfort, and accessibility. Authors concluded that online learning modalities encourage student-centered learning, and they are easily manageable during this lockdown situation [22–25].

Furthermore, the role of e-learning during the outbreak of COVID-19 crisis played a great role during the present crisis. Due to ongoing of COVID-19, there was a significant increase in the number of students who are using the platform and apps of ED-TECH [4, 26]. So, it is suggested that using e-learning is all about learning and taking a step that helps in moving toward higher education level. Learners need to be familiar with technology devices and their use to enhance their education and knowledge. The teachers should use e-learning tools and encourage students to download number of educational Apps to make the journey towards learning pleasing and interesting.

On the other hand, COVID-19 pandemic impacted the education. Jadhav et al. [27] stated in their study that the education issue has become very big and similarly in the future. Exams in all schools and colleges in Maharashtra (India) have been cancelled, mainly on traditional education. With the help of a few questions to know what the students think about online education during the lockdown period and its consequences, hundred undergraduate and post graduate students were the sample of the study. The study sought to know what students think about the impact of COVID-19 on education. The results showed that 90 percent of students stated that lockdown has had a huge impact on education. Additionally, most of the students stated that exams should be done online. In the same regards, Sahu [28] highlighted the potential impact of the terrible COVID-19 outbreak on education and mental health of students and academic staff. The author casted the light on the challenges that universities and learners face due to COVID-19. The results reflected that the greatest challenge was the transition from the traditional method of teaching to the online mode as both teachers and learners should work from their homes. However, it was difficult for them to do so, particularly students of art and music departments, who needed to study practically. Further, many universities did not have enough infrastructure or resources to facilitate online teaching with immediate effort. Additionally, there were students who did not have access to the technological devices and/or internet at their homes [29]. Despite all of those challenges, and even though COVID-19 has impacted education in most of the world countries, online learning became the alternative method of learning, which mitigated the disruption of education and made students linked with their institutions.

Some scholars discussed the challenges of online learning during COVID-19 pandemic. According to Al-Madhagy [30], during COVID-19 pandemic outbreak, education faced many obstacles to continue the education process. The sudden closure created a chaotic situation not only for educational institutions but also for students. Additionally, educators faced challenges and parents had much new burdens upon their shoulders more than ever. Teachers had to deliver the lectures online and the problem resided in many institutions and teachers as well. However, education entities have to update their policies, procedures and practices to cope up with the emergent changes in life. The quality of education should be maintained and the emerging technologies should be updated. Thus, through Taufiq's study it was clear that, there is a need to focus more on challenges that online learning facing to cope up with the 21st century educational systems. However, in this regard, [22], in his study about COVID-19 and online teaching in higher education, focused on a case of Peking University's online education, and suggested five high-impact principles for online education: first, high relevance between online instructional design and student learning; second, effective delivery on online instructional information; third, adequate support provided by faculty and teaching assistants to students; fourth, high-quality participation to improve the breadth and depth of student's learning; and finally, contingency plan to deal with unexpected incidents of online education platforms.

2.2 Online Testing in Higher Education Institutions During the Outbreak of COVID-19

During the outbreak of COVID-19, there are many things have affected the people and increased their fears from this novel virus and forced them to stay at home and leave everything behind them including education. One of those things was the social media posts. Arpaci et al. [31] analyzed in their study about 43 million tweets collected from social media to know the effect of social media posts of public. The results suggested that social media posts may affect human psychology and behavior and increase his/her fears of death due to being sick with COVID-19 disease. The results of such study may help governments, health organizations and ministry of education to better understand the psychology of the public, and thereby, taking necessary decisions to communicate with people to prevent and manage the panic as well as to help the educational process to be continued. Thus, one of those decisions was to shift to the online learning and online testing.

Online testing, during the pandemic of COVID-19, has become a solution to help the continuation of educational process and save the teachers and learners from being affected with the disease. However, as online learning has positive effects, it has also negative effects. Arnold [32], stated that online testing increases the academic dishonesty because it takes place in un-proctored environment [14]. It was argued that students' cheating at formative tests forsake the opportunity to enhance their learning and they may suffer the consequences in subsequent proctored summative tests. The author calculated a score of investigations that have been done in a large school of economics in the Netherlands and found that the likelihood of cheating, based on unexpected grade patterns is negatively related to academic progress. On the other hand, Arnold [33] stated that online test is one of the widespread methods of higher education in the 21st century. It can be valuable to the current technological era if used effectively. It was concluded that even though the online testing might be poorly designed, it is still an effective tool during the outbreak of COVID-19 pandemic.

Therefore, limitations appeared in online testing are not inherent features of online tests, but are a result of poorly conceived design, development, and deployment of online tests. Moreover, some universities worldwide have already suspended the final exams while some other universities made them online. In fact, it is clear that online testing is still in need for further investigation due to the need of HEIs for online testing currently and in the future. That is, online testing is still in need for further investigation in order to highlight its challenges and opportunities, for the benefit of Higher Education Institutions, educators, and students.

3   Methodology

3.1   Research Design

The research design used in the present study is the qualitative design. This design is suitable for the research, in terms of getting sufficient data, since the data were collected through semi-structured interviews. This qualitative data describes the experience of students while attending the external online testing, so the study employed the phenomenological approach. Phenomenological approach is suitable because it focuses on the commonality of a lived experience within a group of participants. The fundamental goal of this approach is to arrive at a description of the nature of the particular phenomenon [34]. In this study, the experience is attending external online testing. The phenomenological approach is used because the purpose is to study the attitudes of participants towards online testing, as well as the challenges they faced and opportunities they got.

The researchers used semi-structured interviews method as it is suitable for the research design, since it helps in getting an in-depth investigation for the experience of the participants in the online testing. "Semi-structured interviews may be conducted in various modes: face-to-face, by telephone, videophone" [35]. So, the researchers use videophone and telephone to make the interview because of the lockdown due to COVID-19, and the existence of the participants in remote places.

3.2 Sample

Semi-structured interviews were conducted with six international students from various universities in state of Maharashtra, India. The researchers followed the purposive sampling method to select six undergraduate and post-graduate participants from various universities to be as the study participants. The selected participants' information is shown as in Table 1.

3.3 Data Analysis

The researchers analysed the data from the interpretation of recordings. Before starting the interviews, the researchers have taken permission from the participants to record the interviews. The participants also have been informed that the data will be used only for research purposes. The collected data were sent back to interviewees for member-checking, then the interviews were analyzed into codes, which were developed later into themes [34]. Hence, the qualitative data were analyzed in the form of thematic patterns to have a clear picture concerning attitudes of students

Table 1 Participants' profile

No.	Level of study	Major
Participant 1	Postgraduate	Accounting
Participant 2	Postgraduate	Business Administration
Participant 3	Postgraduate	Computer Science
Participant 4	Undergraduate	Business Administration
Participant 5	Undergraduate	Business Administration
Participant 6	Undergraduate	Business Management

toward online testing, and challenges they face. However, to achieve the reliability, one of the authors did the analysis and the others revised it in order to reach an agreement concerning the final themes taken from the interviews.

3.4 Findings

The findings of the study showed that the attitudes of students towards online testing can be divided into four categories: (a) psychology, (b) ICT skills, (c) advantages of online testing to traditional testing, and (d) challenges of online testing.

3.5 Psychological Effects of Online Testing During COVID-19

This category is divided into three themes. The first theme is that students' self-confidence. Some participants were self-confident while doing the online testing while others were not. They stated that they were relaxed because the university has provided them with a summary for each subject a month before the exam and thirty questions from the summary are to be included in the exam. Participant 1 stated:

> **P1**: "In fact I feel comfortable while doing the online testing because the university provided us with one hundred twenty questions for every subject a month before the exam. I answered those questions properly that helped me a lot to be self-confident. Thirty questions were included in the exam and all were MCQ, so when I read the question and choices. I could recognize the correct answer"

On the other hand, some participants were very anxious and unconfident. Attending online testing for the first time played a major role in making students worried about various things such as the style of the questions and time. Such anxiety affected the psyche of students especially in the first exam. Regarding this anxiety, participant 2 stated:

> **P2**: "When they inform me that the exam will be conducted online, I was thinking of how to deal with it, how the questions will be and how much time we will get. However, the first exam was difficult because I don't know the style of the questions."

The second theme under this category is motivation. Some participants had some worries before and while the online testing, but they tried to motivate themselves in order to pass the exam. Motivation was considered to be one of the most important factors to overcome difficulties and anxiety while doing the online testing. In this regards, participant 4 stated:

P4: "Even though I was restless, I tried my best to motivate myself. Thinking of graduation motivates me as I am in the last semester and I should score good marks. I took it as a challenge of a self to overcome all worries."

The third theme is time pressure. Time in online testing was different from the traditional testing. In online testing, students have to give more focus and divide time according to the questions. Some participants considered the time as an obstacle that affected their achievement. Regarding the time pressure, participant 2 stated:

P2: "There were thirty questions and thirty minutes. In fact, time was not that much enough. There is no time to review my answers, so that makes me confused and worried about my results".

3.6 Effects of ICT Skills on Online Testing

This section is about the students' skills in using ICT tools during the online testing. It is essential that students of all specializations should have knowledge in using computer. All participants of the present study stated that it was easy for them to use computer devices in online testing. The analysis showed that students did not face any problem in using computer during the online testing. In this regard, participant 6 stated:

P6: "Actually, I'm not perfect in using computer. I can use it for basic things like Word processing, Google search, and YouTube, etc. So, I could deal with computer while the online testing. The task was easy because the university sent us a link that we can log in into the exam directly.

3.7 Advantages of Online Testing to Traditional Testing

In this section the researchers discuss some advantages of online testing. Participants of the study argued that there were advantages for online testing. One of those advantages as the participant 1 stated is that the discussion of the final viva. Throughout the analysis it was shown that doing the final viva online is easier than presenting at university. Participant 1 stated:

P1: "I discussed my project with my supervisor via zoom programme. We were five students and we discussed one by one. In fact, it was easier because standing in front of students, teachers and professors in the classroom is fearful."

Another advantage the participants stated was that the style of questions is easier. The analysis showed that students prefer MCQ style because it is easy to identify the correct answer and also faster than typing. In this regard, participant 3 said:

P3: "I prefer online testing to traditional because it is in MCQ style that depends on understanding the subject more than memorizing. Moreover, if the online testing in WHQ style, we will need too much time for typing the answer and that is a difficult task."

However, some participants preferred the traditional testing to online testing. They argued that in traditional testing they could convey the answer much better, but with the present situation of COVID-19 pandemic, online testing is better. The analysis showed the satisfaction of participants in using online testing. Participant 4 mentioned that instead of wasting time, the choice of online testing is a very good solution which enables students to continue their study in a safer way and place. Participant 4 stated:

P4: "Because of the present situation in which Corona Virus is spreading in such a way, I'm satisfied with the online testing. It is a good way to continue our study rather than wasting time till unknown date. On the other hand, if the situation gets better and I have to options either to choose online testing or traditional, I will of course go for traditional".

3.8   Challenges of Online Testing

This section is about the challenges that the students faced during the online testing. The participants argued that there were some challenges and obstacles related to internet connection. Participant 4 argued that he had a backlog in one subject and the test was orally. They used zoom program to conduct the exam where the connection was really poor, so after many failed attempts, the teacher used a WhatsApp video call as an alternative for zoom. Additionally, participant 2 said that if they lost the internet connection, they would have to restart the test again with new questions. Participant 2 commented:

P2: "In the first exam, it was raining so the connection got poor. I lost the connection when I was in question number eight. When I did refresh the page, I got shocked because I have to restart again from the first question".

In the same regard, participant 5 noted that the students who were not studying computer science faced challenges because they used to use mobile phones to search the information related to their studies. He stated:

P5: "Of course as a student of B.B.A., I use the computer about once a month only. The major use is to watch movies in YouTube. So, when the university sent us the link of online testing, I couldn't log in into the exam page then I asked my roommate to help me".

To sum up, the analysis of the qualitative data showed different themes, including anxiety, worry and low-self-confidence, for which the students used to motivate themselves to overcome such psychological impacts of online testing during COVID-19 outbreak. Besides, ICT skills are not considered a major problem for the students; instead, online testing was easier for the students, especially that most of the questions are multiple choice questions, which made students prefer online testing to traditional testing. However, internet connection and other matters related to online testing tool are still a challenge that might have a negative influence on online testing among students.

4 Discussion

This study aimed to investigate the attitude of the students towards online testing during the outbreak of COVID-19. The findings showed four categories of themes, namely psychology, ICT skills, advantages of online testing to traditional testing, and challenges of online testing. In terms of psychological effects, students experienced anxiety during online testing, because it was their first experience with the online testing mode. Novelty effect has been reported in previous studies as a factor that increases anxiety [36–38]. Besides, time pressure was a major challenge for students, which was attributed to the importance of online tests for the students' academic achievement and grades, which might also has a relationship with self-confidence and learning anxiety [38]. That is, the students could overcome online testing anxiety and improve their self-confidence through motivating themselves to overcome such learning obstacle, which is in line with previous literature [39–41].

Furthermore, the attitudes of students towards online testing as well as online learning was positive, which supports the findings of [16, 17]. These studies showed that online learning is rapidly used, which makes the students have positive attitudes towards the use of technology in HEIs. This might be explained by that students were obliged to take online tests, and after their previous experience, they found it easier and more effective. The students commented that the questions in online tests were clear and easy to deal with, and this increased their positive attitude towards online testing.

In addition, online testing was not difficult for the participants students, since they have the ICT skills to deal with online tools, needed for informal and formal online learning, which supports previous studies [1, 42]. This can be noticed in the comments of the students who stated that online learning and testing do not require professional skills, if it is arranged well by institutions. Hence, it is clear that ICT skills were a challenge for students in past, but the rapid advancement of technology and its influence on education led students to improve such skills required for education in the 21st century.

In addition, COVID-19 pandemic affected education in all over the world so online testing has become the alternative and must solution. Participants in this study also faced online testing challenges due to the transmission from traditional learning into online learning, and due to poor internet connection. These findings are supported by Tamim [43], that is because IT infrastructure of online learning and unprofessional arrangement for online testing are important for students' performance. Academic achievement is valued by students according to Abdul-Ghafour and Alrefaee [44]; however, lack of internet connections and weak online learning systems might increase the students' concern concerning their academic achievements. Therefore, the sudden change to online testing due to the outbreak of COVID-19 shows that infrastructure requirement is still a challenge for students [30].

Nevertheless, the outbreak of COVID-19 and the mandatory online testing shows that students prefer this type of learning to traditional learning. This attitude among students is due the advantages of online testing since it saves time and also enables students to do their exams on time. Hence, online testing encourages students-centred learning, especially that is easily manageable during the lockdown [21]. Further, Arnold [32] argued that online testing negatively affected higher education because it increases dishonesty, especially that online testing takes place in un-proctored environment; however, Boitshwarelo et al. [33] stated that online testing can be valuable in critical situations if used effectively. The outbreak of COVID-19 made online testing a main trend worldwide, and a solution rather than an alternative practice, and students found it a practical, effective and easier method.

Finally, even though students face some challenges in online testing due to poor internet connection and lack of up-to-date tools that help them to carry out online tests perfectly, the attitude of the students shows that positive aspects of online testing outweigh its challenges. Accordingly, it is a good alternative method as it helps in continuing the educational process during lockdowns and critical situations such as the outbreak of COVID-19 pandemic.

5 Conclusion

Online testing became the alternative method during the outbreak of COVID-19 pandemic. Institutions in most of the countries all over the world arranged online testing for students in order to continue the educational process. Teachers and learners have been excited by the transmission from traditional learning into online mode. Universities started preparing for online learning as well as testing; however, there is a need for improving online learning and assessment in HIEs to meet the challenges of learning in digital era. Besides, there is a need to motivate students regarding online testing to take online learning seriously, especially students from least developed countries. This motivation could enable students to overcome the challenges and obstacles they face so that they could get more self-confidence in dealing with online testing, because it was found to be a fast and efficient solution during the outbreak of COVID-19 pandemic.

6 Implications

Based on the findings of the present study, it is recommended that online testing should have more focus on proctored environment. Institutions should encourage students and inform them in advance before the exam to be ready and make sure that their computers are connected to the internet. In term of time pressure, it is recommended that educational institutions should give five minutes extra as an exception to enable the students to revise their answers. Thus, online testing becomes an alternative way to traditional testing, it should be improved by governments as well as institutions.

Moreover, lecturers and teachers should be adequate for the online testing by improving their skill in how to deal with online testing, because some of them might not have experience in designing online exams. Besides, HEIs and lectures need to take internet disruption into consideration, because some students might not submit their online tests due to electricity cut off or internet sudden disconnection. One more important point is that some students might not have the required tools to carry out online tests such as laptops or suitable hand tools. Therefore, the financial situation of students, especially in poor countries, need to be considered.

7 Directions for Future Research

Online testing during the outbreak of COVID-19 was carried out using simple online testing tools, which were dependent on lectures through the university systems. Accordingly, there is a need to investigate online testing that has been carried out through smart systems, which might make exams more reliable. That is, there is a need to investigate how online exams can be carried out without students' cheating or manipulating during the process of exams. Besides, future studies might explore how technology can be used to improve the security and reliability of online exams instead of depending on lecturers' use of traditional ways of monitoring such as using videos to monitor the process of online testing.

References

1. Gudmundsdottir, G.B., Gassó, H.H., Rubio, J.C.C., Hatlevik, O.E.: Student teachers' responsible use of ICT: examining two samples in Spain and Norway. Comput. Educ. 103877 (2020)
2. Basilaia, G., Kvavadze, D.: Transition to online education in schools during a SARS-CoV-2 coronavirus (COVID-19) pandemic in Georgia. Pedag. Res. 5, 1–9 (2020)
3. Lestari, P.A.S., Gunawan, G.: The impact of Covid-19 pandemic on learning implementation of primary and secondary school levels. Ind. J. Elem. Child. Educ. 1, 58–63 (2020)
4. Zhong, B.-L., Luo, W., Li, H.-M., Zhang, Q.-Q., Liu, X.-G., Li, W.-T., et al.: Knowledge, attitudes, and practices towards COVID-19 among Chinese residents during the rapid rise period of the COVID-19 outbreak: a quick online cross-sectional survey. Int. J. Biol. Sci. 16, 1745 (2020)
5. Shah, K., Abdeljawad, T., Mahariq, I., Jarad, F.: Qualitative analysis of a mathematical model in the time of COVID-19. BioMed Res. Int. 2020 (2020)
6. Arpaci, I., Karataş, K., Baloğlu, K.: The development and initial tests for the psychometric properties of the COVID-19 Phobia Scale (C19P-S). Personal. Individ. Diff. 110108 (2020)

7. Sun, L., Tang, Y., Zuo, W.: Coronavirus pushes education online. Nat. Mater. **19**, 687 (2020)
8. Arpaci, I.: A hybrid modeling approach for predicting the educational use of mobile cloud computing services in higher education. Comput. Hum. Behav. **90**, 181–187 (2019)
9. Arpaci, I., Al-Emran, M., Al-Sharafi, M.A., Shaalan, K.: A novel approach for predicting the adoption of smartwatches using machine learning algorithms. In: Recent advances in intelligent systems and smart applications, pp. 185–195. Springer (2020)
10. Means, B., Toyama, Y., Murphy, R., Bakia, M., Jones, K.: Evaluation of evidence-based practices in online learning: a meta-analysis and review of online learning studies (2009). https://repository.alt.ac.uk/629/1/US_DepEdu_Final_report_2009.pdf
11. Keis, O., Grab, C., Schneider, A., Öchsner, W.: Online or face-to-face instruction? A qualitative study on the electrocardiogram course at the University of Ulm to examine why students choose a particular format. BMC Med. Educ. **17**, 194 (2017)
12. Thanji, M., Vasantha, S.: ICT factors influencing consumer adoption of ecommerce offerings for education. Indian J. Sci. Technol. **9**, 1–6 (2016)
13. Arpaci, I.: What drives students' online self-disclosure behaviour on social media? A hybrid SEM and artificial intelligence approach. Int. J. Mob. Commun. **18**, 229–241 (2020)
14. Khan, R.A., Jawaid, M.: Technology enhanced assessment (TEA) in COVID 19 pandemic. Pak. J. Med. Sci. **36**, S108 (2020)
15. Clark, T.M., Callam, C.S., Paul, N.M., Stoltzfus, M.W., Turner, D.: Testing in the time of COVID-19: a sudden transition to unproctored online exams. J. Chem. Educ. (2020)
16. Zhu, Y., Au, W., Yates, G.C.: University students' attitudes toward online learning in a blended course. Aust. Assoc. Res. Educ. (2013)
17. Thakkar, S.R., Joshi, H.D.: Impact of technology availability and self-efficacy on e-learning usage. Int. J. Res. Appl. Sci. Eng. Technol. (IJRASET) **6**(4), 2956–2960 (2018)
18. Herrador-Alcaide, T.C., Hernández-Solís, M., Hontoria, J.F.: Online learning tools in the era of m-learning: utility and attitudes in accounting college students. Sustainability **12**, 5171 (2020)
19. Garip, G., Seneviratne, S.R., Iacovou, S.: Learners' perceptions and experiences of studying psychology online. J. Comput. Educ. 1–21 (2020)
20. Kumar, S.C.: Awareness, benefits and challenges of e-learning among the students of Kurukshetra University Kurukshetra: a study. Int. J. Inf. Dissem. Technol. **8**, 227–230 (2019)
21. Mukhtar, K., Javed, K., Arooj, M., Sethi, A.: Advantages, limitations and recommendations for online learning during COVID-19 pandemic era. Pak. J. Med. Sci. **36** (2020)
22. Bao, W.: COVID-19 and online teaching in higher education: a case study of Peking University. Hum. Behav. Emerg. Technol. **2**, 113–115 (2020)
23. Demuyakor, J.: Coronavirus (COVID-19) and online learning in higher institutions of education: a survey of the perceptions of Ghanaian international students in China. Online J. Commun. Media Technol. **10**, e202018 (2020)
24. Toquero, C.: Challenges and opportunities for higher education amid the COVID-19 pandemic: the Philippine context. Pedagog. Res. **5** (2020)
25. Yassin, A.A., Razak, N.A., Maasum, N.R.M.: Investigating the need for computer assisted cooperative learning to improve reading skills among Yemeni university EFL students: a needs analysis study. Int. J. Virtual Pers. Learn. Environ. (IJVPLE) **9**, 15–31 (2019)
26. Sandars, J., Correia, R., Dankbaar, M., de Jong, P., Goh, P.S., Hege, I., et al.: Twelve tips for rapidly migrating to online learning during the COVID-19 pandemic, vol. 9. MedEdPublish (2020)
27. Jadhav, V.R., Bagul, T.D., Aswale, S.R.: COVID-19 era: students' role to look at problems in education system during lockdown issues in Maharashtra, India
28. Sahu, P.: Closure of universities due to Coronavirus Disease, (COVID-19): impact on education and mental health of students and academic staff. Cureus **12**, 2020 (2019)
29. Meng, L., Hua, F., Bian, Z.: Coronavirus disease 2019 (COVID-19): emerging and future challenges for dental and oral medicine. J. Dental Res. **99**, 481–487 (2020)
30. Al-Madhagy, R.T.H.G.: Educational learning and teaching methods' challenges during COVID-19 outbreak and a sudden transformation towards totally digitizing education

31. Arpaci, I., Alshehabi, S., Al-Emran, M., Khasawneh, M., Mahariq, I., Abdeljawad, T., et al.: Analysis of twitter data using evolutionary clustering during the COVID-19 pandemic. CMC-Comput. Mater. Contin. **65**, 193–204 (2020)
32. Arnold, I.J.: Cheating at online formative tests: does it pay off? Internet Higher Educ. **29**, 98–106 (2016)
33. Boitshwarelo, B., Reedy, A.K., Billany, T.: Envisioning the use of online tests in assessing twenty-first century learning: a literature review. Res. Pract. Technol. Enhan. Learn. **12**, 16 (2017)
34. Creswell, J.W., Poth, C.N.: Qualitative inquiry and research design: choosing among five approaches. Sage publications (2016)
35. Woods, M.: Interviewing for research and analysing qualitative data: an overview, pp. 67–80. Massey University (2011)
36. Yassin, A.A., Razak, N.A.: Investigating the relationship between foreign language anxiety in the four skills and year of study among Yemeni University EFL learners. 3L Lang. Linguist., Lit.® **23** (2017)
37. Jeno, L.M., Vandvik, V., Eliassen, S., Grytnes, J.-A.: Testing the novelty effect of an m-learning tool on internalization and achievement: a self-determination theory approach. Comput. Educ. **128**, 398–413 (2019)
38. Tsay, C.H.H., Kofinas, A.K., Trivedi, S.K., Yang, Y.: Overcoming the novelty effect in online gamified learning systems: an empirical evaluation of student engagement and performance. J. Comput. Assist. Learn. **36**, 128–146 (2020)
39. Dunn, K.: Why wait? The influence of academic self-regulation, intrinsic motivation, and statistics anxiety on procrastination in online statistics. Innov. Higher Educ. **39**, 33–44 (2014)
40. Razak, N.A., Yassin, A.A., Moqbel, M.S.S.: Investigating foreign language reading anxiety among Yemeni international students in Malaysian Universities. Int. J. Engl. Linguist **9** (2019)
41. Mukhlis, H., Triaristina, A., Wahyudi, D.A., Putri, R.H.: Anxiety confronts practice exam reviewed from optimism, emotional intelligence, and social support on student of STIKES. Talent Dev. Excell. **12** (2020)
42. Razak, N.A., Yassin, A.A., Maasum, T.N.R.T.M.: Formalizing informal CALL in learning english language skills. In: Enhancements and limitations to ICT-based informal language learning: emerging research and opportunities, pp. 161–182. IGI Global (2020)
43. Tamim, S.R.: Analyzing the complexities of online education systems: a systems thinking perspective, pp. 1–11. TechTrends (2020)
44. Abdul-Ghafour, A.-Q.K.M., Alrefaee, Y.: The relationship between language learning strategies and achievement among EFL University students. Appl. Linguist. Res. J. **3**, 64–83

ICT-Based Distance Higher Education: A Necessity During the Era of COVID-19 Outbreak

Murat Tahir Çaldağ, Ebru Gökalp, and Nurcan Alkış

Abstract The COVID-19 outbreak, declared as a public health emergency of international concern in March 2020, has forced governments across the world to take action for satisfying social isolation requirements of the pandemic to decrease the spread of coronavirus. The first action higher institutions made is closing universities, suspending face-to-face classes, and switching to distance education. ICT-based distance education for higher institutions seems suitable to alleviate the COVID-19 outbreak since it does not require face-to-face communication; however, it should be planned carefully to get benefit from its advantages. One of the most challenging issues of distance higher education is choosing the appropriate technology. In this scope, this chapter aims to evaluate ICT-based distance higher education technologies, grouped under five categories: learning management systems, massive open online course platforms, platforms for video conferencing, social media platforms, and digital learning content tools. The evaluation of these technologies showed that although different platforms are used for different purposes, there is a lack of an integrated platform that supports all functionalities from a holistic perspective for improving the efficiency of distance higher education offered and supporting shifting to student-centered distance education model. To this aim, the necessities for the future are also described in this study.

Keywords Distance education · Emerging technologies · COVID-19 · Evaluation

M. T. Çaldağ (✉) · E. Gökalp · N. Alkış
Department of Technology and Knowledge Management, Baskent University,
Ankara, Turkey
e-mail: mtcaldag@baskent.edu.tr

N. Alkış
e-mail: nalkis@baskent.edu.tr

E. Gökalp
Institute for Manufacturing Cambridge, University of Cambridge, Cambridge,
United Kingdom
e-mail: eg590@cam.ac.uk

I. Arpaci et al. (eds.), *Emerging Technologies During the Era of COVID-19
Pandemic*, Studies in Systems, Decision and Control 348,
https://doi.org/10.1007/978-3-030-67716-9_23

1 Introduction

The COVID-19 outbreak, firstly reported in December 2019, in Wuhan, Hubei, China, has spread to the world in a short amount of period. It was declared as a pandemic by the World Health Organization (WHO) in March 2020. As of August 11, 2020, confirmed COVID-19 cases amount is up to 20 million as well as the global death toll is over 735,000 [1].

This kind of pandemic crisis can cause a problem of an overwhelming healthcare system if the number of cases increases sharply in a short amount of time. In order to reduce the spread rate of the coronavirus, countries have to apply some restrictions, as quarantine and lockdown, which can have a significant challenge on the education system. Since one of the crowded places, which causes spreading the virus so quickly, is the classroom. Thus, the first decision governments made after learning the first cases in their country is closing schools, suspending face-to-face classes, and switching to distance education. Students living in 186 countries have been affected by this pandemic crisis.

Distance education can be defined as *"education that takes place when a teacher and student(s) are separated by physical distance, and technology is used to bridge the instructional gap"* [2]. Distance education is an effective way of tackling the COVID-19 outbreak since the teachers and learners/students do not have to come together physically at the same place, they can stay at their home being isolated from other people while involving a distance education program. Although the distance education can provide reducing the spread rate of the COVID-19 and it is inevitable to use it to alleviate the COVID-19, determination of the best technology to be used in the distance education brings a challenging issue for the education providers [3].

Countries have taken different actions for their schools and education programs. Most governments temporarily closed all of their educational institutions, while some governments closed their schools locally [4]. Some countries reopened their schools; for example, the children up to 11 years old have returned their school after a break in Denmark. Although distance education may result in some problems for educating younger students, especially for children up to 11, it provides the highest degree of effectiveness in higher institutions. Thus, higher education institutions across the world have changed their school-based education programs to distance education programs for the continuation of education to meet COVID-19 social-isolation requirements. However, the sociological and technological readiness level of the universities for this urgent transition has not been evaluated.

Distance education programs totally depend on the use of technology. The utilization of emerging information and communication technologies (ICT), including different kinds of ICT applications, has the potential of providing increased efficiency and the satisfaction level of the student and teacher for distance education activities [5, 6]. As a result of the evolution of ICT, the technology used for distance education has changed a lot from radio and television to computers and interactive applications [7]. These technological improvements positively affected the acceptance and effectiveness of distance education. However, which technology is the best for distance education is still a debate. In this scope, this study aims to analyze ICT tools and applications used for the purpose of distance education in higher institutions and identify the necessities for more effective applications for distance higher education, especially for urgent transitions from formal education to distance education. To accomplished the aim of the study, ICT based distance education literature and the tools used in distance education have been *reviewed* systematically and synthesized.

The remainder of this chapter has four main sections: first, the result of the literature review of related studies are given, then emerging ICT applications used for distance higher education are described, followed by findings and future directions for distance higher education; lastly, the chapter is concluded with final remarks in the conclusion section.

2 Related Studies

2.1 Distance Education

The developments and improvements in ICT have changed how educational activities are delivered. Traditional education environments have been replaced with electronic environments providing synchronous or asynchronous communication over the internet. The critical characteristic of distance education is removing the necessity of face-to-face communication [3, 8]. However, it should not be thought of as delivering teaching materials at a distance; it also requires the establishment of effective two-way communication [3].

When we look at the history of distance education, it is observed that it has started to be evolved worldwide since the 1980s [2]. Different terms have been used to describe it, such as distance learning, distance teaching, e-learning. Distance education has three generations of evolution in the literature [9]:

- **First-generation distance education**: It covers using only a single technology, which can be tv channels, radio channels, and print-based materials. The drawback of this first-generation distance education is the lack of direct communication between teachers and students.
- **Second-generation distance education**: It includes usage of multiple media applications, including broadcasting and print media. The third person guides communication between parties in a limited manner.
- **Third-generation distance education**: It includes two-way communication between teachers and students via videoconferencing technologies. Bates [10] stated that an efficient learning environment needs to provide interaction between teachers and learners, learners and learners, and learners and learning

materials. Thus, it can be asserted that the third-generation distance education can only satisfy this need. It is needed to utilize ICT tools that provide two-way communication in an effective manner.

Although distance education is not suitable for all face-to-face higher education programs, for example, distance education in the field of dental will be challenging, cause of lack of practice-based learning from a distance. It has many advantages, such as reaching a vast number of students at the same time without having a constraint of class capacity [3], saving time, reducing costs, allowing students to learn anywhere at any time, providing active learner-centered learning instead of teacher-centered one [11].

Another term used with ICT-based distance education is mobile learning (m-learning). M-learning refers to use of mobile devives to access learning content. Researches showed that students are willing to use m-learning and their use is affected from different factors such as perceived ease of use, attitude, perceived behavioral control, and subjective norms [12]. Also, m-learning adoption can be affected from cultural differences [13]. M-learning can be a considered a part of distance education.

2.2   Studies Related to ICT-Based Distance Education

The utilization of ICT tools that support third-generation distance education brings many different concepts in higher education institutions, including e-learning, blended learning, online learning, web-based learning, and distance learning [8, 14]. While e-learning, distance learning, web-based learning, online learning, and distance education concepts sometimes are used synonymously, blended learning refers to the combination of classroom learning and e-learning [15]. Though the differences between online learning, e-learning, and distance learning are a debate, distance education is considered the overall and inclusive term used in higher education; thus, we used this term in this study.

One of the ICT-based distance education tools is the Learning Management System (LMS), which is the base of blended learning as well as e-learning. One of the widely used ICT-based LMS used is the Modular Object-Oriented Dynamic Learning Environment (Moodle) [15], while there are other LMSs, such as Blackboard [16] and various homegrown LMSs. Moodle allows educators to share documents with the students, to give assignments, to conduct exams and quizzes, to communicate with learners over chat and forum [17]. Moodle depends on Social Constructionist Pedagogy, in which students construct their own learning packages from the materials delivered to them [17, 18]. Al-Ajlan and Zedan [17] compared Moodle with different Virtual Learning Environments and identified it as the best platform in terms of functionalities, technical features, and aspects for their university. Another study conducted by Oproiu [19] concluded that the Moodle platform could be used to support students learning and increase the quality of online courses. Flippped classrooms, a strategy of blended learning applications, combines different technologies to improve learning including LMSs, video sharing services, and MOOC [20].

There are many studies related to the evaluation of existing LMSs in the literature. They investigated LMSs based on different criteria. Stewart et al. [21] evaluated LMSs based on systems administration, cost, instructional design, and teaching & learning capabilities. Kim and Lee [22] assessed LMSs based upon instruction management, screen design, technology, interaction, and student evaluation. According to Ozkan et al. [23], it is necessary to provide quality of system, information, and service. One of the latest studies by Arh and Blažič [24], evaluation criteria of LMSs are functional environment, ease of use, course analysis, tutoring & didactics, assessment, and standards supports. Another study conducted by Arpaci [25] examined the effects of perceived usefulness, ease of use and self-efficacy factors on LMS use.

WebCT (Web Course Tools) developed in 1995 are used to deliver courses to students integrating instructional tools, including a glossary, references, quiz module, self-test, and communication tools, including a chat room, e-mail, bulletin board and calendar [26]. Studies showed that students find WebCT as useful to access lecture notes and course information [27]. The studies examining the effectiveness of course websites showed that easy to use websites having communication tools are more successful [28].

Apart from LMSs and WebCT, the ICT tools and applications used for communication in the distance higher education could be classified based on the different characteristics, as synchronous or asynchronous applications. While synchronous tools could be audio graphics, audio conferencing as in a telephone conference, broadcast radio and television, teleconferencing, computer conferencing such as chat and internet telephony [6], asynchronous media includes audio and videotapes CDs, e-mail, computer files transfers, virtual conferences, multimedia products, offline, web-based learning formats [6]. Studies in the literature related to the evaluation of these technologies by UNESCO, World Bank, and OECD [29–31] assess them based on connectivity type, interaction, main functionality, platform, conditions of use, language, target group, subject, format, costs, and offline option.

Also cloud computing services have been used in distance education. They enable students to access their learning materials from anywhere and any device. Integrating these services to educational setting may increase students academic performance and success [32] and universities could develop cloud-based application and encourage use of them [33]. On the other hand cloud computing services could be used to work on same documents in group projects by the students and for collaboration by educators [34].

3 Emerging ICT Applications Used for Distance Education

The utilization of emerging technologies, like data analytics, cloud computing, everything as a service (XaaS), in the field of distance higher education can provide many innovative solutions, such as the development of a more useful learning environment which enables monitoring the learning process [6], as well as collecting and analyzing feedback from learners to improve the provide education service in a more effective manner.

We categorized Emerging ICT applications used in distance higher education into five groups, as LMSs, Massive Open Online Course (MOOC) Platforms, live-video conferencing platforms, social media platforms, and digital learning content tools, as illustrated in Fig. 1. In the scope of this study, we aim to evaluate these technologies to understand the existing situation and define future directions for ICT-based distance higher education.

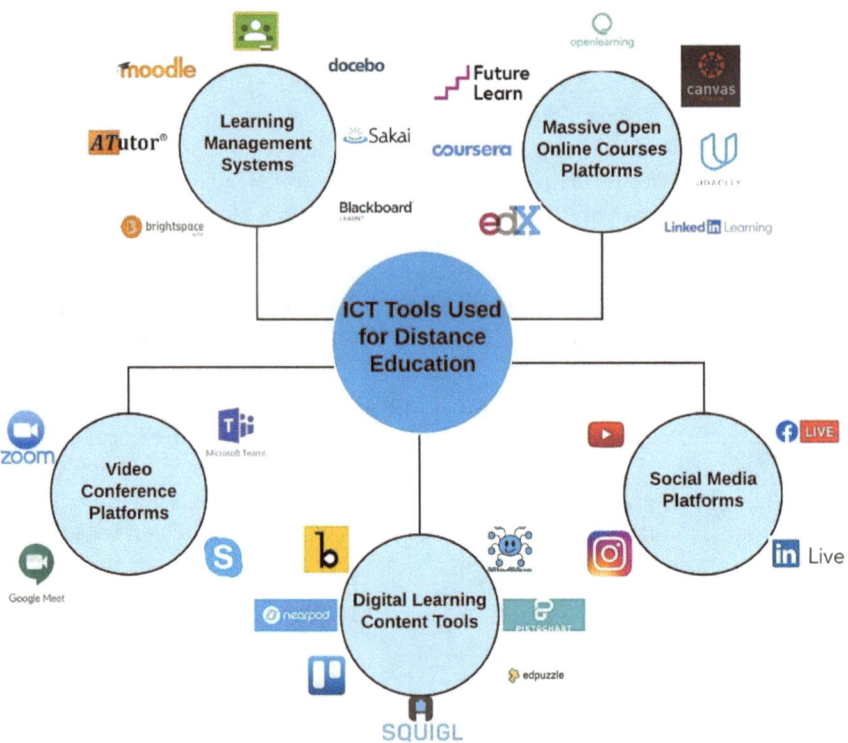

Fig. 1 Emerging ICT applications used in distance education

3.1 Learning Management Systems

LMSs are used to deliver educational course materials, like lecture notes, exams, announcements in a web-based environment. According to Croitoru and Dinu [35], the features and capabilities of LMSs used in higher education are communication, productivity, student involvement, administration, course delivery, curriculum design, hardware/software, and pricing/licensing.

LMS used in higher education can be categorized by considering their purpose, applicability to open source, proprietary, and offering cloud-based solutions [36]. Open-source LMSs provide users to use, change, create and distribute them free of charge. The critical feature of open-source LMSs is the public free license, which offers users to customize their LMS based on their needs and goals [36]. Proprietary LMSs are the platforms that are licensed to their developers or owners on usage and distribution. These systems are required to be installed on the servers and computers of the higher education institution [36]. On the contrary, cloud-based LMSs do not require installation and infrastructure necessities; cloud-based LMSs offer flexibility and a more feasible financial solution by offering it as a service accessible over the internet [36, 37].

Existing LMSs used extensively in higher education can be identified as Moodle, Sakai, ATutor, Google Classroom, Blackboard, D2L Brightspace, and Docebo [38, 39] are evaluated in Table 1. The most widely used LMS is **Moodle**, created by the developer Martin Dougiamas in 1999 and released in 2001 as an open-source platform for educators and learners [40, 41]. Implementation of this open-source software in a web-based platform and customizing it based on the necessities of the higher institution offers interaction between teachers and students on creating dynamic courses accessible from any time and place [41]. Another LMS used in higher education is **Sakai**, developed in 2004 as open-source software to support education with a flexible online-based platform [42]. The LMS focusing on the accessibility of people having disabilities is **ATutor**, which was created in 2002 as an open-source software [43]. **Blackboard**, founded by Matthew Pittinsky and Michael Chasen in 1997 as a web-based proprietary LMS, aims to offer educators to share their course-related materials with students through the web [34, 35]. Eventually, it evolved to a greater scale, which offers services from K-12 to higher education and beyond [44, 46]. **Desire2Learn**, founded by John Baker in 1999, offers services to healthcare, government, corporate, and education sectors [47]. **D2L Brightspace** provides online collaboration tools for educators and students. **Docebo** is a cloud-based LMS with artificial intelligence integration with suggestions and improvements for the users [48]. **Google Classroom** is a cloud-based platform providing free to use service for educators and students for offering communication with collaborative tools [49].

All of the LMSs offer similar features on synchronous & asynchronous communication, customization options, video conference integrations, mobile learning support, and basic user and content tracking tools. The primary difference between open source and proprietary LMSs is the higher institutions' choice of use. Moodle, Sakai, and ATutor systems offer freemium options with a public license that promotes the systems for a more comprehensive and varied user base. Google Classroom offers a different approach as a free to use without a public license. Although there is not any dedicated technical support, the information and problems that are encountered on these systems are answered on community forums or web pages. Blackboard Learn, D2L Brightspace and Docebo offers only premium options with no general use public licenses and dedicated technical support for their customers. Although these systems offer a solution for distance education, the more integration options and detailed user analytics tracking can help educators and students for an improved learning environment.

3.2 Massive Open Online Course Platforms

Massive online open courses (MOOCs) are online courses aimed at providing open, interactive, collaborative learning for massive participation via the web [50–52]. MOOC provide students access to numerous resources at anywhere and anytime [53]. Since innovative solutions providing more accessibility and flexibility can be obtained with MOOC platforms, they have a disruptive effect on the traditional teaching and learning methods. The popular MOOC platforms generally have been spinoffs or a joint effort with universities having an excellent reputation across the world. To illustrate, One of the well-known MOOC platforms of Edx has a joint effort with universities of Harvard and MIT; other examples are Coursera and Udacity, which are collaborated with Stanford University.

There are several challenges for administrating MOOCs, such as accreditation, monetization, recognition, quality, and evaluation mechanisms [54]. Although MOOCs offer a variety of learning subjects, it faces accreditation problems which have a consequence of the lack of evaluation of courses and materials. The most popular MOOCs used for higher education, as Coursera, EdX, Udacity, Canvas Network, OpenLearning, FutureLearn, LinkedinLearning, are evaluated in Table 2. As seen in the table, MOOC having the highest number of participants is Coursera comparing to others. MOOC platforms differentiate on business models rather than the features they provide. Although the primary goal of MOOCs was to offer open access and scalable education for masses when they were firstly launched [55], nowadays, many MOOC providers execute the strategy of offering courses for people who have a premium subscription. While Coursera, EdX, Udacity, and FeatureLearn have hybrid business models with free of charge and paid courses, OpenLearn and LinkedinLearn implement the premium business model with free trial versions. Only Canvas Network offers free access to courses

Table 1 The evaluation of LMSs used in distance higher education

	Moodle	Sakai	ATutor	Google Classroom	Blackboard Learn	D2L Brightspace	Docebo
Business model	Freemium	Freemium	Freemium	Free	Premium	Premium	Premium
Open source	Y	Y	Y	N	N	N	N
License	GPL	ECL	GPL	N/A	N/A	N/A	GPL
Video conference integration	Y	Y	Y	Y	Y	Y	Y
Countries	234	20	N/A	N/A	50+	N/A	10+
Language	139	20	20+	N/A	20+	10+	30+
User tracking analytics	Y	Y	Y	Y	Y	Y	Y
Asynchronous communication	Y	Y	Y	Y	Y	Y	Y
Synchronous communication	Y	Y	Y	Y	Y	Y	Y
Customization option	Y	Y	Y	Y	Y	Y	Y
Mobile learning support	Y	Y	Y	Y	Y	Y	Y
Dedicated technical support	N	N	N	N	Y	Y	Y

Y yes, *N* no, *N/A* not applicable
GPL general public license, *ECL* educational community license

Table 2 The evaluation of MOOCs used in distance higher education

	Coursera	EdX	Udacity	Canvas network	Open learning	Future learn	Linkedin learning
Business model	Freemium	Freemium	Freemium	Free	Premium	Freemium	Premium
Global rank*	358	1.897	5.668	37.183	36.047	4.903	N/A
Total visits (M)*	70.13	21.85	6.47	1.13	1.25	7.8	N/A
Avg. time on platform (min)*	15	8	17	17	20	6	N/A
User tracking analytics	Y	Y	Y	Y	Y	Y	Y
Mobile compatibility	Y	Y	Y	Y	Y	Y	Y
Total No. of learners (mil)	66	24	11.5	–	2.17	10	17

*July 2020 www.similarweb.com Desktop and Mobile Web Statistics

3.3 Platforms for Video Conferencing

Videoconferencing in the context of higher education is defined as synchronous video and audio communication between participants, including educators and students in different geographical places [56, 57]. Because of the COVID-19 pandemic, there has been an emerging shift in digital communication for higher education. Even though videoconferencing technology dates back to 1964, current tools and platforms have not been analyzed in the literature [57]. In our study, we included popular video conference platforms enabling tools, like screen sharing and whiteboard options for higher education. These platforms are Zoom, Google Meets, Microsoft Teams, and Skype.

The COVID-19 outbreak has a massive effect on the usage of Zoom. The peak usage from December 2019 to July 2020 has increased from 10 to 300 million. It offers some limited functionalities for free of up to 100 participants, with a 40 min time limit. The premium option providing unlimited videoconference duration with more features is available for paid subscriptions.

As shown in Table 3, business models of the videoconference platforms have an impact on the capabilities they offered. The full functionality of the tools is locked behind a paywall. The premium options of Zoom, Google Meet, Microsoft Teams, and Skype have almost the same features, except for the lack of whiteboard option in Google Meet. Zoom has the best free to use features to other platforms with limitations on time and participants. Although Microsoft Teams offers no time limit and 250 max participants, the absence of a recording option is a critical factor. From the aspect of the required bandwidth for live video communication, Zoom and Microsoft Teams have the lowest requirements. All free-to-use videoconference tools do not have an analytics tracking option, which is needed for a more efficient evaluation of distance learning.

Table 3 The evaluation of MOOCs used in distance higher education

	Zoom		Google Meet	Microsoft Teams		Skype
	Zoom-Free	Zoom-Premium		Teams-Free	Teams-Premium	
Business Model	Free	Premium	Preemium	Free	Premium	Freemium
Max no of participants	100	500	100–250	250	250	25
Time limit	40 min.	None	None	None	None	None
Document sharing	Y	Y	Y	Y	Y	Y
Whiteboard	Y	Y	N	Y	Y	Y
Share screen	Y	Y	Y	Y	Y	Y
Record option	Y	Y	Y	N	Y	Y
Analytics tracking	N	Y	Y	N	Y	N
Mobile compatibility	Y	Y	Y	Y	Y	Y
Bandwidth requirements*	1.2/1.2	1.2/1.2	3.2/2.6	1.2/1.2	1.2/1.2	1.5/1.5

*720p quality video (upload/download) Mbps

3.4   Social Media Platforms

The usage of social networking platforms, including Facebook, Twitter, YouTube, and blogs, can be used for communication and sharing course materials in the distance higher education [58]. These tools are suitable for improving the capabilities of knowledge sharing and interactivity, since social networking platforms support interaction among learners, especially in different geographical areas [59]. Streaming media platforms offers flexibility, high student engagement, equity of access, the ability for tracking, cost efficiency, and creation of communities [60]. The steep learning curve, low attendance, and communication infrastructure are limitations of these platforms for distance higher education [60]. In this study, we evaluated the most widely used and free of charge social media streaming platforms as Youtube Live, Instagram, Facebook Live, Linkedin Live, as given in Table 4.

All social media platforms for streaming in higher education provide detailed analytics tracking and free to use options. That creates an opportunity for detailed evaluation and improvement of the education offered. The best streaming social media platform is Youtube, with its higher time limit, resolution, and frame per second. Although other platforms have the necessary specifications for streaming lessons, it is absolute that Youtube offers the highest quality among all social media platforms.

3.5   Digital Learning Content Tools

The task of creating digital content for educators used to deliver courses in the traditional teaching environment can be challenging [61]. In order to compensate for the lack of visual and physical connection, different variety of instructional materials can be used in distance education [62]. There are various digital learning

Table 4 The evaluation of social media platforms used in distance higher education

	Youtube	Instagram	Facebook Live	LinkedIn Live
Business model	Free	Free	Free	Free
Live stream	Y	Y	Y	Y
Time limit (h)	12	1	4	4
Resolution	4 K UHD	1080p	720p	1080p
Detailed analytics tracking	Y	Y	Y	Y
Platform	All platforms	All platforms	Smartphones and tablets	All platforms
Frame per second	60	30	30	30

content tools for educators and learners for the purpose of content creation as Piktochart, Squigl, Nearpod, Edpuzzle, Buncee, Mindmup, and Trello [31]. **Piktochart** is a collaborative design platform that offers students and educators to easily create and design visually appealing presentations, infographics, and reports [63]. Another tool used for digital content creation is **Squigl** offering speech and text transformation to animated videos with artificial intelligence [64]. While **Nearpod** is a content creation platform that aims to help educators to create interactive lessons by adding interactive elements, links, and video integrations to existing lesson materials, and it also provides tracking with synchronous features [65]. **Edpuzzle** is a video lesson creation tool with student tracking systems [66]. Another tool, **Buncee** is a creation and presentation tool on visualizing designs [67]. There are also some interactive tools for creating mindmaps online, as **Mindmup,** which offers a design tool for creating mind maps with cloud storage options for improved accessibility [68]. The final tool, **Trello** provides a visual collaboration, organization, and planning tool used for lessons as well as faculty collaborations [69].

As shown in Table 5, all of digital learning content tools offer similar capabilities for the aspect of collaboration, technical support, and mobile compatibility. Some of them, as MindMup, Buncee, Edpuzzle, Trello, Nearpod, offer freemium business models for achieving broader accessibility. They also provide integration with LMSs or other software to increase collaboration and sharing. One of the most critical challenges on distance learning content tools is the limited features on free of charge option. Another challenge can be defined as the storage size.

4 Findings and Future Directions for Distance Higher Education

As a result of analyzing existing ICT tools used in distance higher education, it was observed that although there are different tools used for different purposes, as managing course materials, synchronous videoconferences, asynchronous course delivery for massive people and digital content creation. There is not any integrated platform offering functionalities from a holistic perspective for improving the efficiency of distance higher education offered and supporting shifting from instructor-centered education to student-centered distance education model. In order to satisfy this gap, the necessities for the future are described in this section. We classified these necessities into five main phases of distance higher education; as design and development, administration, delivery, assessment of learning experience, and evaluation of distance education courses.

Table 5 The evaluation of digital learning content tools in distance higher education

	MindMup	Buncee	Edpuzzle	Trello	Nearpod	Squigl	Piktochart
Business model	Freemium	Freemium	Freemium	Freemium	Freemium	Preemium	Preemium
Storage	100 KB–100 MB	N/A	20 videos+	N/A	1–5 GB	10–50 GB	1 GB
Integration with LMS and other software	Y	Y	Y	Y	Y	N	N
Collaboration Tools	Y	Y	Y	Y	Y	Y	Y
Mobile compatibility	Y	Y	Y	Y	Y	Y	Y
Technical support	Y	Y	Y	Y	Y	Y	Y

4.1 Design and Development of Distance Education Courses

One of the most critical keys for offering successful distance higher education is designing and developing course materials before the course begins. As a result of the unexpected COVID-19 pandemic crisis, which forced higher education providers to shift to emergency remote delivery of courses they give, the course materials they had to use were designed for face-to-face education and not appropriate for distance education. That caused challenges for learners and resulted in inefficient course delivery. In order to prevent this problem, the course materials, such as lecture notes, homework, projects, exams for the course should be prepared for distance education. The distance education provider needs to be creative and innovative in the design and structure of the course. A critical issue that should be taken into consideration while designing a distance education course is planning activities encouraging interaction among learners as well as learners and the instructor. Adaptation of successful interactive learning experiences working in the traditional class environment into the distance learning environment could be an approach. Another approach, the visualization of concepts and ideas in lecture notes, could have a significant impact on the efficiency of learning and teaching for the lecture delivered online. Improving visualization of ideas and concepts as a result of the utilization of emergent ICT tools is needed for the future to improve the effectiveness of distance higher education.

4.2 Administration of Distance Education Courses

Delivering course materials from a single channel and administrating this channel is another critical issue for distance education. The requirements of the platform offering distance education can be defined as accessibility from anywhere anytime, reliability, usability, interoperability with other tools, reusability, manageability, capabilities of mixing media and methods, supporting personalized education by considering individual differences, supporting learners' control and involvement, creating group collaboration. While an existing ICT tool, Moodle provides administrating course materials in an effective manner by offering functionalities of content delivery, announcement, e-mail, grade books, forums, chat rooms, exams, and assignment, it does not have features as synchronous videoconferencing, video analytics, prompt messaging via a mobile device, and personalized education. Also, Current Intellectual Property policies and copyright laws should be adapted to overcome problems related to the complexity of protecting ownership rights for the future.

4.3 Delivery of Distance Education Courses

Frequency and quality of interaction have the highest impact on the success of distance higher education. Interaction between the instructor and students, so critical for the distant learner, can be classified as synchronous and asynchronous interaction. While synchronous interaction covers live-stream videoconferencing, online discussion in chat rooms, making the quiz while videoconferencing, asynchronous interaction covers quizzes, exams, homework, projects, lecture notes, posts on forums. During the delivery of distance education courses, communication, and collaboration among students, achieving an active learning experience rather than the passive one, obtaining prompt feedback are vital issues for efficient distance education. Although synchronous interactions are desirable, they could present problems when ensuring equal access for all learners. Time zones differences or overlapping some other responsibilities (e.g., work, other classes, etc.) may pose complications. The most appropriate alternative should be determined for the course while designing it. The ICT tool should satisfy both synchronous and asynchronous video conferencing options with no privacy concern and low bandwidth necessity.

4.4 Assessment of Learning Experience

The process design of the assessment of learning experience is also a vital issue for the successful distance higher education. Alternative instruments for assessing learning experience in an objective manner can be defined as quizzes, exams, assignments, projects, and take-homes. The most appropriate instrument(s) should be selected based on the performance objectives of the course. For example, for a course given through a MOOC platform with asynchronous videos, the assessment instruments could be objective testing or peer-reviewed comments. The selected assessment instrument should satisfy the requirements of transparency, validity, reliability, fairness, and clarity of expectations. The selected assessment instrument should support practicing newly acquired skills, result in suggestive feedback, and provide motivation to succeed for learners.

4.5 Evaluation of Distance Education Courses

We predict that the term of data-driven distance higher education will be used frequently in the future. The utilization of an emergent technology of data analytics [70–72] provides us to measure, collect, analyze, and report data about the progress of the learner as well as the effectiveness and efficiency of the course materials. Tracking learning experience as a result of the utilization of data analytics and visualization of the progress of the learner is a necessity to improve the effectiveness of distance education. Data-based evaluation of the effectiveness of the course materials provides to identify what works and what needs to be improved in the next revised version. Evaluation of courses should be used for continuous improvement.

5 Discussion

This chapter identified the necessities for the direction of future distance education for better implementations. First of all, the findings and the literature showed that distance education courses should be designed and developed before course delivery. COVID-19 outbreak forced rapid transition from formal education to distance education and caused ineffective courses, since courses designed for formal education have been delivered in an improver way, online. Second, the administration of online courses over a single platform is critical issue for distance education. For example, one of widely used LMS Moodle is effective in course delivery including content sharing, announcements, grade book, e-mail tool, forums, chat rooms, quiz module and assignment module, it does not have features like synchronous video-conferencing, video analytics, prompt messaging via a mobile device, and personalized education. Third, the interaction between students and educators, students and students are important for an effective distance education. The distance education technology used should support both synchronous and asynchronous interaction between the participants for an active learning experience, which is required for better learning experience. Fourthly, an important part of distance education is assessment of students' learning activities. Appropriate assessment tools should be selected according to course objectives. Lastly, the evaluation of distance education via data analytics technology will be used for improvement of distance education by measuring, collecting and analyzing data of learner and course materials in the future. As a conclusion a single integrated system serving all needs of distance education is needed for better implementations in the future.

6 Conclusion

COVID-19 outbreak forced many higher education institutions to shift to emergent distance delivery of courses with no face-to-face interactions. The fast switch from formal education to distance education has many challenges, including choosing the most appropriate technology. While it is questioned which technology is best for distance education, with the improvement in ICTs, there are many tools directly or indirectly serving distance education for different purposes like sharing course materials and synchronous video conferencing. These tools categorized under five groups as Learning Management Systems, Massive Open Online Course Platforms, Platforms for Video Conferencing, Social Media Platforms, and Digital Learning Content Tools are evaluated in this chapter. As a result of the evaluation, it is observed that there is a lack of an integrated platform offering functionalities from a holistic perspective for improving the efficiency of distance higher education. In order to satisfy this gap, necessities for the future are defined to improve the effectiveness and user satisfaction of distance higher education in this chapter.

The COVID-19 outbreak forced rapid transition from formal education to distance education and caused ineffective courses since courses designed for formal education have to be delivered in distance education environments. The aim of this study is to prevent problems as delivering ineffective courses and support to take advantage of distance higher education, as saving time, reducing costs, allowing students to learn anywhere at any time, and providing active learner-centered learning.

Contributions of this study are to analyze the current situation by evaluating existing ICT tools used for the higher distance education, and to identify necessities for the future to be able to provide more effective distance education. Educators and higher institutions can benefit from this study for selecting the best technology they need, and ICT providers can also use the findings of this study while developing ICT applications for the higher distance education.

References

1. WHO: WHO coronavirus disease (COVID-19) dashboard (2020)
2. Willis, B.: Distance education: a practical guide. Educational Technology Publications, Englewood Cliffs, New Jersey (1993)
3. Perry, W., Rumble, G.: A short guide to distance education. International Extension College, Cambridge (1987)
4. Education: from disruption to recovery. Available at https://en.unesco.org/covid19/educationresponse
5. Agrawal, A.K. Mittal, G.K.: The role of ICT in higher education for the 21st century: ICT as a change agent for education. In: Multidisciplinary higher education, research, dynamics & concepts: opportunities & challenges for sustainable development, pp. 76–83 (2018)
6. Oyovwe-Tinuoye, G., Adogbeji, B.O.: Information communication technologies (ICT) as an enhancing tool in quality education for transformation of individual and the nation. Int. J. Acad. Res. Bus. Soc. Sci. 3, 21–32 (2013)
7. Casey, D.M.: A journey to legitimacy: the historical development of distance education through technology. TechTrends 52, 45–51 (2008)
8. Moore, J.L., Dickson-Deane, C., Galyen, K.: e-Learning, online learning, and distance learning environments: are they the same? Internet High. Educ. 14, 129–135 (2011)
9. (Tony) Bates, A.W.: Technology, e-Learning and distance education, 2nd edn. EdiRoutledge, New York (2005)
10. (Tony) Bates, A.W.: Technology, open learning and distance education. Routledge, New York, NY (1995)
11. Cowan, J.: The advantages and disadvantages of distance education. In: Howard, R., McGrath, I. (eds.) Distance education for language teachers: a UK perspective, pp. 14–20. Multilingual Matters, Clevedon (1995)

12. Al-Emran, M., Arpaci, I., Salloum, S.A.: An empirical examination of continuous intention to use m-learning: an integrated model. Educ. Inf. Technol. 2899–2918 (2020)
13. Arpaci, I.: A comparative study of the effects of cultural differences on the adoption of mobile learning. Br. J. Educ. Technol. **46**, 699–712 (2015)
14. Kumar, R.: Convergence of ICT and education. World Acad. Sci. Eng. Technol. **40**, 556–559 (2009)
15. Patel, D., Patel, H.I.: Blended learning in higher education using MOODLE open source learning management tool. Int. J. Adv. Res. Comput. Sci. **8**, 439–442 (2017)
16. Andrews, T., Tynan, B.: Distance learners: connected, mobile and resourceful individuals, Australas. J. Educ. Technol. **28** (2012)
17. Al-Ajlan, A., Zedan, H.: Why moodle. In: Proceedings of the IEEE computer society workshop on future trends of distributed computing systems, pp. 58–64 (2008)
18. Aranda, A.D.: Moodle for distance education. Distance Learn. **8**, 25–28 (2011)
19. Oproiu, G.C.: A Study about using e-learning platform (Moodle) in University teaching process. In: Proc. Soc. Behav. Sci. 426–432 (2015)
20. Arpaci, I., Basol, G.: The impact of preservice teachers' cognitive and technological perceptions on their continuous intention to use flipped classroom. Educ. Inf. Technol. **25**, 3503–3514 (2020)
21. Stewart, B., Briton, D., Gismondi, M., Heller, B., Kennepohl, D., McGreal, R., et al.: Choosing moodle: an evaluation of learning management systems at Athabasca, in methods and applications for advancing distance education technologies: international issues and solutions, pp. 167–173. IGI Global (2009)
22. Kim, S.W., Lee, M.G.: Validation of an evaluation model for learning management systems: original article. J. Comput. Assist. Learn. **24**, 284–294 (2008)
23. Ozkan, S., Koseler, R., Baykal, N.: Evaluating learning management systems: hexagonal e-learning assessment model (HELAM). Proc. Eur. Mediterr. Conf. Inf. Syst. EMCIS **2008** (3), 111–130 (2008)
24. Arh, T., Blažič, B.J.: Application of multi-attribute decision making approach to learning management systems evaluation. J. Comput. **2**, 28–37 (2007)
25. Arpaci, I.: The role of self-efficacy in predicting use of distance education tools and learning management systems. Turkish Online J. Distance Educ. **18**, 52–62 (2017)
26. Burgess, L.A.: WebCT as an e-learning tool: a study of technology students' perceptions. J. Technol. Educ. **15** (2003)
27. Devi, P.: An ICT-based distance education model an evaluation of ICT-based modes at the university of the South Pacific. Victoria University of Wellington (2006)
28. Elicker, J.D., O'malley, A.L., Williams, C.M.: Does an interactive WebCT site help students learn? Teach. Psychol. **35**, 126–131 (2008)
29. Reimers, F., Schleicher, A., Saavedra, J., Tuominen, S.: Supporting the continuation of teaching and learning during the COVID-19 Pandemic, pp. 1–38 (2020)
30. World Bank: Remote learning, distance education and online learning during the COVID19 pandemic, remote learning, distance education and online learning during the COVID19 pandemic (2020)
31. Distance Learning Solutions. Available at https://en.unesco.org/covid19/educationresponse/solutions
32. Arpaci, I.: Antecedents and consequences of cloud computing adoption in education to achieve knowledge management. Comput. Hum. Behav. **70**, 382–390 (2017)
33. Arpaci, I.: A hybrid modeling approach for predicting the educational use of mobile cloud computing services in higher education. Comput. Hum. Behav. **90**, 181–187 (2019)
34. Arpaci, I.: Understanding and predicting students' intention to use mobile cloud storage services. Comput. Hum. Behav. **58**, 150–157 (2016)
35. Croitoru, M., Dinu, C.-N.: A critical analysis of learning management systems in higher education. Econ. Inf. **16**, 5–18 (2016)
36. Dobre, I.: Learning management systems for higher education—an overview of available options for higher education organizations. Proc. Soc. Behav. Sci. **180**, 313–320 (2015)

37. Faisal, H., Ubaidullah, M., Alammari, A.: Overview of cloud-based learning management system. Int. J. Comput. Appl. **162**, 41–46 (2017)
38. Keles, M.K., Özel, S.A.: A review of distance learning and learning management systems. In: Virtual learn (2016)
39. Kaya, M.: Distance education systems used in universities of Turkey and Northern Cyprus. Proc. Soc. Behav. Sci. **31**, 676–680 (2012)
40. Moodle History: Available at https://docs.moodle.org/37/en/History
41. Moodle LMS: Open source online learning|moodle. Available at https://moodle.com/lms/
42. Sakai. Available at https://www.sakailms.org/feature-details
43. ATutor: Learning management system. Available at https://atutor.github.io/atutor/index.html
44. Falvo, D.A., Johnson, B.F.: The use of learning management systems in the United States. TechTrends **51**, 40–45 (2007)
45. Bradford, P., Porciello, M., Balkon, N., Backus, D.: The blackboard learning system: the Be All and End All in educational instruction? J. Educ. Technol. Syst. **35**, 301–314 (2007)
46. Blackboard learning management system. Available at https://www.blackboard.com/teaching-learning/learning-management
47. Desire2Learn. Available at https://www.d2l.com/en-mea/about/
48. Docebo Learn. Available at https://www.docebo.com/learning-management-system-lms/
49. Classroom Help. Available at https://support.google.com/edu/classroom
50. Li, Y.: MOOCs in higher education: opportunities and challenges. In: 2019 5th international conference on humanities and social science research (ICHSSR 2019) (2019)
51. Status report on the adoption of MOOCs in higher education in Latin America and Europe. Ecuador (2016)
52. De Freitas, S.I., Morgan, J., Gibson, D.: Will MOOCs transform learning and teaching in higher education? Engagement and course retention in online learning provision. Br. J. Educ. Technol. **46**, 455–471 (2015)
53. Arpaci, I., Al-Emran, M., Al-Sharafi, M.A.: The impact of knowledge management practices on the acceptance of Massive Open Online Courses (MOOCs) by engineering students: a cross-cultural comparison. Telemat. Informatics in press (2020)
54. Zheng, Q., Chen, L., Burgos, D.: The international comparison and trend analysis of the development of MOOCs in higher education. In: Lecture notes in educational technology, pp. 1–9. Springer (2018)
55. Yuan, L., Powell, S.: MOOCs and disruptive innovation: Implications for higher education. eLearning Pap, pp. 1–8 (2013)
56. Krutka, D.G., Carano, K.T.: Videoconferencing for global citizenship education: wise practices for social studies educators. J. Soc. Stud. Educ. Res. **7**, 109–136 (2016)
57. Al-Samarraie, H.: A scoping review of videoconferencing systems in higher education: Learning paradigms, opportunities, and challenges. Int. Rev. Res. Open Distance Learn. **20**, 121–140 (2019)
58. Griffith, S., Liyanage, L.: An introduction to the potential of social networking sites in education. Sites J. 20th Century Contemp. French Stud. (2008)
59. Hung, H.T., Yuen, S.C.Y.: Educational use of social networking technology in higher education. Teach. High. Educ. **15**, 703–714 (2010)
60. Osteen, B., Basu, A., Allan, M.: In the current or swimming upstream? Instructors' perceptions of teaching with streaming media in higher education. In: Streaming media delivery in higher education: methods and outcomes, pp. 136–157. IGI Global (2011)
61. Kebritchi, M., Lipschuetz, A., Santiague, L.: Issues and challenges for teaching successful online courses in higher education. J. Educ. Technol. Syst. **46**, 4–29 (2017)
62. Davis, N.L., Gough, M., Taylor, L.L.: Online teaching: advantages, obstacles and tools for getting it right. J. Teach. Travel Tour. **19**, 256–263 (2019)
63. Piktochart. Available at https://piktochart.com/
64. Squigl. Available at https://squiglit.com/
65. Nearpod. Available at https://nearpod.com/
66. Edpuzzle. Available at https://edpuzzle.com/

67. Buncee. Available at https://app.edu.buncee.com/home
68. MindMup. Available at https://www.mindmup.com/
69. Trello. Available at https://trello.com/
70. Gokalp, M.O., Kayabay, K., Akyol, M.A., Eren, P.E., Koçyiğit, A.: Big data for industry 4.0: A conceptual framework. In 2016 International Conference on Computational Science and Computational Intelligence (CSCI) Dec 15 (pp. 431-434). IEEE (2016)
71. Çoban, S., Gökalp, M.O., Gökalp, E., Eren, P.E., Koçyiğit, A.: [WiP] Predictive maintenance in healthcare services with big data technologies. In 2018 IEEE 11th Conference on Service-Oriented Computing and Applications (SOCA) Nov 20 (pp. 93-98). IEEE (2018)
72. Gökalp, M.O., Kayabay, K., Gökalp, E., Koçyiğit, A., Eren, P.E.: Towards a model based process assessment for data analytics: An exploratory case study. . In European Conference on Software Process Improvement Sep 9 (pp. 617-628). Springer, Cham (2020)